DATE DUE

Demco, Inc. 38-293

Handbook of Sensory Physiology

Volume III/1

Editorial Board
H. Autrum · R. Jung · W. R. Loewenstein
D. M. MacKay · H. L. Teuber

Enteroceptors

By

B. Andersson · M. Fillenz · R. F. Hellon · A. Howe · B. F. Leek
E. Neil · A. S. Paintal · J. G. Widdicombe

Edited by

E. Neil

With 91 Figures

Springer-Verlag Berlin · Heidelberg · New York 1972

ISBN 3-540-05523-1 Springer-Verlag Berlin · Heidelberg · New York
ISBN 0-387-05523-1 Springer-Verlag New York · Heidelberg · Berlin

This work is subject to copyright. All rights are reserved, whether the whole or part of the material is concerned, specifically those of translation, reprinting, re-use of illustrations, broadcasting, reproduction by photocopying machine or similar means, and storage in data banks.

Under § 54 of the German Copyright Law where copies are made for other than private use, a fee is payable to the publisher, the amount of the fee to be determined by agreement with the publisher. © by Springer-Verlag Berlin · Heidelberg 1972. Printed in Germany. Library of Congress Catalog Card Number 78-24187.

The use of general descriptive names, trade names, trade marks, etc. in this publication, even if the former are not especially identified, is not to be taken as a sign that such names, as understood by the Trade Marks and Merchandise Marks Act, may accordingly be used freely by anyone.

Typesetting, printing and binding: Universitätsdruckerei H. Stürtz AG, Würzburg

Preface

This series of concise essays on Enteroceptors is designed to interest the graduate student and to stimulate research.

Even before the advent of electrophysiological studies, classical physiological techniques had shown the essence of the role of many of the enteroceptors. Thus the monitoring influence of the cardiovascular mechanoreceptors on the heart and on the systemic vascular resistance, the role of the arterial chemoreceptors in hypoxia and the influence of the so-called Hering Breuer stretch receptors on breathing had all been documented. The pioneering work of ADRIAN, BRONK, ZOTTERMAN and others using electroneurographic methods gave a remarkable impetus to the study of the enteroceptors themselves. Nowhere is this better exemplified than in the case of the afferent end organs of the heart, the respiratory tract and the abdominal and pelvic viscera. The remarkable development of our knowledge of the multiplicity of types of nerve endings from the thoracic and abdominal viscera acquired from electrophysiological studies has refocussed our attention on the histological details of the sites of such receptors. Once more research on the structural side has been accelerated by the question raised by evidence obtained from functional studies. This is well illustrated in the case of the carotid body, where the long cherished belief that the innervated epithelioid cells constitute the chemoreceptor complex is now under attack.

The detailed consideration of the functional characteristics of each enteroceptor considered has not occupied our whole attention. In so far as possible the influence of each of these afferent end organs on the organism as a whole has been considered.

Electrophysiological studies of hypothalamic neurones responsive to temperature have contributed much to our further appreciation of the complexity of thermoregulation. Techniques of electrical stimulation and/or ablation have indicated that the hypothalamic complex is concerned with relaying the sensations and behavioural activity associated with hunger and thirst. Again, electrophysiological studies have revealed the presence of anterior hypothalamic neurones responsive to changes in blood glucose levels and blood osmolality respectively, but much more evidence is required before these "central receptors" of hunger and thirst can be ascribed a firm role.

ERIC NEIL

Contents

Chapter 1	Cardiovascular Receptors. By A. S. PAINTAL. With 19 Figures	1
Chapter 2	Arterial Chemoreceptors. By E. NEIL and A. HOWE. With 14 Figures	47
Chapter 3	Receptors of the Lungs and Airways. By J. G. WIDDICOMBE and M. FILLENZ. With 23 Figures	81
Chapter 4	Abdominal Visceral Receptors. By B. F. LEEK. With 16 Figures	113
Chapter 5	Central Thermoreceptors and Thermoregulation. By R. F. HELLON. With 11 Figures	161
Chapter 6	Receptors Subserving Hunger and Thirst. By B. ANDERSSON. With 8 Figures	187
Author Index	. . . .	217
Subject Index	. . . .	229

List of Contributors

ANDERSSON, Bengt
    Department of Physiology, Veterinärhögskolan,
    10405 Stockholm 50, Sweden

FILLENZ, Marianne
    University Laboratory of Physiology, Parks Road, Oxford OX1 3PT,
    Great Britain

HELLON, Richard F.
    National Institute for Medical Research, The Ridgeway, Mill Hill,
    London NW7, Great Britain

HOWE, Alan
    Department of Physiology, University of London,
    Chelsea College of Science & Technology, Manresa Road, London SW3,
    Great Britain

LEEK, Barry F.
    Department of Veterinary Physiology,
    Royal Dick School of Veterinary Studies, University of Edinburgh,
    Summerhall, Edinburgh EH9 1QH, Great Britain

NEIL, Eric
    Department of Physiology, Windeyer Building,
    Middlesex Hospital Medical School, Cleveland Street, London W1,
    Great Britain

PAINTAL, A. S.
    Vallabhbhai Patel Chest Institute, University of Delhi, Delhi 7, India

WIDDICOMBE, J. G.
    University Laboratory of Physiology, Parks Road, Oxford OX1 3PT,
    Great Britain

Chapter 1

Cardiovascular Receptors

By

A. S. Paintal, Delhi (India)

With 19 Figures

Contents

I. Peripheral Mechanisms	1
II. Nerve Fibres	3
III. Blocking Temperatures	6
IV. Arterial Baroreceptors	7
A. Carotid Baroreceptors	7
B. Aortic and Brachiocephalic Baroreceptors	10
V. Pseudo Baroreceptors	15
VI. Atrial Receptors	16
A. Type B Atrial Receptors	17
B. Type A Atrial Receptors	27
VII. Ventricular Receptors	31
A. Ventricular Pressure Receptors	32
B. Epicardial Receptors	36
VIII. Pericardial Receptors	38
IX. Pulmonary Arterial Baroreceptors	39
References	40

I. Peripheral Mechanisms

Like all the other visceral receptors, the cardiovascular receptors have been studied by recording impulses in their sensory fibres most of which travel either in the glossopharyngeal or vagus nerves. So far no attempts have been made to study the precise mechanisms of impulse initiation at the ending (except in the case of the mesenteric Pacinian corpuscle described in this section), since it is practically impossible to isolate a single cardiovascular receptor and study the generator potentials produced by it. It will be necessary to await the development of techniques that will enable recording of such potentials (e.g. with microelectrodes) before one can attempt to study such processes in these endings. However, for practical purposes it is not imperative to do this because the information already available in the case of other mechanoreceptors can be applied to the cardiovascular receptors as well. For example as in the case of the muscle spindle (Katz, 1950) and the Pacinian corpuscle (Alvarez-Buylla and Ramirez de Arellano, 1953; Gray and Sato, 1953; Loewenstein and Rathkamp, 1958;

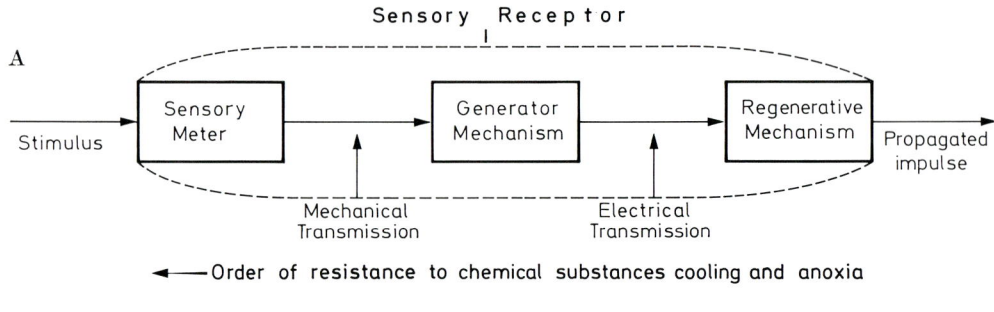

Fig. 1 A and B. Mechanisms involved in the excitation of sensory receptors and their relative resistance to chemical substances, cooling and anoxia (PAINTAL, 1971). B Schematic diagram of sensory endings of medullated and non-medullated nerve fibres showing the two parts of an ending and the probable site of action of drugs at the regenerative region, where there is no diffusion barrier. A greater variety of drugs effect the endings of non-medullated fibres because the fibres themselves are more susceptible to these drugs (PAINTAL, 1964)

for other references see PAINTAL, 1964, 1971) it may be safely assumed that impulses in cardiovascular receptors must be generated when the generator potential reaches a critical threshold for firing an impulse, that the frequency of discharge will depend on the amplitude of the generator potential, that the generator potential will be of a rhythmic nature and that as in the case of the muscle spindle it will have a dynamic off effect (KATZ, 1950) at the end of the pulsatile stimulus.

Most of the cardiovascular receptors are stimulated by certain chemical substances (see PAINTAL, 1964) or are depressed by local anaesthetics (ZIPF, 1966). It has been postulated that all these excitatory drugs and local anaesthetics produce their effects by acting on the regenerative region of the sensory apparatus as shown in Fig. 1B (see PAINTAL, 1964). This hypothesis has received substantial support from recent experiments on muscle spindles and the crayfish stretch receptors (see PAINTAL, 1971). And so for the present it can be safely assumed that in the case of cardiovascular receptors also, the various chemical substances produce their effects by an action on the regenerative region.

It has also been brought out that the regenerative region is much more sensitive to cooling and anoxia and that there is an increasing order of resistance to chemical substances, cooling and anoxia as one proceeds backwards from the regenerative region to the stimulus (Fig. 1 A). The most resistant structures appear to be the sensory meters which in the case of cardiovascular receptors must be the fibro-elastic tissue in which the endings lie (PAINTAL, 1971). It needs to be stressed that the actual meters which transmit, by causing mechanical deformation of the generator region, the intensity of the stimulus, are the non-nervous elements i.e. the fibro-elastic tissue as far as the cardiovascular receptors are concerned. Clearly, if the stress/strain properties of this tissue changes, so will the degree of deformation of the generator region. Under experimental conditions any agent that is strong enough to put out of action the regenerative and generator regions apparently leaves the sensory meter practically unaffected (PAINTAL, 1971). However in certain diseases, e.g. TAKAYASHU's disease (see HEYMANS and NEIL, 1958, p. 84) and possibly mitral stenosis in the case of left atrial receptors it is conceivable that the sensory meter may be affected before any changes occur at the regenerative region. For such studies i.e. alterations in the properties of the sensory meters the cardiovascular receptors may prove more useful than the Pacinian corpuscle or muscle spindle—receptors that have been the source of valuable information about generator processes.

As in the case of other mechanoreceptors (PAINTAL, 1964) there is no firm evidence that a chemical transmitter is involved in the process of impulse initiation at any stage as shown in Fig. 1A (PAINTAL, 1971).

II. Nerve Fibres

Most of the known cardiovascular receptors are connected to medullated nerve fibres as shown in Table 1. Such receptors are therefore unaffected by chemical

Table 1. *Conduction velocities of cardiovascular afferent fibres* (PAINTAL, *1953c, 1963a and unpublished observations*)

Type of receptor	Nature of fibre (i.e. medullated or non-medullated)	Conduction velocity		
		Range (m/sec)	Mean (m/sec)	S.D. (m/sec)
Carotid baroreceptors	med.	[a]	[a]	[a]
	non-med.	0.5–2.0	—	—[b]
Aortic baroreceptors	med.	12–53	33	11
Right atrial type A	med.	13–27	20	4
Left atrial type A	med.	12–19	16	2
Right atrial type B	med.	8–29	17	6
Left atrial type B	med.	11–26	18	3
Ventricular pressure receptors	med.	8–19	13	—
Epicardial receptors	non-med.	1.2–1.9	—	—[c]
		0.4–5.0	1.5	2.3 (SE)[d]
Pericardial receptors	med.	2.5–7.0	5.3	—[e]

[a] Not known.   [b] FIDONE and SATO (1969).   [c] COLERIDGE *et al.* (1964b).
[d] SLEIGHT and WIDDICOMBE (1965a).
[e] SLEIGHT and WIDDICOMBE (1965b) (see text also).

substances that stimulate the endings of non-medullated fibres e.g. acetylcholine, nicotine, phenyl diguanide (PAINTAL, 1964, 1971) because their regenerative region (first node) is much less sensitive to drugs than the regenerative region of endings

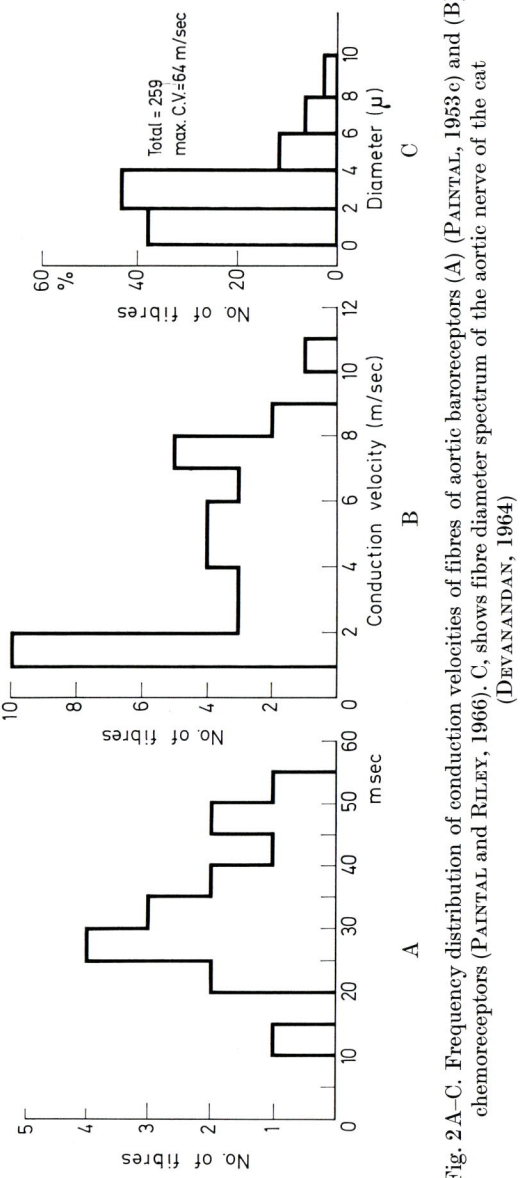

Fig. 2A—C. Frequency distribution of conduction velocities of fibres of aortic baroreceptors (A) (PAINTAL, 1953c) and (B) chemoreceptors (PAINTAL and RILEY, 1966). C, shows fibre diameter spectrum of the aortic nerve of the cat (DEVANANDAN, 1964)

of non-medullated fibres (Fig. 1B). It is therefore not surprising to find that the carotid baroreceptors with non-medullated fibres and the epicardial endings are the only ones that are stimulated by acetylcholine or nicotine (SLEIGHT and WIDDICOMBE, 1965a; FIDONE and SATO, 1969).

The conduction velocities of cardiovascular afferent fibres are given in Table 1. These are the velocities in the cervical vagus (except in the case of epicardial and pericardial receptors) and in some fibres this velocity remains unchanged right up to the ending (PAINTAL, 1962a). However, in most fibres there is a reduction in conduction rate amounting to 10–40% in the intrathoracic stretch of the nerve fibres and there are a few (say 10%) in which the rate is half that of the velocity of the same fibre in the cervical vagus (PAINTAL, 1962a). These conclusions relate to fibres with conduction velocities greater than 10 m/sec, no information is available regarding slower fibres. However Table 1 indicates that there are likely to be very few medullated fibres with conduction velocities less than 10 m/sec. It is also noteworthy that the conduction velocities of all cardiac fibres (of rhythmically active receptors) ranges from 8 to 29 m/sec and that the mean conduction velocities of all of them is approximately the same.

The distribution of the conduction velocities of fibres of aortic baroreceptors and chemoreceptors and the fibre diameter distribution in the aortic nerve are shown in Fig. 2A–C. If one takes the 3 figures together one finds that the diameter distribution is consistent with the distribution of conduction velocities. Indeed, one can say safely, whether one assumes that the conduction velocity: fibre diameter ratio is 6 (HURSH, 1939) or 5 (BOYD, 1964, 1965), that the fibres below $3\,\mu$ in diameter mostly consist of chemoreceptor fibres and those above $3\,\mu$ of baroreceptor fibres. DEVANANDAN (1964) found that the conversion factor of 6 fits the electrophysiological data fairly well. Fig. 2A shows that about 20% of the baroreceptor fibres conduct at rates greater than 40 m/sec (corresponding to a fibre diameter greater than 7 or $8\,\mu$). This is a little greater than what one would expect from the results of DEVANANDAN in general although it would fit in with Fig. 2B taken from his paper. The histological observations of SCHMIDT and STROMBERG (1967) on the aortic nerve of swine agree with those of DAVANANDAN (1964). These results therefore clear up certain discrepancies between electrophysiological and histological observations pointed out earlier (PAINTAL, 1963a).

On the other hand in the case of carotid baro- and chemoreceptor fibres of the cat one is forced to conclude from the results of FIDONE and SATO (1969) taken in conjunction with those of EYZAGUIRRE and UCHIZONO (1961) that all the fastest and slowest medullated fibres are chemoreceptor ones and that the baroreceptor fibres conduct only within the range of 15 to 25 m/sec. Clearly this is quite different from what obtains in the case of aortic chemo- and baroreceptor fibres (Fig. 2). The findings of FIDONE and SATO (1969) are also in conflict with those of DE CASTRO (1951) who concluded from his histological findings that all the large and some medium diameter fibres are baroreceptor fibres, the remainder being chemoreceptor fibres. However, it should be pointed out that the results of FIDONE and SATO were obtained using an unusual method of measuring conduction velocities of medullated fibres, i.e. a system of two differential inputs with a common distal electrode so called "monotopic" (see Fig. 1B of their paper). Measurements of conduction velocities of fibres using their technique has yielded values of 20–30 m/sec in non-medullated fibres. And so there is a strong possibility that some of their medullated chemoreceptor fibres were in fact non-medullated (PAINTAL, 1971).

Finally, there is the important finding by FIDONE and SATO (1969) that about 30% of the non-medullated fibres in the carotid nerve are made up of baroreceptor fibres with conduction velocities ranging from 0.5 to 2.0 m/sec.

III. Blocking Temperatures

The relation between diameter of nerve fibres and the temperature at which conduction in them is blocked was discussed in an earlier review (PAINTAL, 1963a). Since then evidence has been obtained to show that all medullated nerve fibres are blocked at about the same temperature i.e. about 7.6°C regardless of their

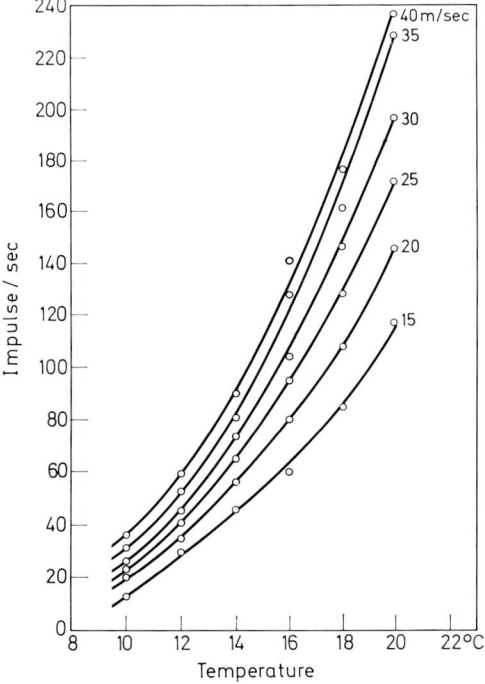

Fig. 3. Maximum transmissible frequency of impulses between 10 and 20°C in fibres with different conduction velocities measured at 37°C (PAINTAL, 1966b)

conduction rates (PAINTAL, 1965a; FRANZ and IGGO, 1968) and so it is not possible to use cold block as a method of blocking the medullated fibres differentially. However it is possible to use low temperatures to reduce the frequency of discharge in nerve fibres since the maximum frequency of impulses that a nerve fibre can conduct varies with the temperature and with its conduction velocity (Fig. 3) (PAINTAL, 1965b, 1966b; FRANZ and IGGO, 1968). It follows that above about 8°C, i.e. above the actual blocking temperature of the nerve fibre, cardiovascular afferent fibres will be able to conduct the first impulse of each burst but the passage of subsequent impulses of the burst will be determined by the temperature of the cold block and the frequency of discharge. Thus the relatively higher frequency of discharges in aortic baroreceptors and type A atrial receptors (PAINTAL, 1953c) would drop out at temperatures which will leave the somewhat lower frequency discharges in some type B atrial receptors largely unaffected. This will explain the observations of TORRANCE and WHITTERIDGE (1948) and WHITTERIDGE (1948) who found that the activity of baroreceptor fibres was blocked at a higher temperature than that needed to block the fibres of atrial type B receptors. It would also be expected that baroreceptor reflexes at low mean blood pressures (e.g. 70 mm Hg) should survive at lower blocking temperatures since the frequency of discharge in the afferent fibres will be lower than normal. Further it should be possible to block practically all impulses of cardiovascular receptors (i.e. baroreceptor, atrial type A and B, ventricular pressure receptors) at about 8°C and leave the impulses of non-medullated chemoreceptor fibres largely unaffected since the frequency of discharge in the latter averages only 9 impulses/sec when they are stimulated during intense hypoxia (PAINTAL, 1967a). At this temperature the relatively high frequency grouped discharges of chemoreceptors will be blocked, but one would not expect the irregularly spaced impulses to be blocked.

Since non-medullated fibres are blocked at a lower temperature than medullated ones (PAINTAL, 1967b; FRANZ and IGGO, 1968) one would expect at least the first impulse of each rhythmic burst to pass through a region of nerve cooled to about 5°C, at which temperature there is total block of conduction in medullated nerve fibres. This would apply particularly to the non-medullated fibres of carotid baroreceptors (SATO et al., 1968; FIDONE and SATO, 1969) and also to the fibres of epicardial receptors (COLERIDGE et al., 1964b; SLEIGHT and WIDDICOMBE, 1965a).

The frequency of discharge produced by excitatory substances e.g. veratridine is relatively high (Figs. 12, 18) (PAINTAL, 1957, 1964). It follows that the discharges so produced by excitatory substances will be blocked at a higher temperature than that needed to block lower frequency natural activity. However, the degree of block will depend on the local experimental conditions and the actual frequency of discharge. Thus under some conditions it is possible for some impulses to pass through a cold block even though the discharge is much above the maximum transmissible frequency. In other cases a steady frequency of discharge only a little higher than the maximum transmissible frequency is completely blocked (PAINTAL, 1965b). This is apparently due to the fact that each impulse arriving at the cooled stretch of nerve produces abortive spikes and although these are of insufficient amplitude for propagation, nevertheless leave behind a refractory period that affects the conduction of subsequent impulses (PAINTAL, 1966a). Such a situation arises from the absolute refractory period being much greater than the absolute refractory period for the initiation of an impulse (ARP_i) at lower temperatures (PAINTAL, 1966a).

IV. Arterial Baroreceptors
A. Carotid Baroreceptors

Of all the cardiovascular receptors, the carotid baroreceptors have been studied most extensively because of the technical advantages available for their study, namely, easy isolation of the nerve fibres, isolation of the sinus region for perfusion and application of controlled stimuli. As clearly shown by BRONK and STELLA (1932, 1935) the carotid baroreceptors are slowly adapting stretch receptors and they therefore respond like the muscle spindle in many ways. Thus the threshold varies from ending to ending, the peak frequency of discharge is attained during the rising phase of the stimulus, there is variable amount of adaptation and there is post-excitatory depression. This particular aspect was studied in more detail by LANDGREN (1952a) who related it to the dynamic off effect seen in the generator potential of the muscle spindle (KATZ, 1950). BRONK and STELLA (1935) also found that the activity of the endings was linearly related to blood pressure in many endings until a "saturation" level was reached above which there was no further increase in activity on further increasing the pressure; this pressure level for most receptors was about 140–180 mm Hg. However, there were other receptors in which the ending responded linearly throughout the range of pressure up to 200 mm Hg. These pressures are actually transmural pressures that are responsible for stretching the endings. When the endings are prevented from being stretched by the application of a rigid cast around the artery, they are not stimulated as indicated by the absence of reflex fall in blood pressure on raising the pressure in the carotid sinus (HAUSS et al., 1949).

Although it is recognized that the natural stimulus for the carotid baroreceptors is the pulsatile variation in pressure, it has nevertheless been of interest to find out whether a steady pressure is as effective as pulsatile pressure (at the same level of mean pressure) in stimulating baroreceptors and eliciting reflex effects. EAD, GREEN and NEIL (1952) studied this problem systematically and they found that the frequency of discharge during application of steady pressure was less than that attained during pulsatile stimulation. Further, application of pulsatile pressure recruited more units into activity. These results were consistent with their observations on reflex effects of pulsatile and steady pressure.

On application of a rectangular stimulus the frequency of discharge rises to a peak and then falls off to a steady level (LANDGREN, 1952a) as in the case of the muscle spindle (MATTHEWS, 1931, 1933). The peak frequency of discharge attained in large baroreceptor fibres varies from 250–350 impulses/sec; the maximum adapted frequency varies from 40–70/sec. On the other hand the peak frequency of discharge attained in the case of fibres with small spikes ranges from 50 to 150/sec and the maximum adapted frequency varies from 20–30 impulses/sec (LANDGREN, 1952a). Such frequencies approximate the maximum frequencies recorded in non-medullated fibres of other sensory receptors e.g. gastric stretch receptors (PAINTAL, 1954) aortic chemoreceptors (PAINTAL, 1967a) (see also PAINTAL, 1964). It is therefore important to consider whether these small-spike fibres were non-medullated since it is now known from the work of FIDONE and SATO (1969) that there are many baroreceptors with non-medullated fibres.

Although in a whole nerve that has not been injured (or in thick filaments) it is reasonable to expect that the size of the spikes will vary with the diameter of the fibres (GASSER and GRUNDFEST, 1939), it is important to realise that this relation does not hold consistently in the case of thin filaments. Indeed not infrequently the opposite obtains (PAINTAL, 1953c) owing to variable experimental conditions, an important variable being injury to nerve fibres, leading to the reduction of the membrane potential near the recording electrodes. Perhaps in some cases it is this factor that is responsible for the spikes of non-medullated fibres being larger than those of medullated fibres conducting faster than 20 m/sec (IGGO, 1958).

However, even allowing for all these variables it seems reasonable to assume that all the medullated fibres of carotid baroreceptors will have, more or less, the same spike height because according to DE CASTRO's observations one would expect the majority of such fibres to range between 4 to 7 μ in diameter (see

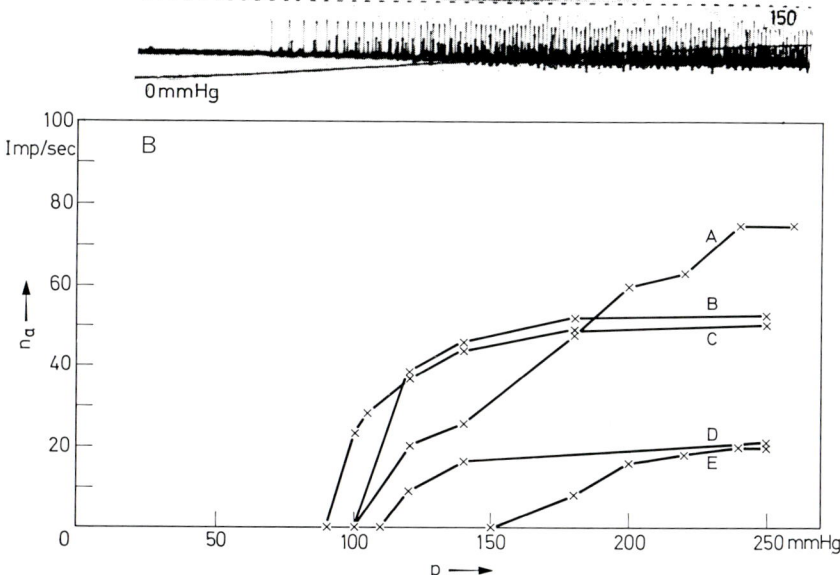

Fig. 4A and B. Impulses in medullated (large spikes) and presumably non-medullated (small spikes) fibres of carotid baroreceptors produced on raising the pressure in the sinus. B, shows the adapted frequency of discharge (ordinate) in fibres with large, medium and small spikes (presumably non-medullated) at various pressures (abscissa). Note that the frequency of discharge in the non-medullated fibres is much less than that in medullated fibres. (From LANDGREN, 1952a)

DE CASTRO, 1951). These fibres probably yield the large and medium size spikes. Assuming this to be the case it follows that the small spikes, which are much smaller than the large and medium size spikes (Fig. 4A), can be presumed to originate in non-medullated fibres. In this connection FIDONE and SATO (1969) have reported that the baroreceptors with C fibres have a lower frequency of discharge and perhaps a higher threshold than their A fibre counterparts i.e. precisely what LANDGREN (1952a) found in the case of fibres with small spikes relative to those with large and medium size spikes (Fig. 4B). Thus for the present it would be reasonable to assume that the small spikes recorded by earlier workers (particularly those recorded in the whole nerve e.g. LANDGREN et al., 1951) arose from non-medullated fibres. These are by far the most numerous (EULER et al., 1941; LANDGREN et al., 1952; HEYMANS and NEIL, 1958, pp. 28 and 75).

It is now possible to try to reconcile histological observations such as those of De Castro (1928) and Rees (1967) showing that the baroreceptor endings are all located in the adventitia and none in the media on the one hand and those of Landgren et al. (1951) and Landgren (1952b) showing that topical application of adrenaline stimulates endings of fibres with small spikes and not those with large spikes on the other. This point has been rather puzzling hitherto (see pp. 73–76 in Heymans and Neil, 1958) because application of adrenaline causes a reduction in the diameter of the vessel and this must presumably reduce the tension in the wall (for the same pressure) in accordance with Laplace's law (see Burton, 1951). As expected the endings with large spike fibres were not stimulated since they are known to lie in the adventitia and hence in parallel with the muscle fibres. On the other hand, the activity of endings with small-spike fibres is increased which implies that such endings must lie in series with the contractile elements. In this connection it is noteworthy that Rees (1967) has found that there are nerve endings in association with smooth muscle cells located in the adventitia of the carotid sinus. In fact some of these presumptive baroreceptor terminals are actually in series with the smooth muscle cells. It is possible that these nerve endings are terminations of non-medullated fibres.

Palme (1943) and Kezdi (1954) observed that stimulation of the local sympathetic branches yields reflex effects typical of baroreceptors. This would fit in with Rees's observation that the smooth muscle cells in the adventitia receive sympathetic nerve endings (Rees, 1967). However Floyd and Neil (1952) could not confirm the observations of Palme. Moreover in agreement with their negative results on reflex effects of sympathetic stimulation they found that stimulation of the local sympathetic branches failed to increase the activity of nearly all the baroreceptor units they tested; only one unit was stimulated. Floyd and Neil (1952) suggested that the effect on carotid baroreceptors could be due to distortion of the sinus region consequent on contraction of the smooth muscle in the adjacent parts of the vessel. The effect of longitudinal tension in the wall of the vessel has also been noted by Angell James (1971) in the case of aortic baroreceptors.

B. Aortic and Brachiocephalic Baroreceptors

The responses of aortic and brachiocephalic baroreceptors are basically similar to those of the carotid ones—the pulsatile discharge being closely related to the aortic pressure curve (Fig. 5) (Whitteridge, 1948; Neil, 1954; Bloor, 1967). The baroreceptors with fibres running in the right aortic nerve of the cat are distributed in several areas of the brachiocephalic trunk (Boss and Green, 1956; Bianconi and Green, 1959a). Those in the arch of the aorta run in the left aortic nerve.

With the chest intact, care has to be exercised in identifying aortic baroreceptors because they could be confused with left atrial type B receptors (Table 3). Such difficulty would not arise if one were to isolate fibres from the aortic nerve in which case one could be more or less confident that one were dealing with aortic baroreceptors. However, when fibres are isolated from the main vagal trunk (Paintal, 1953c) one has to exclude other cardiovascular endings. Table 3 can be of help in identifying easily the different cardiovascular receptors from one another. The chief characteristic of aortic baroreceptors is the prominent systolic burst of impulses beginning within about 80 msec after the Q wave of the e.c.g. This activity is greatly reduced on positive pressure inflation of the lungs and it returns after a few cycles following the release of the inflation (WHITTERIDGE, 1948). The activity of left type B receptors is affected in the same way (Table 3) (PAINTAL, 1953a) but it is usually easy to distinguish the type B receptors because

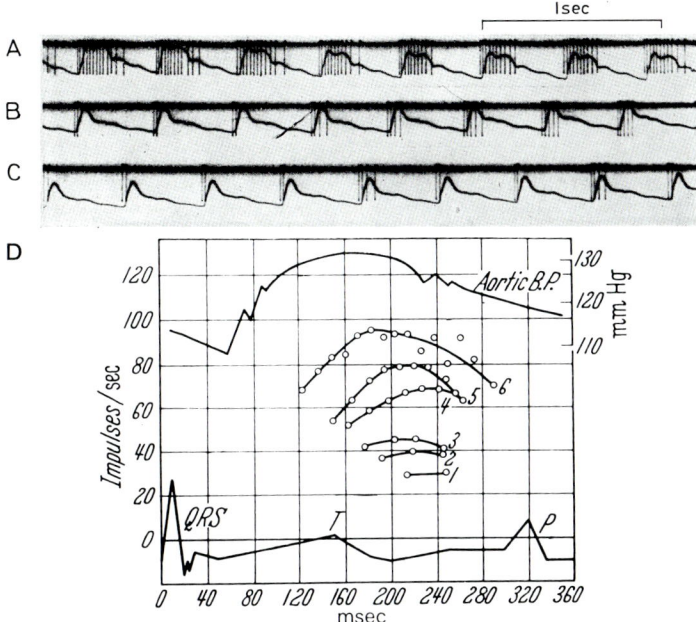

Fig. 5A–D. Activity in an aortic baroreceptor at different levels of blood pressure (A = 125 mm Hg; B = 80 mm Hg and C = 62 mm Hg). Note, the relation of the first impulse to the foot of the pressure pulse is fixed. The pressure pulse is delayed as it is recorded in the carotid artery (NEIL, 1954). D Activity in a type B atrial receptor showing how the first impulse advances earlier with increase in activity during successive stages of inspiration, 1–5. This can be regarded as a typical response of type B Atrial receptors. (From WHITTERIDGE, 1948)

their burst of impulses appears later in systole. However confusion will arise in the case of those type B atrial receptors that have marked activity and in which the burst of impulses starts earlier than usual. In such cases identification is facilitated if one examines the first impulse of each burst during respiratory fluctuations (Fig. 5D). In the case of type B atrial receptors, it will be found that as inspiration progresses (and venous return increases) the first impulse appears earlier and earlier in the cardiac cycle (Fig. 5D). On the other hand in the case of aortic baroreceptors the position of the first impulse does not change visibly relative to the Q wave of the e.c.g. because it is produced during the rising phase of the aortic pressure pulse (Fig. 5) which remains relatively constant in the cardiac cycle. Finally as shown in Table 3 ectopic stimulation of the right atrium will provide a decisive answer.

The difficult task of studying the activity of baroreceptors of the isolated aortic arch of the rabbit perfused with Krebs-Henseleit solution or blood at 38° C has been accomplished by Angell James (1968, 1971). Although judging from the responses of aortic baroreceptors *in vivo* it was expected that their responses *in vitro* would be qualitatively similar to those of carotid baroreceptors, her experiments have shown beyond doubt that this is in fact so. The threshold pressure for different endings varied from 0–188 mm Hg (mean 51 mm Hg), the impulse frequency varied linearly with the pressure until a "saturation" pressure was attained beyond which the increase in frequency for equal increments of pressure became less and less (levelling off); this pressure varied from 63 to 158 mm Hg. One must keep in mind that these values pertain to the isolated perfused preparation. Another important finding was that as in the case of carotid baroreceptors (Ead et al., 1952), although pulsatile pressure produced a higher frequency of discharge than static pressure, the total number of impulses per cycle remained the same. Moreover, pulsatile pressure caused recruitment of other baroreceptors during the rising phase of the pressure curve (Angell James, 1968). However an important difference is that although the reflex effects of pulsatile carotid pressure are more pronounced than those of steady pressure, this is not so in the case of aortic baroreceptors examined in the same dog (Angell James and Daly, 1970). It is possible that the observations of Angell James (1968, 1971) on the baroreceptors of the rabbit may not be applicable in the dog. In the rabbit it is significant that fluctuations in "intrathoracic" pressure can vary the activity of the endings (Angell James, 1968). This means that it is important to use effective aortic pressure i.e. transmural pressure as a measure of the stimulus particularly at lower levels of mean pressure when the fluctuations in intrathoracic pressure will become significant.

The aortic baroreceptors of the new-born respond in qualitatively the same way as those of adult rabbits (Bloor, 1967). There is a linear relation between peak discharge frequency and the mean arterial pressure but the range of pressure which stimulates the endings of the new-born is less because the mean blood pressure is low (about 40 mm Hg) in the new-born.

Are there Aortic Baroreceptors with Non-Medullated Fibres? Since it is known that there are carotid baroreceptors with non-medullated fibres (Sato et al., 1968; Fidone and Sato, 1969) one must now consider whether there are similar non-medullated fibres in the aortic nerve or the vagus. Assuming that such fibres exist then one would expect that, allowing for conduction time in fibres of 1 to 2 m/sec conduction velocity, the first impulse would appear about 130 to 180 msec after the Q wave of the e.c.g. if the impulses are recorded near the nodose ganglion of the cat. This would give the discharge the appearance of a type B atrial receptor and such endings would therefore have been lumped with the type B atrial fibre population in a random sample. However, the lowest conduction velocity of type B atrial fibres is 8 m/sec (Paintal, 1953c) (Table 1). It can therefore be concluded that either there are no non-medullated baroreceptor fibres in the aortic nerve or vagus, or if there are any, their number is insignificant.

Effect of Temperature. There are two main studies on the effects of temperature on baroreceptors, one by Diamond (1955) on carotid baroreceptors of the cat and the other by Angell James (1968, 1971) on aortic baroreceptors of the rabbit. As expected, both studies have shown that the frequency of discharge falls with fall in temperature, the temperature coefficient (Q_{10}) being about 2 in the case of carotid baroreceptors (DIAMOND, 1955). Although there is no qualitative change in the stimulus-response relationship, there is, in the case of carotid baroreceptors, a slight increase in the threshold (DIAMOND, 1955). On the other hand in the case of aortic baroreceptors ANGELL JAMES (1971) observed a small reduction in threshold on lowering the temperature. How far these differences between aortic and carotid baroreceptors are real or are due to differences in experimental techniques remains to be determined.

A Q_{10} of 2 in carotid baroreceptors is similar to the value obtained by MATTHEWS in the frog's muscle spindle (MATTHEWS, 1931). OTTOSON (1965) also

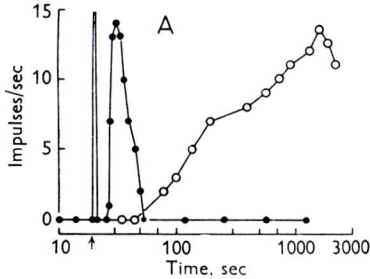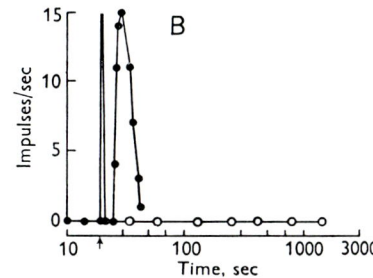

Fig. 6A and B. Excitation of a carotid baroreceptor (with presumably non-medullated fibre—see text) following injection at arrows of ACh (filled circles) and adrenaline (open circles). Ordinate, average frequency of discharge; abscissa, time on logarithmic scale. In A, 30 μg ACh and 1 mg adrenaline were injected using normal perfusion fluid. In B the perfusion fluid contained 2×10^{-6} g/ml dihydroergotamine; ACh 10 μg and Adrenaline 1 mg were injected. Note delayed stimulation following adrenaline in A (DIAMOND, 1955)

obtained a Q_{10} of 2 for the frequency of discharge. However, significantly he found no reduction in the generator potential on cooling and he therefore attributed the fall in frequency of discharge to the increase in the refractory period. This would seem reasonable because the Q_{10} for the ARP in frog nerve fibres is about 3 [as indicated by the results of SCHOEPFLE and ERLANGER (1941)] and the Q_{10} for the relative refractory period is about the same as that for the ARP (PAINTAL, 1966a). Thus the reduction in the frequency of discharge on cooling can be attributed to the increase in the relative refractory period. It can be safely assumed that this also applies to carotid (and aortic) baroreceptors since the Q_{10} is also the same i.e. about 2.

Although the amplitude of the generator potential of the muscle spindle does not fall on cooling (OTTOSON, 1965) it should be remembered that there *is* a reduction of the generator potential in the Pacinian corpuscle—the Q_{10} being about 1.5 in the range of 25–40°C (INMAN and PERUZZI, 1961; ISHIKO and LOEWENSTEIN, 1961). Further till direct evidence is obtained it would be reasonable to assume that the sensory meter has an insignificant temperature coefficient (PAINTAL, 1971) as in the case of the Pacinian corpuscle (ISHIKO and LOEWENSTEIN, 1961).

Effects of Chemical Substances. A large number of observations on several endings has revealed that ACh does not have a primary excitatory effect on sensory receptors with medullated fibres (PAINTAL, 1964). The only well known exception has been the carotid baroreceptors (Fig. 6). This exception has been puzzling because it has been assumed that all carotid baroreceptors have medullated fibres and so ACh would not be expected to stimulate such endings (PAINTAL, 1964). Secondly, in conformity with this expectation, aortic baroreceptors (known to have medullated fibres) (Table 1) are definitely not stimulated by relatively large doses of ACh (100–300 µg) injected into the aorta (PAINTAL, 1971). This exception is no longer puzzling because it is now known that there are many carotid baroreceptors with non-medullated fibres (Table 1) and that these are stimulated by ACh (FIDONE and SATO, 1969). The situation becomes even more clear if one accepts the view put forward above that the small spikes, recorded by various investigators including DIAMOND, originated in non-medullated fibres particularly because the baroreceptors with large spike fibres (i.e. medullated) were apparently not stimulated by ACh in the same way as the endings with small spike fibres (DIAMOND, 1955). It is clear from DIAMOND's records and his conclusion that ACh mainly stimulated the endings with small spike fibres; so did nicotine (DIAMOND, 1955).

An important observation of DIAMOND which indicates that the small spike fibres stimulated by ACh in fact belonged to the same class as those studied by LANDGREN et al. (1952) and LANDGREN (1952b) (i.e. non-medullated fibres as argued above) is that the endings that were stimulated by ACh were also stimulated by adrenaline (Fig. 6) with a time course that was of the same order observed by LANDGREN et al. (1952). The marked difference in the rate of development of excitation between that produced by ACh in seconds and that by adrenaline in minutes (Fig. 6) led DIAMOND to conclude justifiably that ACh acted directly on the endings unlike adrenaline whose excitatory effect is secondary to the contraction of the smooth muscle of the artery (LANDGREN, 1952b). This conclusion is further supported by the fact that the effects of ACh survive after application of atropine and also after dihydroergotamine which abolishes the effect of adrenaline but not that of ACh (Fig. 6B). This observation also indicates that ACh does not act by stimulating the synapses at the isolated ganglion cells of the sympathetic system.

Stimulation of baroreceptors by local application of noradrenaline and pitressin is similar to that following adrenaline (LANDGREN et al., 1952) i.e. it is secondary stimulation. Sodium nitrite has the opposite effect (LANDGREN et al., 1952; LANDGREN, 1952b) i.e. it reduces the activity of baroreceptors secondary to relaxation of the smooth muscle fibres.

Considering that ACh must apparently act like nicotine in a non-specific manner (cf. PAINTAL, 1964), it is noteworthy that the excitatory effects of ACh are blocked by hexamethonium (10^{-3}) and d-tubocurarine (1 ml of 10^{-3} solution). ACh also seems to have a blocking effect on a second dose of ACh in some cases (DIAMOND, 1955).

Primary excitation of baroreceptors is also produced by two other groups of excitants namely, veratrum alkaloids (e.g. see JARISCH et al., 1952; WITZLEB, 1952) and by volatile anaesthetics (ROBERTSON et al., 1956; PRICE and WIDDI-

COMBE, 1962). The veratrum alkaloids stimulate the carotid baroreceptors but it is not known whether they also desensitize them as in the case of other endings (PAINTAL, 1964). As expected, the volatile anaesthetics—ether, chloroform and trichlorethylene do not stimulate the aortic and carotid baroreceptors but they sensitize them as they do the pulmonary stretch receptors (WHITTERIDGE and BÜLBRING, 1944; WHITTERIDGE, 1958; PAINTAL, 1957, 1964). On prolonged exposure the endings are blocked (ROBERTSON et al., 1956). An interesting point is that in the case of the baroreceptors the threshold of the endings is reduced; this is not a feature of the pulmonary stretch receptors (see PAINTAL, 1964). It appears that the baroreceptors of both cats and dogs are also sensitized by about 25 to 33% cyclopropane but this effect requires at least 3 min of administration before the effect sets in—this being a possible reason why ROBERTSON et al. (1956) did not observe any effects since they apparently made observations up to 2 min only (PRICE and WIDDICOMBE, 1962). An interesting point is that the effects of the anaesthetics were observed in "large-spike" fibres of the carotid nerve (ROBERTSON et al., 1956; PRICE and WIDDICOMBE, 1962). Moreover ROBERTSON et al. have shown that the effect of the anaesthetics is a primary one and is not secondary to an action on the smooth muscles as is the case with topical application of adrenaline (see above).

As discussed elsewhere (PAINTAL, 1964) the evidence points to an action of the anaesthetics on the regenerative region of the endings i.e. the first node (Fig. 1 B).

V. Pseudo Baroreceptors

From time to time investigators have come across endings whose main function is other than to signal changes in blood pressure but which yield bursts of impulses engendered by the arterial pulse. Some pulmonary stretch receptors are notable for such responses (ADRIAN, 1933; WHITTERIDGE, 1948; BIANCONI and GREEN, 1959c). In some fibres the pattern of discharge may resemble that of aortic baroreceptors closely but it is easy to recognise such fibres as pulmonary stretch receptors by inflating the lungs when, unlike the baroreceptors, a continuous discharge is produced in them.

The mesenteric Pacinian corpuscles are the best known receptors of this group. These corpuscles are known to serve as vibration receptors (HUNT and McINTYRE, 1960; HUNT, 1961; SATO, 1961). Indeed the corpuscles in the leg are so sensitive to vibratory stimuli that HUNT and McINTYRE stated that their sensitivity approached that of a seismograph. It is therefore not surprising that they should be stimulated by the arterial pulse as first reported by GAMMON and BRONK (1935) who made a systematic study of their responses. Later GERNANDT and ZOTTERMAN (1946) also made some observations on these endings.

GAMMON and BRONK (1935) found that in addition to the responses to the arterial pulse these receptors yielded a prolonged discharge to static pressure when the mesenteric vessels were perfused above threshold pressure. However, GAMMON and BRONK concluded that these endings did not signal changes in mean blood pressure. This has been confirmed by LEITNER and PERL (1964) who studied these endings by recording impulses in the dorsal roots. They found that the receptors that had a cardiac rhythm also responded to sinusoidal vibration at a frequency of 50—500 cycles/sec. They also concluded that the endings did not signal changes in mean blood pressure. Moreover they found that their activity increased after injection of adrenaline intravenously. This is the opposite of what was found by Gammon and Bronk (1935). Presumably these differences can be attributed to variations in experimental techniques e.g. manner of mounting the mesentery leading to changes in local conditions. In this connection it is noteworthy that Leitner and Perl (1964) observed that the sensitivity of the Pacinian corpuscle could be increased by local injections of 0.2 µg of adrenaline.

The conduction velocity of the majority of afferent fibres from these receptors is about 42 to 72 m/sec although there are a few that have conduction rates down to 18 m/sec (Leitner and Perl, 1964).

VI. Atrial Receptors

It is now established that there are two main types of atrial receptors, the type A and the type B which are quite different from each other in their pattern of discharge and their responses to changes in cardiovascular dynamics (Paintal,

Table 2. *Ratio of type A to type B atrial receptors in cats and dogs*

Reference	Ratio in cats	Ratio in dogs
Coleridge et al. (1957)	—	3:14
Langrehr (1960a)	1:1	2:28
Paintal (1963a)	1:1 [a]	—
Arndt et al. (1971a)	1:1	—
Fahim and Gupta (1970)	—	3:47

[a] Random sample.

1953a, 1963a, 1963b). For those who work on cats this distinction between the two is clear cut because in this animal one encounters many pure type A fibres i.e. fibres with only one burst of impulses in the cardiac cycle co-incident with the *a* wave of the atrial pressure curve (Fig. 14). In fact in the cat the ratio of type A to type B receptors was found to be 1:1 in a random sample (Paintal, 1963a) (Table 2). It would appear from the results of Langrehr (1960a) and Arndt et al. (1971a) that they too obtained a ratio of 1:1 (Table 2) although they did not make a special attempt to get a random sample. Obtaining a random sample is important because it helps to decide whether the type A and type B receptors belong to two different populations or whether they belong to the same group as suggested by Langrehr (1960a). In a random sample one would expect to encounter many more intermediate types than pure type A and type B receptors if the type A and type B receptors belonged to one homogenous group. Instead it was found that there are equal numbers of type A and type B receptors but relatively very few intermediate types (Paintal, 1963a). This has been confirmed by Arndt et al. (1971a) who found only one intermediate type receptor and so there is no doubt that in the cat the type A and type B receptors constitute distinctly separate populations.

On the other hand in the dog there are relatively few type A receptors, the ratio being less than 3:14 (Table 2). FAHIM and GUPTA (1970) made a specific attempt to study a random sample and they obtained a ratio of 3:47 (Table 2). There is only one preliminary study on the monkey by CHAPMAN and PEARCE (1959) who found 9 type B and one type A atrial receptors. This gives a ratio of 1:9 which is of the same order as that in the dog.

It is therefore certain that there are marked "species" differences in the relative numbers of type A and type B atrial receptors. Moreover, since there are so few type A receptors in the dog one can now appreciate why those who work only on dogs tend to adopt the view that there is only one type of ending (i.e. type B) and therefore prefer to refer to these endings generally as atrial receptors (e.g. see KIDD et al., 1966; KAPPAGODA et al., 1970). However since such studies often have a bearing on studies of reflex effects (LEDSOME and LINDEN, 1964, 1967) it would be of value to know the actual identity of the ending especially because the reflex effects of type B receptors are likely to be different from those of type A receptors (HAKUMÄKI, 1970; BRAMBRING et al., 1969; ARNDT et al., 1971).

A. Type B Atrial Receptors

Of the two, the type B receptors have been studied more extensively particularly since a number of investigators have studied these endings in dogs in which, as pointed out above, there are relatively few type A receptors. Information about these endings is contained in papers by the following authors: WHITTERIDGE (1948; cats); PEARCE and WHITTERIDGE (1950; cats); PAINTAL (1953a, 1953b, 1953c, 1955, 1957, 1963a, 1963b; all on cats); HENRY and PEARCE (1956; dogs); MÜHL et al. (1956; dogs); COLERIDGE et al. (1957, 1964b; dogs); KRAMER (1959; dogs); CHAPMAN and PEARCE (1959; monkeys); LANGREHR (1960a, 1960b; cats and dogs); NEIL and JOELS (1961; cats); SLEIGHT and WIDDICOMBE (1965a; dogs); GUPTA et al. (1966; dogs); WELLHÖNER and HAFERKORN (1966; cats); HAKUMÄKI (1970; cats); ARNDT et al. (1971a, 1971b; cats).

Identification. Under *normal* conditions with the chest intact the chief characteristic of these endings is that they fire one main burst of impulses (per cardiac cycle) (Figs. 5D, 7 and 8) which begins *most commonly* a little after the beginning of the upstroke of the aortic pressure curve (Fig. 5D) a timing that can be suitably described as late systolic (Table 3). The latency of the burst of impulses from the Q wave of the e.c.g. is about 70 to 210 msec in the cat (PAINTAL, 1953a); in most endings it is about 140 msec. Table 3 provides useful information for identification of the endings with the chest intact. This table includes two additional criteria for identification which have not been described before. One is that, unlike the aortic baroreceptors in which the position of the first impulse is relatively fixed in relation to the Q wave in spite of marked respiratory fluctuations in the intensity of discharge, the first impulse of the burst in type B receptors varies markedly, and as each cycle is observed on the oscilloscope it can be seen that the first impulse of each subsequent burst appears earlier and earlier as inspiration progresses (Fig. 5D).

As shown in Table 3 the respiratory fluctuations of left atrial receptors are similar in timing to those of the aortic baroreceptors, as are their responses after

Table 3. *Summary of useful criteria for identification of various types of rhythmically active cardiovascular afferent fibres*

Type of receptor	Main volley of impulses	Effect of inspiration on		Effect of inflation of lungs on impulse activity	Return of activity after inflation	Effect of ectopic atrial stimulus	
		Impulse activity	position of 1st impulse			activity during atrial contraction	activity during premature ventricular contraction
Right atrial type B	late systolic	early increase	advances earlier	reduced or abolished	early	present	reduced
Left atrial type B	late systolic	late increase	advances earlier	reduced or abolished	late	present	increased
Aortic baroreceptors	systolic	late increase	fixed	reduced abolished	late	absent	reduced or abolished
Left ventricular pressure receptors	early systolic	little or no change	fixed	little or no change	late	absent	reduced or abolished
Right ventricular pressure receptors	early systolic	little or no change	fixed	little or no change	indefinite	absent	reduced or abolished
Right atrial type A	pre-systolic	slight increase	fixed	slightly reduced	early	present	little or absent
Left atrial type A	pre-systolic	slight increase	fixed	slightly reduced	late	present	little or absent

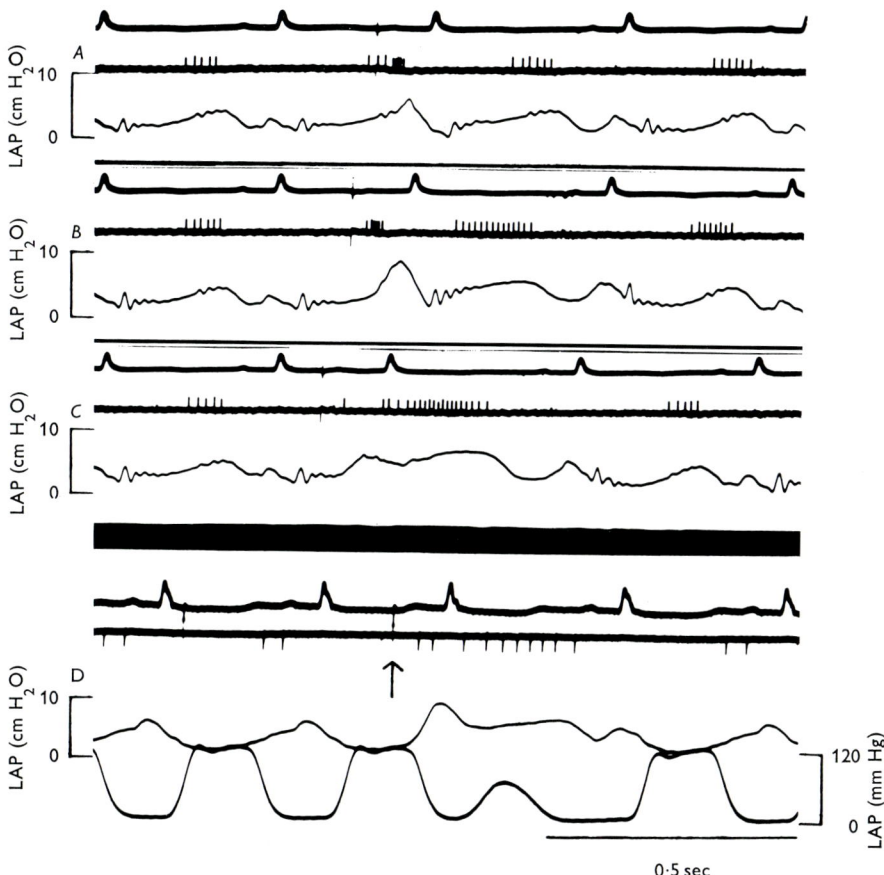

Fig. 7 A–D. Effect of premature ventricular contractions on the activity of two left atrial type B receptors. The first cycles in A–C show the normal pattern of activity in the endingl An electrical stimulus was applied to the atrium in the second cycle of each of the 3 traces. In A and B the resulting atrial contraction stimulated the ending markedly. The three records show that the earlier the application of the stimulus to the atrium in the 2nd cycle, the greater is the increase in activity during premature ventricular contraction; this is related to the rise in left atrial pressure. Total conduction in this fibre was 10.0 msec. D is a record from another fibre along with a left ventricular pressure tracing. Total conduction time in this fibre was 11 msec. From above downwards in each record: e.c.g., impulses in a fibre, left atrial pressure and in D, left ventricular pressure (PAINTAL, 1963b)

releasing a maintained inflation. Consequently some difficulty in identification may arise if the phasic changes in atrial volume are large leading to an earlier onset of the burst of impulses. However a reliable method for differentiating the left atrial endings from the aortic ones is to produce ectopic beats by stimulating the right atrium through an electrode passed down a catheter with its tip in the right atrium. The left atrial endings (like the right atrial ones) will be stimulated by ectopic atrial contractions (Fig. 7). Moreover in the case of left atrial endings the activity is increased during the period of premature ventricular contraction

(PAINTAL, 1963b) (Figs. 7 and 9). On the other hand the activity of aortic baroreceptors is either greatly reduced or abolished. This method also distinguishes the left atrial endings from the right atrial ones because unlike the former, activity of most of the right atrial type B receptors is reduced (Figs. 8 and 9). This technique is equally applicable in the case of dogs because it was found by FAHIM and GUPTA (1970) that endings identified as belonging to a particular atrium with intact chest were in fact located in that chamber, after opening the chest, by known methods (PAINTAL, 1953a; COLERIDGE et al., 1957). Indeed, of all the methods, this one (i.e. ectopic stimulation) is the most convenient. However

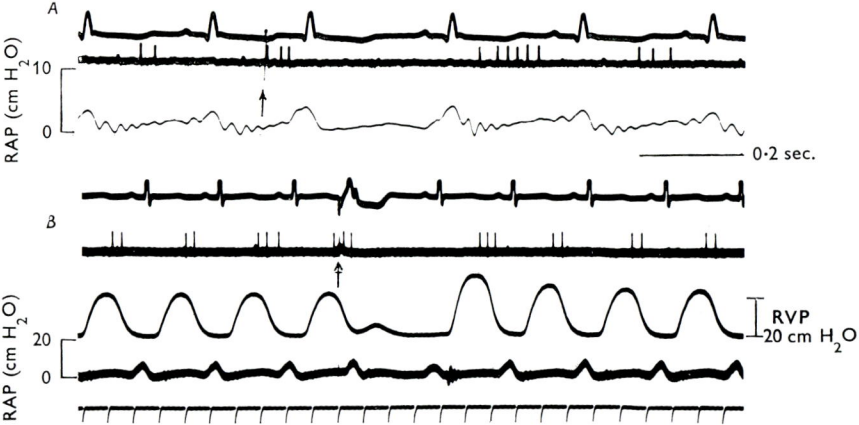

Fig. 8A and B. Records showing that the activity of two right atrial type B receptors (A and B respectively) is abolished during premature ventricular contraction produced by stimulating the right atrium at arrows. Note increased activity in post-extrasystolic beat in A and that right atrial pressure falls during premature ventricular contraction in both A and B. From above downwards in A: e.c.g., impulses in a fibre and right atrial pressure; in B: e.c.g., impulses in a fibre, right ventricular pressure, right atrial pressure and 0.1 sec time marks. Total conduction time in the fibre of A was 9.5 msec and in that of B, 14 msec (PAINTAL, 1963b)

the right atrial endings can also be distinguished from aortic baroreceptors by the relatively rapid return of their activity after releasing a maintained inflation (Table 3). Although as pointed out before (PAINTAL, 1953a, 1963b; COLERIDGE et al., 1957, 1964b) one cannot be certain about the location of an ending unless the chest is opened and the ending is localized to a particular chamber preferably by punctate stimulation (COLERIDGE et al., 1957, 1964b) one can now have considerable confidence about the location of a particular type B receptor by its response to ectopic stimulation along with other evidence (Table 3). It is therefore no longer necessary to open the chest in order to assign an ending to a particular chamber in cats and dogs because the likely possibility of making a mistake is less than 1 in 16 in dogs (FAHIM and GUPTA, 1970).

This expectation is valid only under normal conditions of blood pressure, respiration and cardiac output and in the absence of transfusions, injections of drugs etc.; otherwise difficulties are bound to arise. For instance, imagine the confusion that would result if one were to isolate and identify fibres of type B atrial receptors after their pattern of discharge had been altered as grossly as that shown in Fig. 3B and C in PAINTAL (1953b). As shown in Fig. 4 of that paper (i.e. PAINTAL, 1953b) such changes are due to changes in intra-atrial pressure. The dependence of the discharge on intra-atrial pressure (i.e. the v wave) is illustrated in Fig. 7. Under certain conditions the v wave is altered in such a way that the burst of impulses produced (in type B receptors) by it gives the impression that one is dealing with an aortic baroreceptor.

Pattern of Discharge. Typically the discharge develops in a crescendo fashion (Figs. 5D and 7) unlike that in most aortic baroreceptors, in which the peak

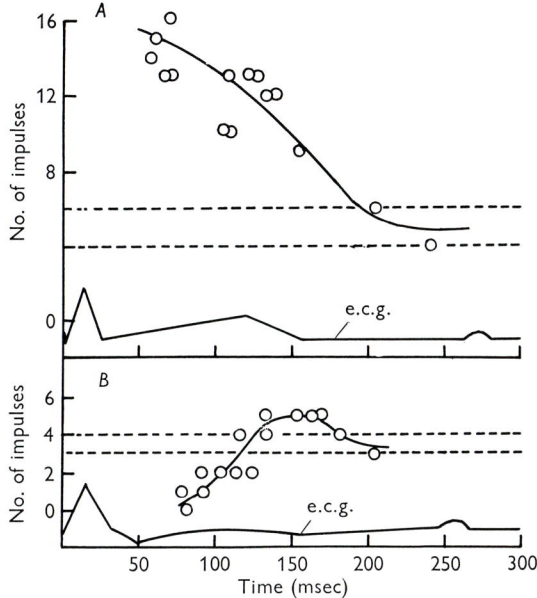

Fig. 9. Graphs showing how the activity of left (A) and right (B) atrial type B receptors during premature ventricular contraction varies with the position of the ectopic stimulus in the cardiac cycle. Abscissa indicates the interval between the Q wave of the e.c.g. and the stimulus to the right atrium. The ordinates indicate the number of impulses during premature ventricular contraction. The interrupted lines represent the range of normal activity in the absence of any stimulation in the two fibres respectively (PAINTAL, 1963b)

frequency is attained initially and this starts to decline soon after the beginning (Fig. 5A). The pattern of discharge is closely related to the v wave of the atrial pressure tracing (Fig. 7) and it retains this characteristic in spite of marked fluctuations in the intensity of discharge. When the heart rate is high the last few impulses of the burst may appear after the p wave of the e.c.g. giving the appearance of the presence of an a burst of impulses. This is due to the v wave not having come to an end before the start of atrial contraction. Reducing the heart rate by stimulating the vagus abolishes these pseudo a impulses. However a true a burst can appear under unusual conditions (see PAINTAL, 1963a) but this, unlike the burst of impulses in type A receptors, is of relatively low frequency. It has been pointed out that such impulses which appear under unusual haemodynamic conditions e.g. severe haemorrhage (PEARCE et al., 1956) (see Fig. 21 in PAINTAL, 1963a) should not be given undue importance for identifying the endings, because a restoration of normal conditions will invariably bring back the normal pattern of activity (PAINTAL, 1963a).

Natural Stimulus. By studying the responses of the left atrial endings in the isolated in situ left atrium it was demonstrated that these endings were slowly adapting stretch receptors that respond to pulsatile changes in atrial filling and that their activity is linearly related to the volume of fluid in the atrium (PAINTAL, 1953a). These observations, indicating that there were volume receptors in the heart, proved to be of considerable interest in connection with studies of reflex mechanisms concerned with regulation of fluid volume (GAUER et al., 1956; HENRY et al., 1956) and the results obtained in the cat were rapidly confirmed in the dog (HENRY and PEARCE, 1956; COLERIDGE et al., 1957; MÜHL et al., 1956; KRAMER, 1959). However, in the absence of a suitable method of measuring changes in atrial volume, it was not possible to correlate the intensity of the natural stimulus with the activity of the endings in normally beating hearts even though it was recognised that the activity was related to the v wave of the atrial pressure curve (PAINTAL, 1953a; HENRY and PEARCE, 1956). Recently HAKUMÄKI (1970) has also demonstrated the close dependence of the development of the v wave and the activity in the ending.

It has been pointed out that the v wave of the atrial pressure curve (provided the pressure is recorded satisfactorily) reflects the gradual increase in atrial volume as a result of atrial filling (PAINTAL, 1963a. 1963b). This assumption is reasonable because (1) the pressure in the atrium is linearly related to atrial volume in isolated or intact heart preparations within normal limits of filling (OPDYKE et al., 1948; LITTLE, 1949, 1960; IRISAWA et al., 1959) and (2) it can be assumed that the compliance of the atrium does not change appreciably after relaxation starts in it following atrial systole (PAINTAL, 1963a, 1963b). Accordingly, the amplitude of the v wave serves as a convenient quantitative index of the amount of blood entering the atrium during ventricular systole. It was therefore not surprising to find that the activity of type B receptors was related to the amplitude of the v wave (Fig. 10B). This relationship held fairly well under different conditions unlike the peak pressure of the v wave or the mean atrial pressure whose relationship to the activity depended on whether atrial pressure was rising or falling (Fig. 10C) (PAINTAL, 1963b). It should be pointed out that actually the activity of any particular ending (i.e. number of impulses per beat) should depend not only on the amplitude of the v wave but also on the rate of rise of pressure [because there is evidence that the endings accommodate to slowly rising stimuli (PAINTAL, 1963b)] and the duration the pressure remains raised (PAINTAL, 1963a). The best relation (i.e. least scatter) therefore exists between peak pressure minus threshold pressure and the activity in the ending (Fig. 10A). This is particularly true in measurements under conditions of equilibrium (FAHIM and GUPTA, 1970).

A striking feature of these endings is that they are silent during atrial contraction and the a wave that accompanies it i.e. at a time when the pressure in the atrium may be the highest (Fig. 11). This behaviour persists even during intense stimulation of the endings (PAINTAL, 1953a). [See also Fig. 1 in KRAMER

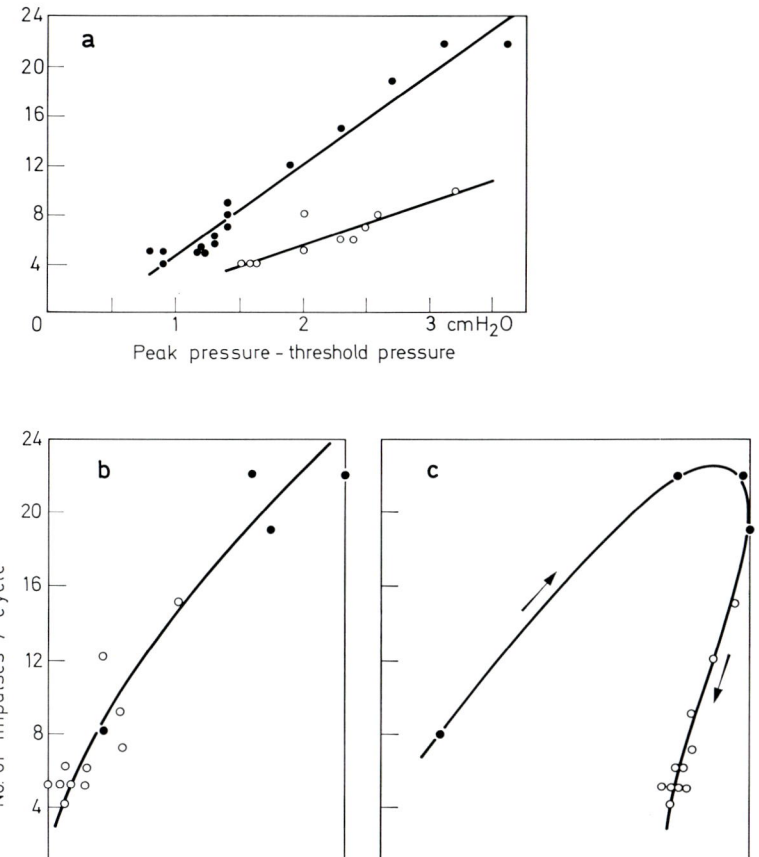

Fig. 10 A–C. Graphs in A show that the discharge in two left atrial type B receptors is related to atrial volume (see text) (PAINTAL, 1963a). B and C show the relation of activity in a type B atrial receptor to the amplitude of the **V** wave (B) and peak pressure of the **V** wave (C). The filled and open circles in B and C are the values obtained when atrial pressure was rising and falling respectively. B and C show that the effective stimulus for the receptor is better represented by the amplitude of the **V** wave. (From PAINTAL, 1963b)

(1959) and Fig. 4 in GUPTA et al. (1966).] It is due to the fact that the volume of the atrium (which is their natural stimulus) is low at this time and it falls still further when the atrium empties (PAINTAL, 1953a). However when there is little or no emptying the endings are stimulated (PAINTAL, 1953a). This is well illustrated in Fig. 11 which shows the effect of rise in intra-atrial pressure with and without emptying of the atrium.

Effects of Ectopic Contractions. Unlike normal atrial systoles, those produced by an ectopic focus yield significant bursts of impulses partly perhaps because the atrium contracts against closed a-v valves (Figs. 7 and 11). However the most

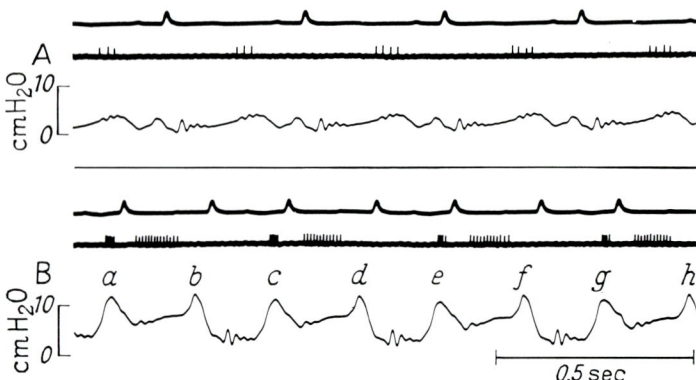

Fig. 11 A and B. Impulses in a fibre, from left atrial type B receptor, with total conduction time of 10 msec during normal heart beats (A) and during arrythmia (B). The chest was open. In B atrial contractions marked a, c, e and g which stimulated the ending markedly were associated with little emptying of the atrium as shown by the raised post-atrial systolic pressure as compared with the pre-systolic pressure. During contractions b, d and f the atrium emptied as shown by the lower post-atrial systolic pressure (PAINTAL, 1963a)

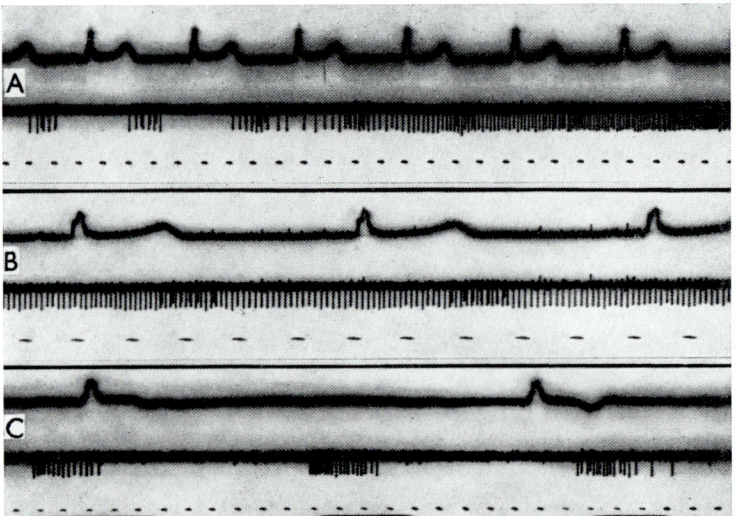

Fig. 12 A–C. Impulses in two fibres from left atrial type B receptors. Veratridine, 22 µg was injected into the right atrium 6.8 sec before beginning of record A. B was recorded 5.6 sec after A. Note that the fibre with the smaller spike was not stimulated by veratridine. This stimulated the ending with the larger spike markedly; C, shows the response of the same receptor to touching a part of the left atrium after cutting away the ventricles
(PAINTAL, 1955)

helpful observation in this connection is that the discharge of the left-sided endings is increased during the period of premature ventricular contraction and that most of the right-sided ones is decreased (Figs. 7 and 8). The increase or decrease depends on when the ectopic stimulus is applied; the earlier it comes, the greater is the change (Fig. 9). These effects can be explained by the changes in atrial filling during premature ventricular contraction (PAINTAL, 1963b). Apart from the fact that the observations on the left-sided receptors have helped to dispose of the suggestion that the activity of type B atrial receptors is produced by ventricular contractions (LANGREHR, 1960a, 1960b), they have provided an apparently reliable method for distinguishing the right atrial type B endings from the left-sided ones since in a random sample of 16 endings in dogs FAHIM and GUPTA (1970) were able to show that the endings assigned to a particular chamber by this method were all located in that chamber after opening the chest. This series of experiments were done after FAHIM and GUPTA had already demonstrated in an earlier series that all left or right atrial endings responded to ectopic stimulation in the same way as those in the cat (PAINTAL, 1963b).

Effect of Anoxia. It is noteworthy that like all other visceral mechanoreceptors e.g. pulmonary stretch receptors (ADRIAN, 1933), carotid baroreceptors (BRONK and STELLA, 1935), gastric stretch receptors (PAINTAL, 1954) these endings are not visibly affected by anoxia. In fact they continue to respond for several minutes after circulatory arrest (PAINTAL, 1953a; COLERIDGE et al., 1957).

Effects of Chemical Substances. Certain substances, notably phenyl diguanide, which stimulate the endings of non-medullated fibres do not, as expected (see PAINTAL, 1964), stimulate the type B or type A atrial receptors (PAINTAL, 1953b). The same is true for 5-HT (MOTT and PAINTAL, 1953). Such substances do not seem to influence the first node i.e. the regenerative region of the ending (Fig. 1B). On the other hand as shown in Fig. 12 some of the endings are markedly stimulated by veratrum alkaloids (PAINTAL, 1955, 1957; KRAMER, 1959; NEIL and JOELS, 1961). From the evidence available (i.e. PAINTAL, 1955, 1957) one can conclude that the endings are stimulated and desensitized as in the case of ventricular receptors. Aconitine has similar effects (WELLHÖNER and HAFERKORN, 1966). In small doses (i.e. about 22 μg veratridine or 200 μg veriloid) enough to produce the BEZOLD-JARISCH effect only about a third of the left atrial endings of the cat and none of the right atrial ones were stimulated (PAINTAL, 1955). In a subsequent study (PAINTAL, 1957) it was reported that the right atrial endings could be stimulated by larger doses of the alkaloids e.g. 175 μg germitrine. NEIL and JOELS (1961) found that 100 μg veratrine which is a mixture of alkaloids (like veriloid) stimulated the left as well as the right atrial receptors. One wonders whether smaller doses of the substance would have brought out the difference in the relative sensitivity of the endings. Alternatively it is possible that the mixture (veratrine) contains alkaloids that are different from veratridine in their effects.

In view of the fact that the endings in the two chambers are identical in every way examined i.e. histologically, conduction velocity of their fibres (Table 1) and their responses to natural stimulation, it was suggested that the differences in their responses to veratridine or veriloid could be due to the fact that the right-sided ones were possibly exposed to a higher pCO_2 resulting from the fact that the endings are subendocardial in location and could therefore be influenced by the higher pCO_2 of the mixed venous blood in the right atrium (PAINTAL, 1964). This is a reasonable expectation because oxygen does reach even the muscle fibres of the isolated atrium placed in a bath (BROOKS et al., 1955).

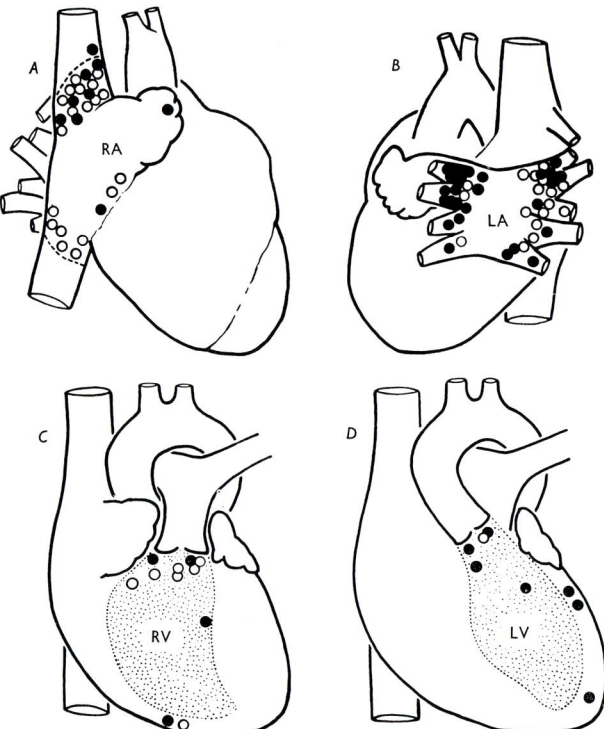

Fig. 13 A–D. Diagrammatic representation of the position of ninety-two cardiac receptors each of which had been located by electrophysiological means. Each receptor is indicated by a circle: open circle, afferent fibre in the right vagus; filled circle, afferent fibre in left vagus. A and B show location of atrial receptors and C and D location of ventricular receptors with a cardiac rhythm (i.e. presumably ventricular pressure receptors—see text). In C and D the stippling represents the cavities of the right and left ventricles respectively (COLERIDGE et al., 1964b)

The effect of the alkaloids can be reversed by the injection of calcium chloride so that the continuous discharge produced by the alkaloids (e.g. germitrine) is reconverted into rhythmic activity (PAINTAL, 1957).

An interesting feature is the long latency before the excitatory effect sets in. In the case of the left atrial receptors it is 19 to 70 sec following injection of veriloid. This is much longer than the latency for excitation of ventricular receptors in which it is 8–15 sec following veriloid (PAINTAL, 1955). Local differences in the circulation would need to be taken into consideration in seeking an explanation for the difference.

Location of the Endings. It is possible to locate the position of an ending by punctate stimulation. Using this method it was found that all the 10 endings (6 right and 4 left atrial) were located in the posterior part of the atrial wall of the cat; none of these endings was located in the auricular appendage (PAINTAL, 1953a). These observations were subsequently confirmed by COLERIDGE et al. (1957) and LANGREHER (1960a) in the dog. In a further study by COLERIDGE et al. (1964b), out of the 73 endings located in the right and left atria only one was located in the auricular appendage; Fig. 13 A and B shows the location of their endings in the dog. This distribution of the endings is in agreement with the earlier histological findings of Nonidez (1937) in the cat.

B. Type A Atrial Receptors

Identification. It is least confusing if the identification of type A receptors is carried out according to the criteria described earlier (Paintal, 1963a). According to this, a type A receptor is one which has only one burst of impulses in each cardiac cycle coincident with the a wave of the atrial pressure curve apart from occasional one or two impulses in any other part of the cycle (Fig. 14). This definition excludes endings with both prominent a and v bursts which are better segregated for working purposes into a separate group- the intermediate type. These probably represent extreme variations of type A or type B receptors. What they actually are would depend on their natural stimulus (Paintal, 1963a).

Because of their characteristic burst of activity just after the p wave of the e.c.g. differentiation from other cardiovascular endings seldom presents a problem except in the case of certain pulmonary stretch receptors with a prominent a burst of impulses (Coleridge et al., 1957; Bianconi and Green, 1959c); such endings can be easily identified by inflating the lungs, upon which the discharge will become continuous. However, unlike the type B receptors it is not easy to differentiate the right atrial type A receptors from the left atrial ones because ectopic stimulation, so convenient for distinguishing right atrial type B endings from left atrial endings is not of much help, for although each stimulus produces a burst of impulses due to atrial contraction (Fig. 14C), the discharge in the case of left atrial endings is not increased (unless the left atrial pressure rises considerably) during the period of ventricular systole (Fig. 14C). In any case, even when activity does increase during premature ventricular contraction in left atrial endings the increase is not conspicuous enough to enable identification of the ending without opening the chest as is possible in the case of type B receptors. The responses following release of maintained inflation although qualitatively similar to the corresponding type B atrial receptors (Table 3) are quantitatively not large enough to enable one to assign ending to a particular chamber with certainty. For certain location it is therefore safest to open the chest and locate the ending by occluding the pulmonary artery, punctate stimulation etc. as described in the case of type B receptors (Paintal, 1953a; Coleridge et al., 1957; Coleridge et al., 1964b).

Natural Stimulus. The natural stimulus for the type A receptors still remains to be established although there are some indications as to what it could be (Paintal, 1963a). It would appear that earlier work in this field by Whitteridge (1948), Dickinson (1950), Struppler and Struppler (1955) has been concerned mainly with studies on intermediate type atrial receptors (i.e. those with both prominent a and v bursts of impulses and possibly a c burst as well (see Paintal, 1963a) Such studies by Whitteridge (1948) showed that the activity in his "venous fibres" was related to the a, c and v waves of the venous pressure curve. Dickinson (1950) found a close correlation between intra-atrial pressure and the frequency of discharge in an experiment in which the heart was in a nodal rhythm.

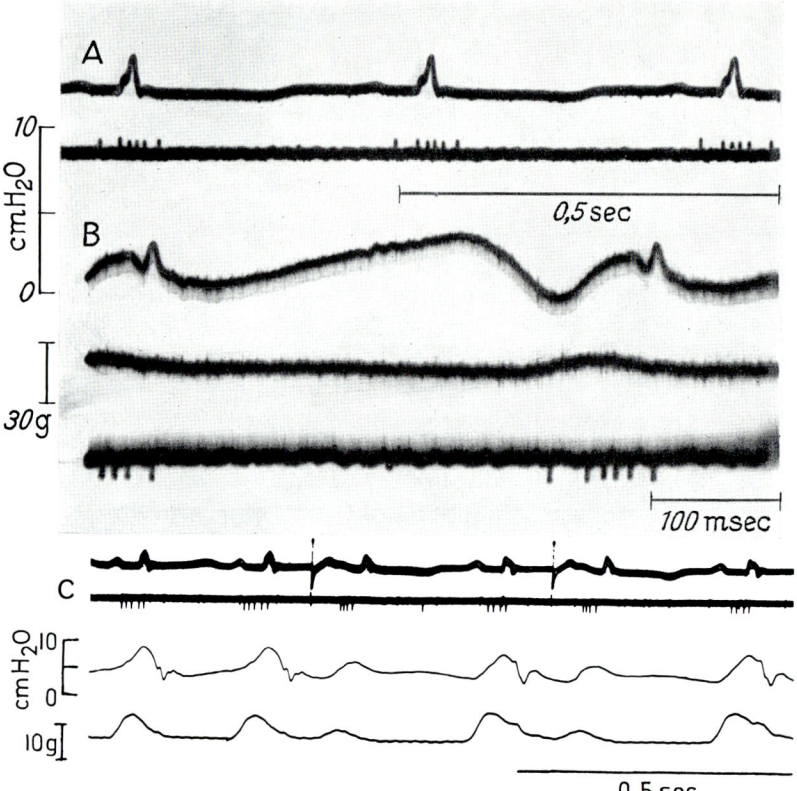

Fig. 14 A–C. Impulses in two fibres from atrial type A receptors. A, is a section of a continuous record with e.c.g. and impulses in a fibre from a left atrial receptor; total conduction time 12 msec. B is a sweep consisting of (from above downwards) left atrial pressure, left atrial myogram indicating contraction of the left atrium) and impulses in the fibre. The first burst of impulses in the sweep (i.e. B) corresponds to the first burst in A. Note the first impulse of each burst is attributable to atrial contraction and not the *a* wave because it appears before the *a* wave (PAINTAL, 1963a). C is a record of activity in a right atrial type A receptor during normal and ectopic beats produced by stimulating the right atrium at the stimulus artifacts. The lower two traces represent right atrial pressure and the myogram. Total conduction time in the fibre was 8 msec. Note increased frequency of discharge during ectopic contractions in spite of smaller tension and *a* wave. Such responses suggest that these contractions occurred against partly closed a.v. valves as revealed by the pressure tracing (PAINTAL, 1962 unpublished observations)

Later STRUPPLER and STRUPPLER (1955) concluded that the activity of the endings was related to the rate of rise of intra-atrial pressure.

Some observations relating to the natural stimulation of type A receptors were reported earlier (PAINTAL, 1963a). Two significant observations were mentioned. One was that if total conduction time in the nerve fibre is taken into account then it is clear that in several endings the first one or two impulses of the burst set in before the *a* wave starts and these impulses therefore cannot be due to the rise in pressure (Fig. 14A). Clearly these impulses generated as a result of atrial systole must have been initiated by some other factor—presumably an increase in tension in the wall during contraction. The second important observation leading to the same conclusion was that a burst of impulses could be produced by each atrial systole in the absence of any pressure in it, since such activity was generated with the a-v junction slit widely and the atrium cut open. However it was also apparent that although a rise in intra-atrial pressure is not indispensable for stimulating the endings it is nevertheless an important stimulus *normally* because the later impulses of the burst are apparently attributable to the *a* wave (PAINTAL, 1963a) (Fig. 14). It is very likely that both atrial systole and rise in pressure (*a* wave) contribute to the stimulation of the ending by increasing the tension and thereby stretching the endings (PAINTAL, 1963a). The dependence of the frequency of discharge on intra-atrial pressure in the isolated *in situ* atrium is clearly revealed by the results of HOMMA and SUZUKI (1966).

The fact that these endings behave like stretch receptors almost exactly like the type B atrial receptors when subjected to a sinusoidal stimulus has been well brought out by the recent experiments by ARNDT *et al.* (1971a). In these experiments ARNDT *et al.* removed the ventricles and after fixing the other atrium, applied a sinusoidal stretch to a strip of the atrium (1.5 cm broad and about 2 to 3 cm long) containing the type A (or type B) receptor earlier identified with intact chest. Fig. 15 shows that there is no difference in the responses of type A and type B receptors when subjected to this type of stimulus and that *under these conditions* the type A do not show greater adaptation rates than the type B receptors. The hyperbolic relation between the frequency of stretch and the number of impulses per cycle is to be expected since the endings behave like slowly adapting stretch receptors under these conditions. In fact the number of impulses per cycle would be even greater if the frequency of stimulation were less than 1 cycle/sec. In view of the close correspondence between the responses of type A and type B endings as assessed in various ways, ARNDT *et al.* (1971a) came to the conclusion that the characteristic differences between type A and type B receptors in the intact animal must be due to differences in the location of the endings in the atria. It is therefore necessary to carry out investigations in the cat along the lines of work by COLERIDGE *et al.* (1957, 1964b) in order to find out whether the location of type A endings differs from that of the type B. The location of what may be regarded as type B endings have been well established by COLERIDGE *et al.* (1964b) in the dog (Fig. 13A and B).

Function of Type A Receptors. In view of the fact that the natural stimulus for these endings is not clearly established, it is not possible to state what the precise function of these endings is. However it is possible to state that they are definitely not volume receptors like the type B endings (PAINTAL, 1963a; ARNDT *et al.*, 1971b). One view is that these endings signal the tension in the atrial wall which will increase when there is atrial contraction or rise in pressure. On this basis the absence of impulses during the *v* wave normally is presumed to be due to the tension in the wall being probably insufficient at that time. However under unusual conditions when the amplitude of the *v* wave is much increased the tension may rise sufficiently to produce a few *v* impulses (PAINTAL, 1963a).

Another view put forward by ARNDT and his co-workers recently is that the function of type A receptors is to signal heart rate (BAMBRING *et al.*, 1969; ARNDT *et al.*, 1971 b) because they found that the activity of these endings (whether measured as number of impulses per beat, peak frequency of discharge, or average frequency of the burst) remained remarkably constant in spite of large changes in the amplitude of the *a* wave, rate of rise of pressure or the pre (atrial) systolic pressure. All their observations were made on spontaneously breathing cats and they varied the atrial pressure by infusion, bleeding, adrenaline or artificial stimulation of the heart. In two cases they observed that when the amplitude of the *a* wave was increased greatly there was marked increase in the number of impulses per burst. However, this occurred during abnormal contractions produced by artificial stimulation of the heart which led to atrial contractions against a-v valves.

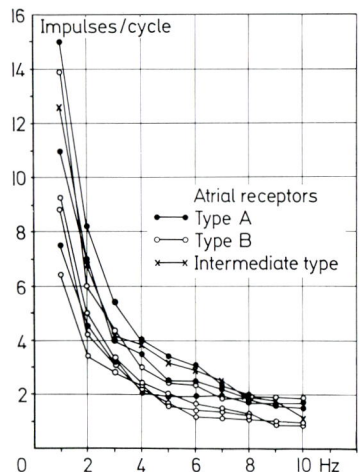

Fig. 15. Relation between frequency of sinusoidal stretch (abscissa) to a strip of atrium containing a type A or type B atrial receptor (or an intermediate type) and the number of impulses generated per cycle (ordinate). The responses of 8 receptors are shown. Each point represents the number of impulses averaged over 10 cycles (ARNDT *et al.*, 1971a)

The type A receptors studied by ARNDT *et al.* (1971b) are noteworthy because in all 16 of them a *v* burst of impulses was not produced by infusion, bleeding or injection of adrenaline. This is not unexpected in the cat but it is important to mention this point because it is sometimes held that conversion of one type of pattern (e.g. A type) in to another is a *common occurrence*.

Since the activity in type A receptors is generated by atrial contraction (whose frequency depends on the heart rate) and since ARNDT *et al.* found that the activity of these endings remained constant in spite of variations in atrial mechanics they came to the conclusion that the type A receptors might signal heart rate (BRAMBING *et al.*, 1969; ARNDT *et al.*, 1971b); clearly the impulses/sec averaged over several cycles would depend on the heart rate if the number of impulses/burst remained constant. This suggestion merits serious consideration. However it should be remembered that unlike the cat there are relatively few type A receptors in the dog (Table 2) so that it would appear that such a mechanism (if specific for signalling heart rate) would not be of much significance in the dog (and perhaps the monkey). This point is particularly relevant in connection with the observation of HAKUMÄKI (1970) (who concluded that the activity of type A receptors causes reflex increase in sympathetic activity) because he has tried to explain the observations of certain investigators (e.g. COLERIDGE and LINDEN, 1955) on the dog in which there are few type A receptors (Table 2) on the basis of his own observations on type A receptors in the cat. In fact from what is already known it would be very surprising if the observations of HAKUMÄKI are confirmed in the dog.

VII. Ventricular Receptors

It is now certain that there are at least two types of sensory receptors in the ventricular wall, the ventricular pressure receptors (Figs. 16, 17) with medullated fibres (PAINTAL, 1955) (Table 1) and epicardial receptors (Fig. 19) with non-

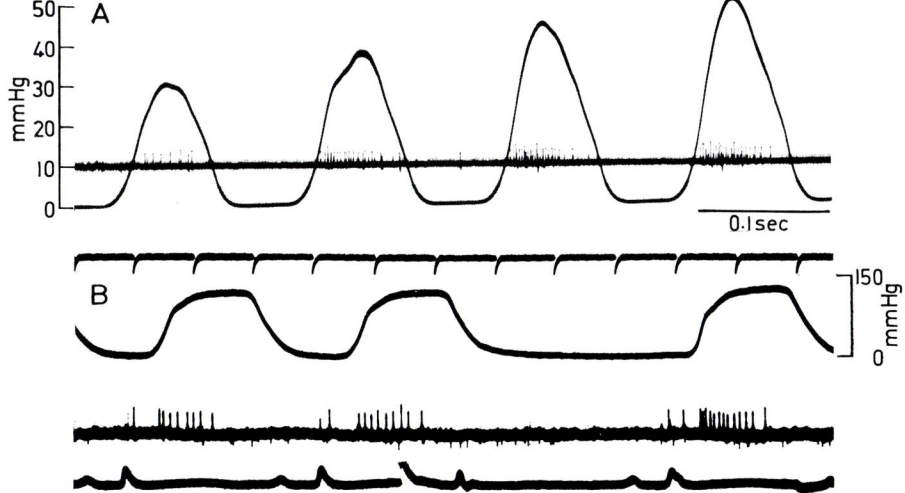

Fig. 16 A and B. Impulses in ventricular pressure receptors. A, shows the impulses in a fibre from a ventricular pressure receptor from a frog's heart following occlusion of the ductus arteriosus. The record shows that the activity is related to the magnitude of the pressure (KOLATAT et al., 1957). B shows the response of a similar ending in the left ventricle of the cat. The total conduction time in this fibre was at least 17 msec. If an allowance for this is made then it is clear that the first 3 impulses of the last early systolic burst (i.e. post extrasystolic beat) must have been initiated by ventricular contraction before the rise in intraventricular pressure. This record shows that the activity of the ending depends on the strength of ventricular contraction or the rise in pressure produced by it. Note the receptor is silenced during premature ventricular contraction following the application of a stimulus indicated by the artifact in the e.c.g. trace (lowest tracing). The premature contraction was too weak to produce a rise in pressure (PAINTAL, 1962 unpublished observations). Uppermost trace in B, 0.1 sec time marks

medullated fibres (COLERIDGE et al., 1964b; SLEIGHT and WIDDICOMBE, 1965a). This knowledge presents a significant advance since writing a previous review on the subject (PAINTAL, 1963a) because at that time it was not known what the nature of the fibres (and the location of their endings) encountered by JARISCH and Zotterman (1948) were (see also Schaefer, 1950). It would now appear that the small-spike impulses recorded by them on application of artificial stimuli such as pinching the ventricle may have arisen from epicardial receptors with non-medullated fibre. However it should be noted that so far there is no report describing the responses of epicardial receptors in single fibres in the cat on which Jarisch and Zotterman (1948) did their experiments; both the studies on epicardial receptors relate to the dog (Coleridge et al., 1964b; Sleight and Widdicombe, 1965a). No doubt such endings must exist in the cat. On the other hand it is known that there are ventricular pressure receptors in the cat (Paintal, 1955; Neil and Joels, 1961), frog (Kolatat et al., 1957) and the dog (Coleridge et al., 1964b; Sleight and Widdicombe, 1965a).

It needs to be pointed out that the earliest work in this field was published by Amann and Schaefer (1943) on multifibre preparations. Subsequently Whitteridge and Dickinson reported observations on single fibres (Whitteridge, 1947, 1948; Dickinson, 1950) but the location of their endings were not established precisely by occlusion of vessels and application of local mechanical stimuli—the value of which technique came to be realized in the subsequent work on type B atrial receptors (Paintal, 1953a). Coleridge et al. have emphasized the importance of this method of localization and indeed they give it overriding importance in deciding the location of an ending in the heart (Coleridge et al., 1964b). While there is justification for this view, in the view of the author it is equally important to take into consideration the actual stimulus that is responsible for generating a particular cyclical discharge (see pp. 39–40). In this connection it is important not to assume that total conduction time is negligible. For example, a burst of impulses co-incident with the a wave of the atrial pulse may be generated by a ventricular receptor that is stimulated by the pre-atrial systolic filling phase of the ventricle, but these impulses appear after a delay owing to the drop in intra-thoracic temperature that occurs after opening the chest-particularly on the exposed surface wheir nerve branches are seen to run.

A. Ventricular Pressure Receptors

Identification. The main characteristic feature of these receptors is that they fire an early systolic burst of impulses before the aortic valves open (Figs. 16 and 17). However since there are occasional atrial receptors with an early systolic discharge (Paintal, 1955) it is necessary to exclude these. This is easily done by stimulating the right atrium electrically through an electrode passed down the atrial catheter. The ectopic atrial contraction will produce a burst of impulses in atrial type A or type B receptors (Figs. 7 and 13) but not in ventricular pressure receptors. Moreover, the expected early systolic discharge during the resulting premature ventricular contraction will be reduced or absent (Fig. 16B) (Table 3). In this respect these endings are quite different from some of those regarded as ventricular receptors by Whitteridge (1947, 1948) because in his endings a burst of impulses was produced during extra systoles. In the cat further support for identification will be available if the ending is stimulated by about 22 μg veratridine (i.v.) since all the ventricular pressure receptors isolated so far are stimulated by veratridine (Paintal, 1955) or veratrine (100 μg; Neil and Joels, 1961).

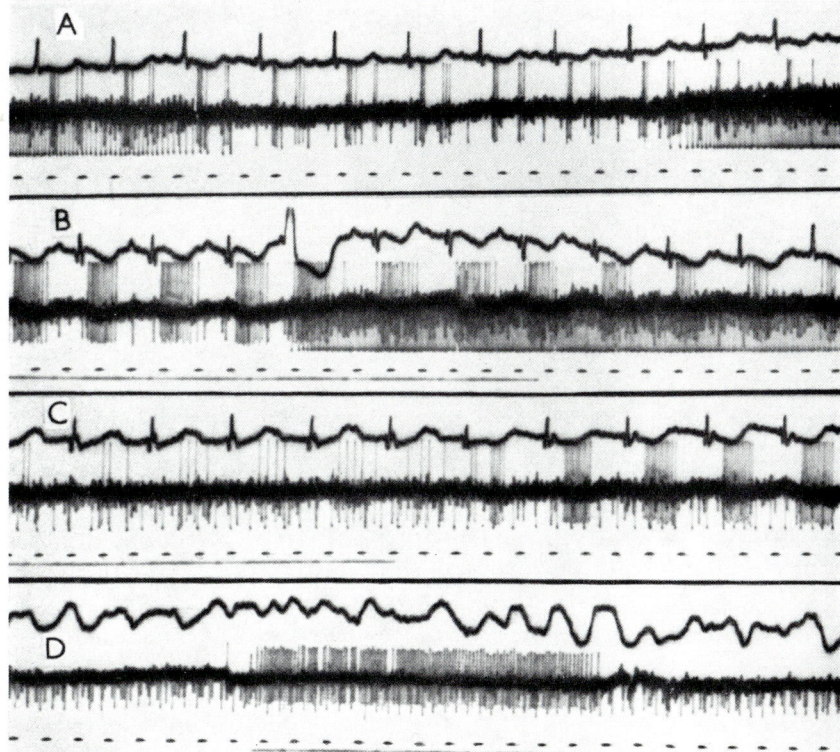

Fig. 17 A–D. Impulses in a right ventricular fibre (large spiked fibre above base line). A, is a normal record with open chest and artificial ventilation. The large spikes below the base line are from a pulmonary stretch fibre. B and C show respectively the effect of occluding the pulmonary artery and right a-v junction; end of signal indicates end of occlusion. In D the right ventricle was pressed during signal after clamping the pulmonary artery and right a-v junction. Responses from this fibre ceased suddenly when the ventricle was incised transversely across its middle. From above downwards in each record; e.c.g., impulses in fibres; 0.1 sec time marks, and signal (PAINTAL, 1955)

After opening the chest the location can be determined by raising the pressure in the ventricle by occluding the aorta or pulmonary artery when the discharge will increase greatly (Fig. 17). However in order to be certain (as emphasized by COLERIDGE et al., 1964b) it is necessary to isolate the segment of the ventricular wall in which the ending is believed to lie and to ascertain that the ending is stimulated by squeezing only this segment. Finally the discharge so produced should cease abruptly when this segment is separated from the rest of the heart (PAINTAL, 1955) (see also Fig. 17).

Although the evidence provided by ectopic stimulation of the right atrium is invaluable for location of a suspected ventricular pressure receptor with an early systolic discharge (Fig. 16B) and makes it almost unnecessary to open the chest for concluding that the ending is located in a ventricle, it nevertheless does not enable one to establish in which particular ventricle the receptor is located.

Moreover, this method of selection would exclude those endings which do not have a prominent early systolic burst.

Natural Stimulus. There are a number of receptors whose activity appears to be directly related to the pressure in the ventricle (PAINTAL, 1955) and whose pattern of discharge seems to follow the ventricular pressure curve (Fig. 16). In the case of some endings the frequency of discharge appears to be directly related to the rate of rise of pressure as was observed by KOLATAT et al. (1957) in the case of ventricular receptors of the frog. Therefore even though the precise natural stimulus (i.e. whether it is contraction of the ventricle or rise in pressure, or both) remains to be established it is reasonable to refer to them as ventricular pressure

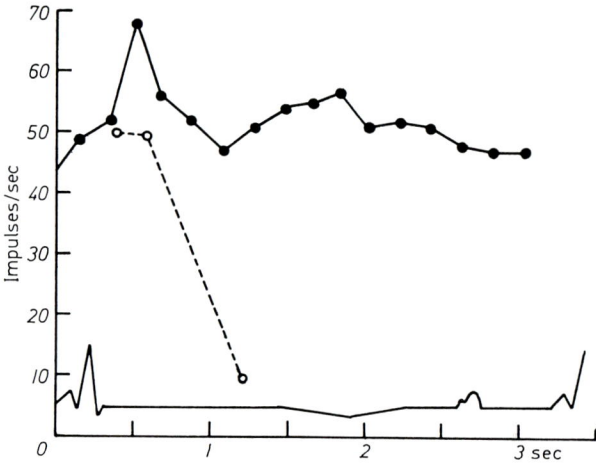

Fig. 18. Pattern of discharge in a left ventricular receptor during one cardiac cycle; time is measured from the beginning of the Q wave of the e.c.g. The open circles represent normal activity, the peak frequency of discharge being about 50 impulses/sec. The filled circles represent the activity following injection of veriloid. Note, the ending is stimulated markedly but that the peak frequency of discharge is not increased much. This response suggests that the ending was stimulated and desensitized by veriloid (PAINTAL, 1955)

(or tension) receptors. It is now certain, after taking into consideration the total conduction time from the ending to the recording electrodes, that the initial impulses of a number of these endings appear with the onset of ventricular contraction at or before the start of rise of intraventricular pressure (e.g. 4th burst in Fig. 16B). Some of the endings show no evidence of being sensitive to distension. For example in the case of the right ventricular receptor of Fig. 16 there are no impulses during atrial systole when the ventricle must be distended to a maximum (during occlusion of the pulmonary artery). On the other hand, there are occasional endings which do yield one or more impulses when blood enters the ventricle during atrial systole (Fig. 16B, 4th burst).

The observation on cats described above have in general been confirmed in dogs (COLERIDGE et al., 1964b). The location of 19 such receptors is shown in Fig. 13C and D.

Responses to Veratrum Alkaloids. In the cat one of the consistent findings is that these endings are all stimulated by veratridine (22 µg), or veriloid (200 µg) or veratrine (NEIL and JOELS, 1961) (Fig. 18). The evidence consisting of evaluating the lowest frequency of discharge with the peak frequency of discharge indicates that the ending is stimulated and simultaneously desensitized as in the case of pulmonary stretch receptors (PAINTAL, 1957, 1964) (Fig. 18). As stated in the beginning such an action of the alkaloids is best explained by an action at the regenerative region (Fig. 1 B) of the ending.

The latency between injection and stimulation of the endings is longer following veriloid (mean, 10 sec); following veratridine the mean latency is 6 sec. The duration of stimulation following veriloid is also much longer, 8 min (mean) following veriloid, and only 8 sec (mean) following veratridine. Apart from these differences there are no obvious differences between veratridine and veriloid, e.g. neither of them appears to sensitize the endings. The most noteworthy point is that veratridine does not appear to stimulate the ventricular pressure receptors of dogs because 3 receptors studied by COLERIDGE et al. (1964b) and one by SLEIGHT and WIDDICOMBE (1965a) were not stimulated by veratridine. Such marked "species" difference between the cat and the dog, although not surprising, is of considerable importance because it shows that the sensory mechanisms underlying the BEZOLD-JARISCH effect (KRAYER, 1961), originally observed in the cat and rabbit, must be different in the cat on the one hand and the dog on the other.

At this point it is appropriate to consider BROWN's conclusion that many of these endings (i.e. ventricular pressure receptors) are in fact coronary artery mechanoreceptors particularly because all his endings (in cats), like the ventricular pressure receptors of the cat studied so far (PAINTAL, 1955; NEIL and JOELS, 1961) were stimulated by veratridine (BROWN, 1965). The fact that there are many features in common between his endings and the ventricular pressure receptors is clear e.g. presence of early systolic discharge, presence of variable patterns not related to the aortic pulse, and excitation by veratridine.

Out of 21 endings BROWN was able to locate 9 endings. These "were probably situated in or near the superficial, main coronary arteries". Eight of the nerve endings were near the main left coronary artery and one was found on the posterior surface of the heart over the left circumflex artery. The remaining 12 receptors which had a diastolic discharge or both a systolic and diastolic discharge could not be located. Perhaps they were located in the rest of the ventricle (see Fig. 13). The location of the 8 endings is not different from the location of the endings encountered by COLERIDGE et al. (1964b) in the dog who found that 12 of them were located at the base of the heart near the origin of the pulmonary artery and aorta, an area close to the coronary arteries.

It is certain that BROWN's endings could not be coronary *baroreceptors* since their discharge was not closely related to the pressure pulse in the coronary artery. In fact BROWN found that the pressure pulse did not seem to be an important factor in producing the bursts of activity, because in several instances there were gross time lags of one or two cycles between change in the coronary artery pressure and the change in the discharge (see Figs. 6 and 7 in BROWN, 1964). He felt that contraction of cardiac muscle was one of the factors responsible particularly in the case of those endings which discharged during the phase of isometric contraction of the ventricle. Such endings continued to be activated even after the pressure in the coronary artery was greatly reduced as shown in Fig. 5 of his paper.

In order to explain this response Brown suggested that each nerve fibre may be connected to two or more receptors one in the coronary artery and the other in the adjacent myocardium, and that the latter continued to be stimulated with each heart beat despite the low pressure. Such a suggestion has not been advanced in the case of any other cardiovascular receptor. Unfortunately intra-ventricular pressure was not recorded in these experiments. Such records would have been helpful since the activity of ventricular pressure receptors clearly depends on the magnitude of intra-ventricular pressure (Paintal, 1955) (Fig. 16) or rate of rise of pressure (Kolatat et al., 1957).

Brown was unable to suggest what the natural stimulus for these endings was, particularly because of the lack of any consistent relation between the coronary pressure pulse and the discharge in these endings. On the other hand since there is a relation between activity in these endings which (Brown thinks were earlier regarded as ventricular pressure receptors) and intra-ventricular pressure and myocardial contraction we may safely regard Brown's endings as ventricular pressure receptors that are located in juxtaposition to the coronary arteries particularly near the base of the heart where such receptors are known to be located in the dog (Coleridge et al., 1964b).

B. Epicardial Receptors

By recording impulses in certain cardiac nerves Kulaev was able to show that the spikes produced following local application of nicotine solutions on the surface of the ventricles, of the cat and rabbit, were due to stimulation of epicardial receptors. He concluded that these impulses were conducted in non-medullated fibres, since their spikes were much smaller than those with prominent cardiac rhythms which are known to be conducted in medullated fibres (Kulaev, 1963). In retrospect one can conclude that Jarisch and Zotterman (1948) (and perhaps also Amann and Schaefer, 1943) had also recorded impulses from such endings. However our detailed knowledge about the location, and stimulation of these endings by physical and chemical stimuli is derived from the subsequent work by Coleridge et al. (1964b) and Sleight and Widdicombe (1965a) who recorded impulses in individual fibres dissected from the cervical vagus (Coleridge et al., 1964b) or cardiac branches of the vagus (Sleight and Widdicombe, 1965a). Both these studies were carried out in dogs.

In order to select fibres from such epicardial receptors Sleight and Widdicombe (1965a) looked for activity (in each filament dissected) produced by palpation of the epicardium of the ventricles or by intrapericardial or intracoronary injection of nicotine. As stated by them it would seem that the endings studied by them were similar to those examined by Coleridge et al. (1964b). Thus in both investigations, measurements of the conduction velocities revealed that the fibres were non-medullated. In both studies, a rise in ventricular pressure, produced by occluding the aorta did not stimulate the endings obviously, nor were the endings stimulated by hypoxia or hypercapnia. The peak frequency of discharge in these

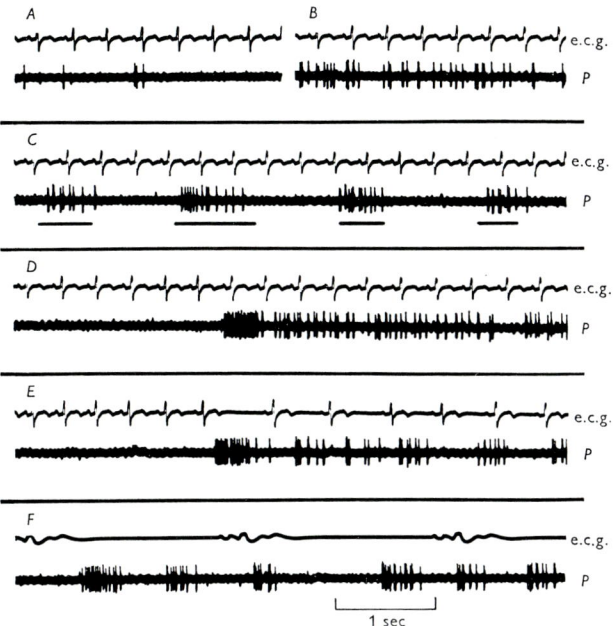

Fig. 19 A–F. Responses in a fibre from an epicardial receptor of the dog. A, control; note sparse discharge and absence of cardiac rhythm. B shows the response after local surface application (over a small area) of capsaicin solution (0.001%) and D 1 sec after injection of Capsaicin 5 μg/kg into the left atrium. E is a record 4 sec after injection of veratridine 7 μg/kg into the left atrium. C, shows the responses to gently stroking the epicardial surface of the intact ventricle and F stroking the same area after the animal had been killed, the heart opened and the endocardium pared away from this part of the ventricular wall (COLERIDGE et al., 1964b)

fibres was found to be less than 60 to 70 impulses/sec when stimulated by veratridine (or other drugs). In both studies, veratridine stimulated most or all the endings. The endings were located in or near the epicardium. The only difference between the two studies is that whereas SLEIGHT and WIDDICOMBE did not locate any of the epicardial receptors in the right ventricle, COLERIDGE et al. found that out of nine endings accurately localized, two were located in the right ventricle, five in the left ventricle and two in the region of the interventricular groove.

The activity in these fibres (with open chest and exposed heart) consists of an irregular discharge with no obvious cardiac rhythm and with a frequency of about 1 to 5 impulses/sec (Fig. 19A); occasionally a few fibres showed a single impulse with each heart beat (COLERIDGE et al., 1964b; SLEIGHT and WIDDICOMBE, 1965a). The most characteristic feature is the discharge that appears on stroking the epicardial surface of the ventricle (Fig. 19C). Such responses could be obtained repeatedly. It should be noted that the response to such local mechanical stimulation was not confined to a small precise localized area as was possible to establish in the case of ventricular pressure receptors, but the responses could be elicited by punctate stimulation at several points within an area of about 1 cm diameter

(Coleridge et al., 1964b). Similar observations were made by Sleight and Widdicombe (1965a).

As shown in Fig. 19 the endings were stimulated by veratridine. Coleridge et al. (1964b) found that they could also be stimulated by injected capsaicine (Fig. 19D) or by applying a 0.001% capsaicine solution over the epicardium (Fig. 19B). The endings could be stimulated by injecting nicotine into the pericardial sac, or the coronary artery (Sleight and Widdicombe, 1965a) or into the left atrium (Coleridge et al., 1964b). This is to be expected because the endings have non-medullated fibres (Paintal, 1964). For the same reason the peak frequency of discharge attained in them following injection of excitatory substances was 50–70 impulses/sec (Coleridge et al., 1964b).

The function of these endings has yet to be determined.

Clearly, these endings have to be kept in mind for finding an explanation for the Bezold-Jarisch type reflexes elicited by veratrum alkaloids in different animals. As stated above the mechanisms responsible for such reflexes may be different in different animals.

VIII. Pericardial Receptors

Sleight and Widdicombe (1965b) have located certain endings that are stimulated by mechanical stimuli and which are located in the rostral third of the parietal pericardium. They observed that the responses of these endings were in general similar to those obtained from endings in the parietal pleura or the pleura over the hilum of the right lung. Such endings with vagal afferent fibres have been referred to as mediastinal receptors and were described by Adrian (1933) and Widdicombe (1954). Holmes and Torrance (1959) found similar receptors with fibres running in the sympathetic nerves.

Measurements of the diameter of the nerve fibres likely to be the ones from which Sleight and Widdicombe (1965b) recorded impulses revealed that they were between 2.6 to 4.0 μ in diameter. This would imply that the conduction velocities of the fibres range from 13 to 24 m/sec [assuming a conversion factor of 5 (Boyd, 1964, 1965) or 6 (Hursh, 1939)]. However the measured conduction velocities were less than 7 m/sec. This discrepancy might be due to the method of measuring conduction velocity used by the authors i.e. local stimulation near the ending. Two main sources of error are the local fall in temperature (see Paintal, 1962a) and that due to indirect stimulation of the ending (and not the fibre itself) (see Paintal, 1962b).

These endings are stimulated when traction is applied to the pericardium or by touching the pericardium. The endings are also stimulated by distension of the pericardial sac by 0.9% NaCl although the amounts needed for doing this seem excessive. For example in one ending the discharge increased from 2.5 impulses/sec to only 5 impulses when 30 ml of fluid was introduced into the pericardial sac. The addition of 20 ml more of fluid raised the discharge to 12.5 impulses/sec.

Unlike the epicardial receptors these endings are not stimulated by intrapericardial injections of nicotine. Nor are they stimulated by injections of veratridine into the right atrium.

IX. Pulmonary Arterial Baroreceptors

The pulmonary arterial baroreceptors are endings in the pulmonary artery or its branches that are supposed to be stimulated by the pulsatile rise in pulmonary artery pressure (COLERIDGE and KIDD, 1961). Accordingly it would be expected that the activity in these endings would bear a close relation to the pressure pulse in the pulmonary artery. Thus it would be expected that the pulsatile discharge would commence with the upstroke of the pressure pulse, peak frequency would be attained at the peak of the pulse (or on its rising phase) and the frequency of discharge would start to decline thereafter, the end of the burst occurring at a higher (certainly not lower) level of pressure than the threshold for stimulating the ending. It would appear that a large number of endings regarded as pulmonary artery baroreceptors do not fulfill one or all of these requirements. Thus in some endings the discharge starts only at the peak of the pressure pulse (e.g. Fig. 1A and Fig. 6A and C in COLERIDGE and KIDD, 1961) while in others the main burst appears after the start of the falling(!) phase of the pressure pulse (Fig. 4 in BIANCONI and GREEN, 1959b). In others there are cycles in which it starts before the upstroke of the pressure pulse (e.g. see second cycle in Fig. 5B of COLERIDGE and KIDD, 1961). In certain others, variation in the pulmonary artery pressure pulse does not produce quantitatively consistent changes in the discharge in these endings (e.g. Fig. 5 in COLERIDGE and KIDD, 1961). On the other hand there are instances in which the discharge disappears in one or more cycles in spite of a practically constant pulmonary artery pressure (e.g. see 3rd cycle of second trace of Fig. 3A in BEVAN and KINNISON, 1965). In some endings the onset of the burst in relation to the upstroke of the pressure pulse is variable (e.g. Fig. 2 in COLERIDGE and KIDD, 1961). Finally there are several endings in which instead of the normal systolic discharge there is a burst that is (as stated by COLERIDGE and KIDD, 1960) typical of type A atrial receptors (e.g. Fig. 8A in COLERIDGE and KIDD, 1960) while in others maximum activity is seen during the P-R interval (Fig. 4B in COLERIDGE and KIDD, 1960). Such endings would, in the opinion of certain investigators, be regarded as atrial receptors. "In some instances the discharge during ventricular diastole was the most prominent" (COLERIDGE and KIDD, 1960). Such endings would be considered to be atrial type B receptors.

The above uncertain position regarding these pulmonary artery baroreceptors has arisen partly because COLERIDGE and KIDD who have done the most extensive work on these endings (COLERIDGE and KIDD, 1960, 1961, 1963; COLERIDGE et al., 1961, 1964a) apparently do not attach over-riding importance to the role of the natural stimulus in deciding which is a pulmonary artery baroreceptor and which is not. They state "These results emphasize that a receptor cannot be assigned with certainty to a particular great vessel or chamber of the heart by inspection of the discharge and its relationship to the e.c.g.; it must be located by appropriate means in the animal with open chest. Such factors may have contributed to the previous paucity of information about receptors in the pulmonary artery". It should be noted that the onset of the natural stimulus (in the case of most endings) is linked with the e.c.g.

Direct localization of an ending in a particular chamber or vessel is an invaluable aid for establishing where an ending is located (see PAINTAL, 1953a).

Punctate stimulation can also be of much help, but the results of such stimulation must be interpreted in conjunction with the results of natural stimulation. Perhaps it may not be desirable to give the results of punctate stimulation over-riding importance in deciding the location of an ending particularly when the sensory receptors are concentrated near the roots of the great vessels and the veins where several types of endings are in abundance. It is therefore not surprising to find that pressure at one point may yield pronounced activity even though the ending is not directly under the point of application of pressure as happens not infrequently.

The position concerning the pulmonary artery baroreceptors therefore needs to be carefully reassessed. In particular it would be desirable to study endings that are dependent for their activity almost entirely (apart from occassional adventitious impulses) on the pressure pulse of the pulmonary artery i.e. endings in which activity begins at (and not before) the upstroke of the pressure pulse (after allowing for conduction time in the sensory fibre) and ends with the end of the pressure pulse. These endings (like the type B atrial receptors or the aortic baroreceptors) should not be stimulated by normal atrial contraction and their activity should be reduced or abolished during premature ventricular contractions depending on when, in the cardiac cycle, the ectopic stimulus is applied i.e. in proportion to the reduction in the amplitude of the pulmonary artery pressure pulse. Occlusion of the pulmonary artery should abolish their activity. Finally it should be possible to stimulate such endings mechanically after removing the pulmonary veins (and left atrium) in order to exclude the type B receptors located in them (see Fig. 5 in PAINTAL, 1953a).

So far, attempts to isolate an ending that satisfies the above requirements have not been successful in the cat (PAINTAL, 1963–70, unpublished observations). This negative observation is only of limited value, but it does indicate that if such endings exist in the cat, there must be very few of them (with afferent fibres in the right vagus) in comparison with the relatively large numbers of type B and type A atrial receptors.

References

ADRIAN, E. D.: Afferent impulses in the vagus and their effect on respiration. J. Physiol. (Lond.) **79**, 332–358 (1933).
ALVAREZ-BUYLLA, R., RAMIREZ DE ARELLANO, J.: Local responses in Pacinian corpuscles. Amer. J. Physiol. **172**, 237–244 (1953).
AMANN, A., SCHAEFER, H.: Über sensible Impulse im Herznerven. Pflügers Arch. ges. Physiol. **246**, 757–789 (1943).
ANGELL JAMES, J. E.: Studies of the impulse activity in baroreceptor fibres from an isolated aortic arch preparation of the rabbit. J. Physiol. (Lond.) **169**, 51–52 P (1968).
— The responses of aortic arch and right subclavian baroreceptors to changes of non-pulsatile pressure and their modification by hypothermia. J. Physiol. (Lond.). **214**, 201–224 (1971).
— DALY, M. DE B.: Reflex vasomotor responses elicited from the carotid sinus and aortic arch baroreceptors: Comparison of pulsatile and non-pulsatile pressure. J. Physiol. (Lond.) **209**, 22–23 P (1970).
ARNDT, J. O., BRAMBRING, P., HINDORF, K., RÖHNELT, M.: Das Verhalten von Vorhof-Afferenzen des A- und B-Typs bei sinusförmiger Dehnung des Vorhofstreifenpräparates der Katze. Pflügers Arch. ges. Physiol. In press (1971a).
— — — — Das Entladungsmuster von Vorhofafferenzen des A-Typs der Katze. Herzfrequenzmessung durch den A-Typ-Rezeptor? Pflügers Arch. ges. Physiol. In press (1971b).

References

Bevan, J. A., Kinnison, G. L.: Action of lobeline on pulmonary artery mechanoreceptors of the cat. Circulat. Res. **17**, 19–29 (1965).
Bianconi, R., Green, J. H.: Baroreceptor innervation of the bifurcation of the brachiocephalic trunk in the cat. Arch. ital. Biol. **97**, 47–52 (1959a).
— — Pulmonary baroreceptors in the cat. Arch. ital. Biol. **97**, 305–315 (1959b).
— — Cardio-respiratory afferent fibres in the vagus of the cat. Arch. Sci. Biol. **43**, 454–463 (1959c).
Bloor, C. M.: Aortic baroreceptor threshold and sensitivity in rabbits at different ages. J. Physiol. (Lond.) **174**, 136–171 (1964).
Boss, J., Green, J. H.: The histology of the common carotid baroreceptor areas of the cat. Circulat. Res. **4**, 12–17 (1956).
Boyd, I. A.: The relation between conduction velocity and diameter for the three groups of efferent fibres in nerves to mammalian skeletal muscle. J. Physiol. (Lond.) **175**, 33–35 P (1964).
— Differences in the diameter and conduction velocity of motor and fusimotor fibres in nerves to different muscles in the hind limb of the cat. In: Studies in physiology, ed. Curtis, D. R., and A. K. McIntyre, pp. 7–12. Berlin-Heidelberg-New York: Springer 1965.
Brambring, P., Röhnelt, M., Hindorf, K., Arndt, J. O.: Das Entladungsmuster von Herzvorhof-Afferenzen des A-Typs unter Volumenänderungen, Noradrenalin und Schrittmacherreiz. Pflügers Arch. **312**, R 24 (1969).
Bronk, D. W., Stella, G.: Afferent impulses in the carotid sinus nerve. I. The relation of the discharge from single end organs to arterial blood pressure. J. cell. comp. Physiol. **1**, 113–130 (1932).
— — The response to steady pressures of single end organs in the isolated carotid sinus. Amer. J. Physiol. **110**, 708–714 (1935).
Brooks, C. McC., Hoffman, B. F., Suckling, E. E.: Excitability of the heart, pp. 104–105. New York: Grune & Stratton 1955.
Brown, A. M.: Mechanoreceptors in or near the coronary arteries. J. Physiol. (Lond.) **177**, 203–214 (1965).
Burton, A. C.: On the physical equilibrium of small blood vessels. Amer. J. Physiol. **164**, 319–329 (1951).
Chapman, K. M., Pearce, J. W.: Vagal afferents in the monkey. Nature (Lond.) **184**, 1237–1238 (1959).
Coleridge, H. M., Coleridge, J. C. G., Kidd, C.: Role of the pulmonary arterial baroreceptors in the effects produced by capsaicin in the dog. J. Physiol. (Lond.) **170**, 272–285 (1964a).
— — — Cardiac receptors in the dog, with particular reference to two types of afferent ending in the ventricular wall. J. Physiol. (Lond.) **174**, 323–339 (1964b).
Coleridge, J. C. G., Hemingway, A., Holmes, R. L., Linden, R. J.: The location of atrial receptors in the dog: A physiological and histological study. J. Physiol. (Lond.) **136**, 174–197 (1957).
— Kidd, C.: Electrophysiological evidence of baroreceptors in the pulmonary artery of the dog. J. Physiol. (Lond.) **150**, 319–331 (1960).
— — Relationship between pulmonary arterial pressure and impulse activity in pulmonary arterial baroreceptor fibres. J. Physiol. (Lond.) **158**, 197–205 (1961).
— — Reflex effects of stimulating baroreceptors in the pulmonary artery. J. Physiol. (Lond.) **166**, 197–210 (1963).
— — Sharp, J. A.: The distribution, connexions and histology of baroreceptors in the pulmonary artery, with some observations of the sensory innervation of the ductus arteriosus. J. Physiol. (Lond.) **156**, 591–602 (1961).
— Linden, R. J.: The effect upon the heart rate of increasing the venous return by opening an arterio-venous fistula in the anaesthetized dog. J. Physiol. (Lond.) **130**, 674–702 (1955).
De Castro, F.: Sur la structure et l'innervation du sinus carotidien. Nouveaux facts sur l'innervation et la fonction du glomus carotidien. Trab. Lab. Invest. biol. Univ. Madrid **25**, 331–380 (1928).
— Sur la structure de la synapse dans les chemorecepteurs: leur mécanisme d'excitation et rôle dans la circulation sanguine locale. Acta physiol. scand. **22**, 14–43 (1951).

Devanandan, M. S.: A study of the myelinated fibres of the aortic nerve of cats. J. Physiol. (Lond.) **171**, 361–367 (1964).

Diamond, J.: Observations on the excitation by acetylcholine and by pressure of sensory receptors in the cat's carotid sinus. J. Physiol. (Lond.) **130**, 513–532 (1955).

Dickinson, C. J.: Afferent nerves from the heart region. J. Physiol. (Lond.) **111**, 399–407 (1950).

Ead, H. W., Green, J. H., Neil, E.: A comparison of the effects of pulsatile and non-pulsatile blood flow through the carotid sinus on the reflexogenic activity of the sinus baroceptors in the cat. J. Physiol. (Lond.) **118**, 509–519 (1952).

Euler, U. S. von, Liljestrand, G., Zotterman, Y.: Baroceptive impulses in the carotid sinus and their relation to the pressure reflex. Acta physiol. scand. **2**, 1–9 (1941).

Eyzaguirre, C., Uchizono, K.: Observations on the fibre content of nerves reaching the carotid body of the cat. J. Physiol. (Lond.) **159**, 268–281 (1961).

Fahim, M., Gupta, P. D.: Personal communication (1970).

Fidone, S. J., Sato, A.: A study of chemoreceptor and baroreceptor A and C-fibres in the cat carotid nerve. J. Physiol. (Lond.) **205**, 527–548 (1969).

Floyd, W. F., Neil, E.: The influence of the sympathetic innervation of the carotid bifurcation on chemoreceptor and baroreceptor activity in the cat. Arch. int. Pharmacodyn. **91**, 230–239 (1952).

Franz, D. N., Iggo, A.: Conduction failure in myelinated and non-myelinated axons at low temperatures. J. Physiol. (Lond.) **199**, 319–345.

Gammon, G. D., Bronk, D. W.: The discharge of impulses from Pacinian corpuscles in the mesentery and its relation to vascular changes. Amer. J. Physiol. **114**, 77–84 (1935).

Gasser, H. S., Grundfest, H.: Axon diameters in relation to the spike dimensions and the conduction velocity in mammalian A fibers. Amer. J. Physiol. **127**, 393–414 (1939).

Gauer, O. H., Henry, J. P.: Beitrag zur Homöostase des extraarteriellen Kreislaufs. Volumenregulation als unabhängiger physiologischer Parameter. Klin. Wschr. **34**, 356–366 (1956).

Gernandt, B., Zotterman, Y.: Intestinal pain: An electrophysiological investigation on mesenteric nerves. Acta physiol. scand. **12**, 56–72 (1946).

Gray, J. A. B., Sato, M.: Properties of the receptor potential in Pacinian corpuscles. J. Physiol. (Lond.) **122**, 610–636 (1953).

Gupta, P. D., Henry, J. P., Sinclair, R., Baumgarten, R. von: Responses of atrial and aortic baroreceptors to nonhypotensive hemorrhage and to transfusion. Amer. J. Physiol. **211**, 1429–1437 (1966).

Hakumäki, M. O. K.: Function of the left atrial receptors. Acta physiol. scand., Suppl. **344**, 79, 1–54 (1970).

Haus, W. H., Kreuziger, H., Asteroth, H.: Über die Reizung der Pressorezeptoren im Sinus caroticus beim Hund. Z. Kreisl.-Forsch. **38**, 28–33 (1949).

Henry, J. P., Gauer, O. H., Reeves, J. L.: Evidence of the atrial location of receptors influencing urine flow. Circulat. Res. **4**, 85–90 (1956).

— Pearce, J. W.: The possible role of cardiac atrial stretch receptors in the induction of changes in urine flow. J. Physiol. (Lond.) **131**, 572–585 (1956).

Heymans, C., Neil, E.: Reflexogenic areas of the cardiovascular system. London: Churchill 1958.

Holmes, R., Torrance, R. W.: Afferent fibres of the stellate ganglion. Quart. J. exp. Physiol. **44**, 271–281 (1959).

Homma, S., Suzuki, S. S.: Phasic properties of aortic and atrial receptors observed from their afferent discharge. Jap. J. Physiol. **16**, 31–41 (1966).

Hunt, C. C.: On the nature of vibration receptors in the hind limb of the cat. J. Physiol. (Lond.) **155**, 175–186 (1961).

— McIntyre, A. K.: Characteristics of responses from receptors from the flexer longus digitorum muscle and the adjoining interosseous region of the cat. J. Physiol. (Lond.) **153**, 74–87 (1960).

Hursh, J. B.: Conduction velocity and diameter of nerve fibers. Amer. J. Physiol. **127**, 131–139 (1939).

Iggo, A.: The electrophysiological identification of single nerve fibres, with particular reference to the slowest-conducting vagal afferent fibres in the cat. J. Physiol. (Lond.) **142**, 110–126 (1958).

INMAN, D. R., PERUZZI, P.: The effects of temperature on the responses of Pacinian corpuscles. J. Physiol. (Lond.) **155**, 280–301 (1961).
IRISAWA, H., GREER, A. P., RUSHMER, R. F.: Changes in the dimensions of the venae cavae. Amer. J. Physiol. **196**, 741–744 (1959).
ISHIKO, N., LOEWENSTEIN, W. R.: Effects of temperature on the generator and action potentials of a sense organ. J. gen. Physiol. **45**, 105–124 (1961).
JARISCH, A., LANDGREN, S., NEIL, E., ZOTTERMAN, Y.: Impulse activity in the carotid sinus nerve following intra-carotid injection of potassium chloride, veratrine, sodium citrate, adenosintriphosphate and α-dinitrophenol. Acta physiol. scand. **25**, 195–211 (1952).
— ZOTTERMAN, Y.: Depressor reflexes from the heart. Acta physiol. scand. **16**, 31–51 (1948).
KAPPAGODA, C. T., LINDEN, R. J., SNOW, H. M.: Further evidence for right atrial receptors affecting heart rate. J. Physiol. (Lond.) **210**, 132–133 P (1970).
KATZ, B.: Depolarization of sensory terminals and the initiation of impulses in the muscle spindle. J. Physiol. (Lond.) **111**, 261–282 (1950).
KEZDI, P.: Control by the superior cervical ganglion of the state of contraction and pulsatile expansion of the carotid sinus arterial wall. Circulat. Res. **2**, 367–371 (1954).
KIDD, C., LEDSOME, J. R., LINDEN, R. J.: Left atrial receptors and the heart rate. J. Physiol. (Lond.) **185**, 78–79 P (1966).
KOLATAT, T., KRAMER, K., MÜHL, N.: Über die Aktivität sensibler Herznerven des Frosches und ihre Beziehungen zur Herzdynamik. Pflügers Arch. ges. Physiol. **264**, 127–144 (1957).
KRAMER, K.: Die afferente Innervation und die Reflexe von Herz und venösem System. Verh. Dtsch. Ges. Kreisl.-Forsch. 25. Tagg, 1959, S. 142–163.
KRAYER, O.: The history of the Bezold-Jarisch effect. Naunyn-Schmiedebergs Arch. exp. Path. Pharmak. **240**, 361–368 (1961).
KULAEV, B. S.: Characteristics of afferent impulses evoked in cardiac nerves by chemical stimulation of epicardial receptors. Fed. Proc. **22**, T 749–T 754 (1963). Translation of paper published in Fiziol. Zh. (Mosk.) **48**, 1350–1362 (1962).
LANDGREN, S.: On the excitation mechanism of the carotid baroreceptors. Acta physiol. scand. **26**, 1–34 (1952a).
— The baroreceptor activity in the carotid sinus nerve and the distensibility of the sinus wall. Acta physiol. scand. **26**, 35–56 (1952b).
— NEIL, E., ZOTTERMAN, Y.: The response of the carotid baroceptors to the local administration of drugs. Acta physiol. scand. **25**, 24–37 (1952).
LANGREHR, D.: Entladungsmuster und allgemeine Reizbedingungen von Vorhofsreceptoren bei Hund und Katze. Pflügers Arch. ges. Physiol. **271**, 257–269 (1960a).
— Beziehungen zwischen Vorhofreceptoraktivitäten und Herzmechanik von Hund und Katze bei verschiedenen Kreislaufzuständen. Pflügers Arch. ges. Physiol. **271**, 270–282 (1960b).
LEDSOME, J. R., LINDEN, R. J.: A reflex increase in heart rate from distension of the pulmonary-vein-atrial junctions. J. Physiol. (Lond.) **170**, 456–473 (1964).
— — The effect of distending a pouch of the left atrium on the heart rate. J. Physiol. (Lond.) **193**, 121–129 (1967).
LEITNER, J.-M., PERL, E. R.: Receptors supplied by spinal nerves which respond to cardiovascular changes and adrenaline. J. Physiol. (Lond.) **175**, 254–274 (1964).
LITTLE, R. C.: Volume elastic properties of the right and left atrium. Amer. J. Physiol. **158**, 237–240 (1949).
— Volume pressure relationships of the pulmonary-left heart vascular segment. Circulat. Res. **8**, 594–599 (1960).
LOEWENSTEIN, W. R., RATHKAMP, R.: The sites of mechano-electric conversion in a Pacinian corpuscle. J. gen. Physiol. **41**, 1245–1265 (1958).
MATHEWS, B. H. C.: The response of a single end organ. J. Physiol. (Lond.) **71**, 64–110 (1931).
— Nerve endings in mammalian muscle. J. Physiol. (Lond.) **78**, 1–53 (1933).
MOTT, J. C., PAINTAL, A. S.: The action of 5-hydroxytryptamine on pulmonary and cardiovascular vagal afferent fibres and its reflex respiratory effects. Brit. J. Pharmacol. **8**, 238–241 (1953).
MÜHL, N., SCHOLDERER, I., KRAMER, K.: Über die Aktivität der intrathorakalen Gefäßrezeptoren und ihre Beziehung zur Herzfrequenz bei Änderung des Blutvolumens. Verh. Dtsch. Ges. Kreisl.-Forsch., 22. Tagg., S. 122–126 (1956).

Neil, E.: The carotid and aortic vasosensory areas. Arch. Middlesex Hosp. **4**, 16–27 (1954).
— Joels, N.: The impulse activity in cardiac afferent vagal fibres. Naunyn-Schmiedebergs Arch. exp. Path. Pharmak. **240**, 453–460 (1961).
Nonidez, J. F.: Identification of the receptor areas in the venae cavae and pulmonary veins which initiate reflex cardiac acceleration (Bainbridge's reflex). Amer. J. Anat. **61**, 203–231 (1937).
Opdyke, D. F., Duomarco, J., Dillon, W. H., Schreiber, H., Little, R. C., Seely, R. D.: Study of simultaneous right and left atrial pressure pulses under normal and experimentally altered conditions. Amer. J. Physiol. **154**, 258–272 (1948).
Ottoson, D.: The effects of temperature on the isolated muscle spindle. J. Physiol. (Lond.) **180**, 636–648 (1965).
Paintal, A. S.: A study of right and left atrial receptors. J. Physiol. (Lond.) **120**, 596–610 (1953a).
— The response of pulmonary and cardiovascular vagal receptors to certain drugs. J. Physiol. (Lond.) **121**, 182–190 (1953b).
— The conduction velocities of respiratory and cardiovascular afferent fibres in the vagus nerve. J. Physiol. (Lond.) **121**, 341–359 (1953c).
— A study of gastric stretch receptors. Their role in the peripheral mechanism of satiation, of hunger and thirst. J. Physiol. (Lond.) **126**, 255–270 (1954).
— The study of ventricular pressure receptors and their role in the Bezold reflex. Quart. J. exp. Physiol. **40**, 348–363 (1955).
— The influence of certain chemical substances on the initiation of sensory discharges in pulmonary and gastric stretch receptors and atrial receptors. J. Physiol. (Lond.) **135**, 486–510 (1957).
— Determination of intrathoracic conduction time in cardiovascular afferent fibres of the vagus nerve. J. Physiol. (Lond.) **163**, 222–238 (1962a).
— Responses and reflex effects of pressure-pain receptors of mammalian muscles. In: Muscle receptors, ed. Barker, D., pp. 133–142. Hong Kong: Hong Kong Univ. press 1962b.
— Vagal afferent fibres. Ergebn. Physiol. **52**, 74–156 (1963a).
— Natural stimulation of type B atrial receptors. J. Physiol. (Lond.) **169**, 116–136 (1963b).
— Effects of drugs on vertebrate mechanoreceptors. Pharmacol. Rev. **16**, 341–380 (1964).
— Block of conduction in mammalian myelinated nerve fibres by low temperatures. J. Physiol. (Lond.) **180**, 1–19 (1965a).
— Effects of temperature on conduction in single vagal and saphenous myelinated nerve fibres of the cat. J. Physiol. (Lond.) **180**, 20–49 (1965b).
— The influence of diameter of medullated nerve fibres of cats on the rising and falling phases of the spike and its recovery. J. Physiol. (Lond.) **184**, 791–811 (1966a).
— Re-evaluation of respiratory reflexes. Quart. J. exp. Physiol. **51**, 151–163 (1966b).
— Mechanism of stimulation of aortic chemoreceptors by natural stimuli and chemical substances. J. Physiol. (Lond.) **189**, 63–84 (1967a).
— A comparison of the nerve impulses of mammalian non-medullated nerve fibres with those of the smallest diameter medullated fibres. J. Physiol. (Lond.) **193**, 523–533 (1967b).
— Mechanism of stimulation of type J pulmonary receptors. J. Physiol. (Lond.) **203**, 511–532 (1969).
— Action of drugs on sensory nerve endings. Ann. Rev. Pharmacol. **11**, 231–240 (1971).
— Riley, R. L.: Responses of aortic chemoreceptors. J. appl. Physiol. **21**, 543–548 (1966).
Palme, F.: Zur Funktion der branchiogenen Reflexzonen für Chemo- und Presso-Reception. Z. ges. exp. Med. **113**, 415–461 (1943–1944).
Pearce, J. W., Henry, J. P., Chapman, K. M.: The behaviour and possible functions of cardiac atrial stretch receptors. Abstr. XX. int. physiol. Congr. 1956, p. 711–712.
— Whitteridge, D.: The relation of pulmonary arterial pressure variations to the activity of afferent pulmonary vascular fibres. Quart. J. exp. Physiol. **36**, 177–188 (1950).
Price, H. L., Widdicombe, J.: Actions of cyclopropane on carotid sinus baroreceptors and carotid body chemoreceptors. J. Pharmacol. exp. Ther. **135**, 233–239 (1962).
Rees, P. M.: Observations on the fine structure and distribution of presumptive baroreceptor nerves at the carotid sinus. J. comp. Neurol. **131**, 517–547 (1967).

ROBERTSON, J. D., SWAN, A. A. B., WHITTERIDGE, D.: Effect of anaesthetics on systemic baroreceptors. J. Physiol. (Lond.) **131**, 463–472 (1956).

SATO, A., FIDONE, S., EYZAGUIRRE, C.: Presence of chemoreceptor and baroreceptor C-fibers in the carotid nerve of the cat. Brain Res. **11**, 459–463 (1968).

SATO, M.: Response of Pacinian corpuscles to sinusoidal vibration. J. Physiol. (Lond.) **159**, 391–409 (1961).

SCHAEFER, H.: Elektrophysiologie der Herznerven. Ergebn. Physiol. **46**, 71–125 (1950).

SCHMIDT, E. M., STROMBERG, M. W.: The myelinated fibers in the aortic nerve of the swine. Anat. Rec. **159**, 41–45 (1967).

SCHOEPFLE, G. M., ERLANGER, J.: The action of temperature on the excitability, spike height and configuration, and the refractory period observed in the responses of single medullated nerve fibers. Amer. J. Physiol. **134**, 694–704 (1941).

SLEIGHT, P., WIDDICOMBE, J. G.: Action potentials in fibres from receptors in the epicardium and myocardium of the dog's left ventricle. J. Physiol. (Lond.) **181**, 235–258 (1965a).

— — Action potentials in afferent fibres from pericardial mechanoreceptors in the dog. J. Physiol. (Lond.) **181**, 259–269 (1965b).

STRUPPLER, A., STRUPPLER, E.: Über spezielle Charakteristika afferenter vagaler Herznervenimpulse und ihre Beziehungen zur Herzdynamik. Acta physiol. scand. **33**, 219–231 (1955).

TORRANCE, R. W., WHITTERIDGE, D.: Technical aids in the study of respiratory reflexes. J. Physiol. (Lond.) **107**, 6–7 P (1948).

WELLHÖNER, H. H., HAFERKORN, D.: Die Wirkung von Aconitin auf die Impulsbildung durch Mechanoreceptoren des Sinus caroticus und des Herzens. Naunyn-Schmiedebergs Arch. Pharmak. exp. Path. **255**, 407–418 (1966).

WHITTERIDGE, D.: Afferent impulses from the heart and lungs. Abstr. XVII. int. physiol. Congr. (1947).

— Afferent nerve fibres from the heart and lungs in the cervical vagus. J. Physiol. (Lond.) **107**, 496–512 (1948).

— Effects of anaesthetics on mechanical receptors. Brit. med. Bull. **14**, 5–7 (1958).

— BÜLBRING, E.: Changes in activity of pulmonary receptors in anaesthesia and their influence on respiratory behaviour. J. Pharmacol. exp. Ther. **81**, 340–359 (1944).

WITZLEB, E.: Über die Wirkung des Veratrins auf die chemo- und pressoreceptorischen Aktionspotentiale im Carotissinusnerven. Pflügers Arch. ges. Physiol. **256**, 234–241 (1952).

ZIPF, H. F.: The pharmacology of viscero-afferent receptors with special reference to endo-anaesthesia. Acta neuroveg. (Wien) **28**, 169–196 (1966).

Chapter 2

Arterial Chemoreceptors

By

ALAN HOWE and ERIC NEIL, London (Great Britain)

With 14 Figures

Contents

I. Definition and History 47
II. Structure 49
   A. Light Microscopy 49
   B. Electron Microscopy 51
   C. Nature of the Terminals on Type I Glomus Cells 57
III. Function 58
   A. The Influence of Hypoxia 58
   B. Blood Flow through the Carotid Body 61
   C. The Influence of the Sympathetic Nerve Supply upon Glomeral Oxygen Usage 64
   D. Possible Significance of A–V Anastomoses 65
   E. The Role of Energy Rich Phosphates in the Chemoreceptors 66
   F. The Effects of CO_2 and pH 67
   G. The Influence of Efferent Components in the Sinus and Aortic Nerves . . 69
   H. The Sensory Unit and its Excitation 71
   I. The Role of the Chemoreceptors in Eupnoeic Respiration 75
References 76

I. Definition and History

The so-called arterial chemoreceptors are located in the vicinity of each carotid bifurcation (carotid body) and the roots of the main thoracic arteries (aortic bodies) as depicted in Figs. 1 and 2. The carotid body is supplied by branches of the occipital and ascending pharyngeal arteries. The aortic bodies are nourished by small branches from the neighbouring systemic vessels (COLERIDGE, COLERIDGE and HOWE 1966, 1967, 1970).

Sensory nerve fibres from the carotid body course in the carotid sinus nerve which joins the glossopharyngeal (IXth) nerve and enters the medulla; the cell bodies of these chemoreceptor fibres are situated in the petrous ganglion of IX. Vagal sensory fibres from the aortic bodies are similarly distributed to the medulla and have their cell bodies in the nodose ganglion of X.

The carotid and aortic bodies serve as chemoreceptors which respond to changes in the chemical composition of their extracellular environment, usually occasioned by alterations of pO_2, pCO_2 or pH of their arterial blood supply.

HEYMANS and HEYMANS (1927) showed that chemical changes in the blood perfusing the aortic arch provoked reflex changes in the activity of the respiratory centre and showed that the vagi provided the afferent limb of this reflex pathway.

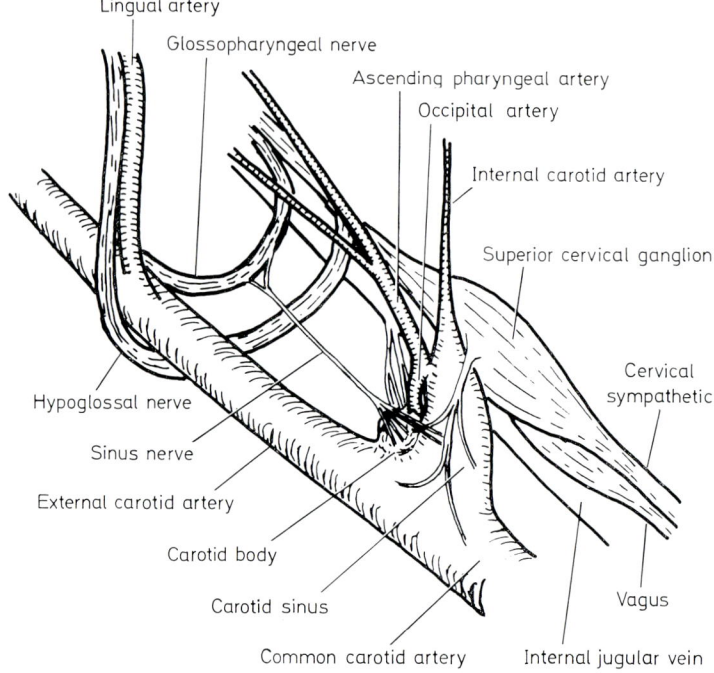

Fig. 1. Diagram of the ventral view of the right carotid bifurcation of the cat. (From N. Joels, 1960)

In 1928 DE CASTRO correctly adumbrated the function of the carotid body from histological and degeneration studies. He proved that its generous innervation was largely sensory and inferred that the organ served to sample the composition of the arterial blood.

Unaware of DE CASTRO's findings, HEYMANS and BOUCKAERT (1930) proved that the carotid bifurcation was the site of a reflexogenic zone responsive to hypoxia, to hypercapnia and to acidaemia. Any of these chemical changes acting at this site provoked hyperpnoea and vasoconstriction; these responses were abolished by cutting the carotid sinus nerve. The identity of the carotid body as the progenitor of these chemoreceptor reflexes was proved by 1933 (for references see HEYMANS, BOUCKAERT and REGNIERS, 1933).

MURATORI (1933, 1935) provided the histological evidence for the homologous aortic bodies and COMROE (1939) showed that stimulation of these structures by chemical means caused reflex changes of respiration and cardiovascular activity qualitatively similar to those evoked from the carotid bodies.

The all important contribution of the carotid and aortic reflexogenic zones to the animal's economy lies in their response to acute hypoxia. Acute hypoxia causes depression of both the respiratory and cardiovascular centres of the medulla leading to death *unless* these reflexogenic areas are intact. Chemoreceptor stimulation by hypoxia causes reflex hyperpnoea which helps to improve the arterial pO_2. Although the carotido-aortic chemoreceptors are responsive to hypercapnia

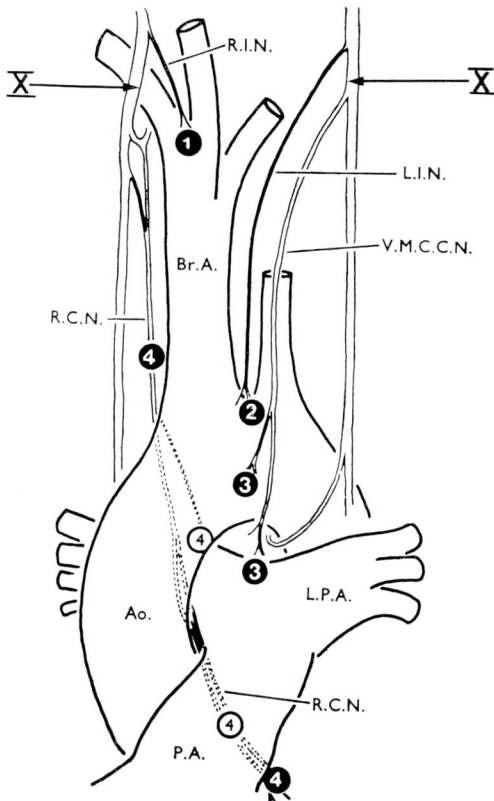

Fig. 2. Diagram of the ventral view of the aortic arch and main thoracic arteries to show the distribution of the various groups of aortic bodies (Groups 1–4) in the cat. Solid circles: bodies on the ventral surface; open circles: bodies on the dorsal surface. *Ao* aorta; *P.A.* pulmonary artery; *L.P.A.* left pulmonary artery; *Br.A.* brachiocephalic artery; *X* left and right vagi; *L.I.N.*, *R.I.N.* left and right innominate nerves; *R.C.N.* recurrent cardiac nerve; *V.M.C.C.N.* ventromedial cervical cardiac nerve. (From COLERIDGE, COLERIDGE, and HOWE, 1967)

and acidaemia and cause reflex hyperpnoea, such chemical changes also stimulate the medullary neurones which increase respiratory and cardio-vascular activity.

The study of the activity of the chemoreceptors by electrophysiological methods began with HEYMANS and RIJLANT (1933), BOGUE and STELLA (1935), and ZOTTERMAN (1935). EULER, LILJESTRAND and ZOTTERMAN (1939) were the first to relate chemoreceptor activity to measured changes of oxygen saturation and to alterations of pCO_2 in the arterial blood.

The serious interpretation of the mode of response of the chemoreceptors, however, presupposes a detailed knowledge of their structure.

II. Structure
A. Light Microscopy

The classical studies of DE CASTRO (1926, 1928) and MURATORI (1933, 1935) linked the basic morphology of the vascular chemoreceptors to a sensory function—

"Nous supposons, comme hypothèse vraisemblable, que le Glomus caroticum représente un organe sensoriel, le seul jusqu'à présent, chargé de recevoir certaines variations qualitatives du sang, ..." (DE CASTRO, 1928).

These small organs (glomera) were seen to be composed essentially of relatively large rounded cells (the so-called epithelioid or glomus cells) often arranged in characteristic whorl-like groups and richly supplied with blood vessels and nerve fibres.

(i) Cell types: With the light microscope it was appreciated that apart from vascular, neural and connective tissue elements more than one type of glomus cell existed (see HOLLINSHEAD, 1943; DE KOCK, 1951). DE KOCK (1951, 1954) employed the terms "Type I glomus cell" and "Type II glomus cell" to distinguish the two main varieties of cell of the carotid body. The rounded principal Type I possessed a very granular cytoplasm and she remarked that the smaller irregular Type II cells, lacking cytoplasmic granulation, were characteristically interposed between the blood sinusoids and groups of Type I cells, or were closely applied to the latter (see also Section B on Electronmicroscopy).

(ii) Vasculature: In the adult animal both carotid and aortic bodies are irrigated by systemic arterial blood (for details see HEYMANS and NEIL, 1958). The developmental changes which occur in the arterial supply of the aortic bodies from foetal to post-natal life are reviewed by COLERIDGE, COLERIDGE, and HOWE (1967, 1970), together with the evidence discounting the physiological importance of a pulmonary chemoreceptor system in the adult animal.

DE CASTRO (1928, 1940) drew attention to the strikingly rich intrinsic vasculature of the carotid body. The glomus cells were described as being in close approximation to capillaries (sinusoids) so situated "... à percevoir quelques modifications qualitatives du sang ..." (DE CASTRO, 1928).

The actual capillary bed, however, may not be very extensive in some species, such as the dog (SERAFINI-FRACASSINI and VOLPIN, 1966). It is seldom displayed completely by injection procedures (DE CASTRO and RUBIO, 1968) and the vascular network is usually demonstrated as well or sometimes better by retrograde venous injection as by arterial injection (HOWE, 1956, 1957). The presence of arteriovenous anastomoses was recognized early as a feature of the carotid and aortic bodies (GOORMAGHTIGH and PANNIER, 1939; DE CASTRO, 1940, 1951, 1962; DE BOISSEZON, 1943; CELESTINO DA COSTA, 1944; SERAFINI-FRACASSINI and VOLPIN, 1966). Most of them are situated around the periphery of the organs and it is possible that they provide a by-pass for some of the blood which reaches the glomus (see p. 65).

(iii) Innervation: The vessels of supply and their branches receive postganglionic fibres from the appropriate sympathetic ganglia. The carotid body vessels are innervated by such fibres from the superior cervical ganglion, whereas stellate ganglion cells furnish the innervation of the aortic body vessels.

Of far more importance is the parasympathetic innervation. Much of this was shown to be sensory by DE CASTRO in degeneration studies of the carotid body using silver staining methods and by HOLLINSHEAD (1939, 1940) in similar experiments on the aortic bodies. The cell bodies of the sensory fibres of the carotid bodies were shown by such experiments to be situated in the petrous ganglion of IX and those of the aortic bodies in the nodose ganglion of X. DE CASTRO

(1928) reported that intracranial section of IX did not lead to any changes in the innervation of the glomus cells and HOLLINSHEAD (1939, 1940) similarly found no alteration in the innervation of the aortic bodies after cutting the vagus nerve above the nodose ganglion. However, a marked degree of degeneration of nerve endings in these bodies followed glossopharyngeal or vagus nerve section below the petrous or nodose ganglia respectively.

DE CASTRO's view that the innervation of the Type I glomus cell was *sensory*, the glomus cell being the chemoreceptive element, interposed between the blood on the one hand and the associated sensory fibre on the other, was widely accepted. DE CASTRO (1951) even went so far as to postulate that the unmyelinated afferent fibre actually penetrated the Type I cell to end within its cytoplasm in a "menisque terminal" in juxta position to the nucleus. This last assertion has received no support whatever from any subsequent study made with either the light or the electron microscope. Nevertheless, the concept that the Type I cell and its innervation constituted the sensory complex of the glomus dominated our ideas on the arterial chemoreceptors for forty years. The application of electron-microscopy coupled with degeneration studies has cast some doubt on DE CASTRO's interpretation of the function of the Type I cell.

B. Electron Microscopy

(i) Cell types: Two main types of cells can be recognized; the principal (parenchymal, epithelioid) or Type I glomus cells and the less abundant sustentacular or Type II glomus cells.

Type I cells may show direct cell-cell contact with each other but more usually are almost completely enveloped by long cytoplasmic processes of the Type II cell (Fig. 3). In such cases the cell body of the Type II cell often lies to one side of a group of Type I cells, its processes ensheathing all the Type I cells in that group (Fig. 3).

Type I cells are generally larger and rounder than Type II cells, and show a large oval nucleus. They have few processes and are characterized by the presence of electron-dense membrane bound cytoplasmic granules (dense-cored vesicles)— see Figs. 3–6. A Golgi complex and endoplasmic reticulum are prominent features; free cytoplasmic ribosomes and various types of inclusion bodies are found.

Some workers (LEVER, LEWIS, and BOYD, 1959; HOGLUND, 1967; DE CASTRO and RUBIO, 1968; CHEN, YATES, and DUNCAN, 1969; ABBOTT, DALY, and HOWE, 1971) have classified Type I cells into two types and MORITA, CHIOCCHIO, and TRAMEZZANI (1969), even have described four types, on the basis of cytoplasmic electron density, ribosome concentration and morphological features of their granules (see Fig. 3). These varieties are not rigidly separated and intermediate forms exist.

Type II glomus cells are flattened and angular; they have a relatively small cell body with dense cytoplasm. Their electron-dense nucleus is smaller than that of the Type I cell and frequently has an irregular profile; their numerous long cytoplasmic processes extend between and surround nearby Type I cells. They appear to act as a supporting tissue for Type I cells, separating them in most places from the extracellular environment. The endoplasmic reticulum is sparse

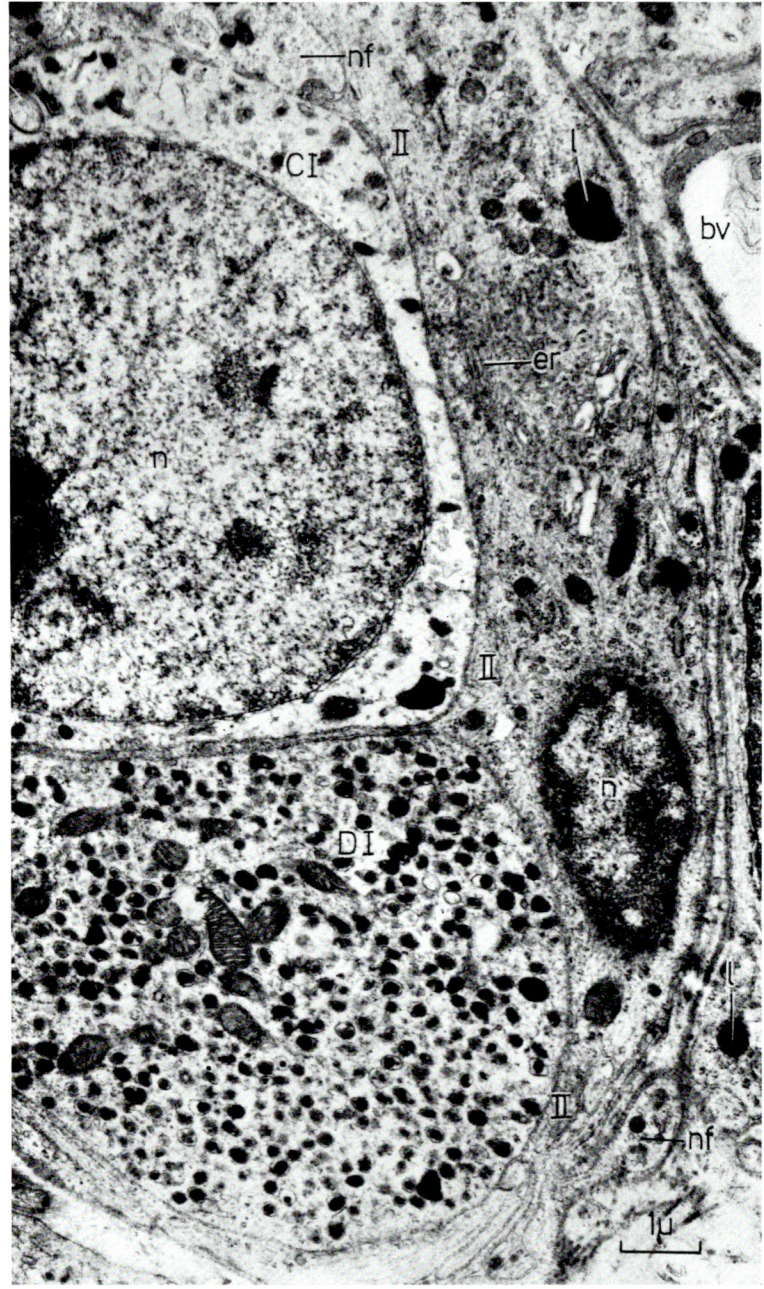

Fig. 3. Carotid body of the cat. Two Type I cells (Dense form, *DI*, exhibiting many granules und mitochondria in this case, and the Clear form, *CI*, with nucleus in profile) enveloped by cytoplasm of a Type II cell (*II*). Nerve fibres (*nf*) in association with Type II cells are visible at top and bottom of the micrograph. Part of a second Type II cell with nucleus, and a blood vessel (*bv*) in section, can be seen at the right hand edge. *n* nucleus; *l* lysosome; *er* endoplasmic reticulum

and usually scattered throughout the cytoplasm, rather than grouped together as is normally the case in the Type I cell. The Golgi complex is often more extensive than that of the Type I cell. A typical Type II cell profile possesses one or more lysosome-like bodies. Glomus granules are never present.

(ii) Granules of Type I cells: In an attempt to elucidate their role, if any, in chemoreceptor function, these granules have been studied in carotid bodies removed from animals subjected to hypoxia or to conditions known to affect the catecholamine content of tissues such as the adrenal medulla.

(1) Reserpine Treatment. LEVER et al. (1959) shewed that reserpine depleted the granules of Type I cells and FILLENZ (1968) reported that the fluorescence (due to catecholamines) of the Type I cells was reduced by reserpine. Both LEVER et al. and FILLENZ used rabbits in which the main catecholamine in the Type I cells is dopamine (DEARNALEY, FILLENZ, and WOODS, 1968). HESS (1968), DE CASTRO and RUBIO (1968) and ABBOTT, HOWE, and JOELS (unpublished) using cats could find no change in granule content of Type I cells after reserpine. CHEN and YATES (1969) and CHEN, YATES, and DUNCAN (1969), used specific sensitive cytochemical methods involving EM autoradiography and labelled precursors of catechol and indol-amines. They showed that reserpine treatment reduced the electron opacity of the granules but did not reduce their number. They concluded that reserpine caused an intracellular discharge of catecholamines. Probably their use of glutaraldehyde-dichromate as a specific stain for these amines, accounts for their positive findings. HESS, DE CASTRO and RUBIO and ABBOTT et al. all used orthodox EM preparative methods which render the granules electronopaque on the basis of their protein content rather than that of their amines.

(2) Hypoxia. Hypoxia causes no change either in number or electron density of the granules (AL-LAMI and MURRAY, 1968; CHEN et al., 1969; ABBOTT et al., unpublished). CHEN et al. conclude that the monoamines of these granules are not involved in the chemoreceptor response to hypoxia.

(3) Sinus Nerve Efferent Stimulation. YATES, CHEN and DUNCAN (1970) reported that efferent stimulation of the sinus nerve decreased the electron density of the granules; atropine prevented this response. As sinus nerve efferents directly innervate Type I cells and influence chemoreceptor discharge (see p. 69), this important finding merits further investigation.

YATES et al. showed that sympathetic stimulation did not alter the electron density of the granules.

(iii) Blood vessels: Little new information has accrued from EM studies. The capillaries are fenestrated as are those of many tissues. Surprisingly, no reports of arterio-venous anastomoses have yet been made, despite the possible importance of these structures (see p. 65).

(iv) Innervation: An analysis of cross sections of the nerves as they enter the glomus wrapped in Schwann cell cytoplasm shows that most are unmyelinated at this point and are approximately 0.2–0.5 μ in diameter. After losing their Schwann sheath, they course in the intralobular connective tissue in mesaxon-like folds of Type II cell cytoplasm. Many of the nerve fibres establish a "synapse" with Type I cells; these terminals are bulbous or flat and vary in size from 0.5 μ even to 9 μ. At this point the Type I cell is usually free from investing Type II cytoplasm (Fig. 5). The "synaptic" region is characterised by the presence of one

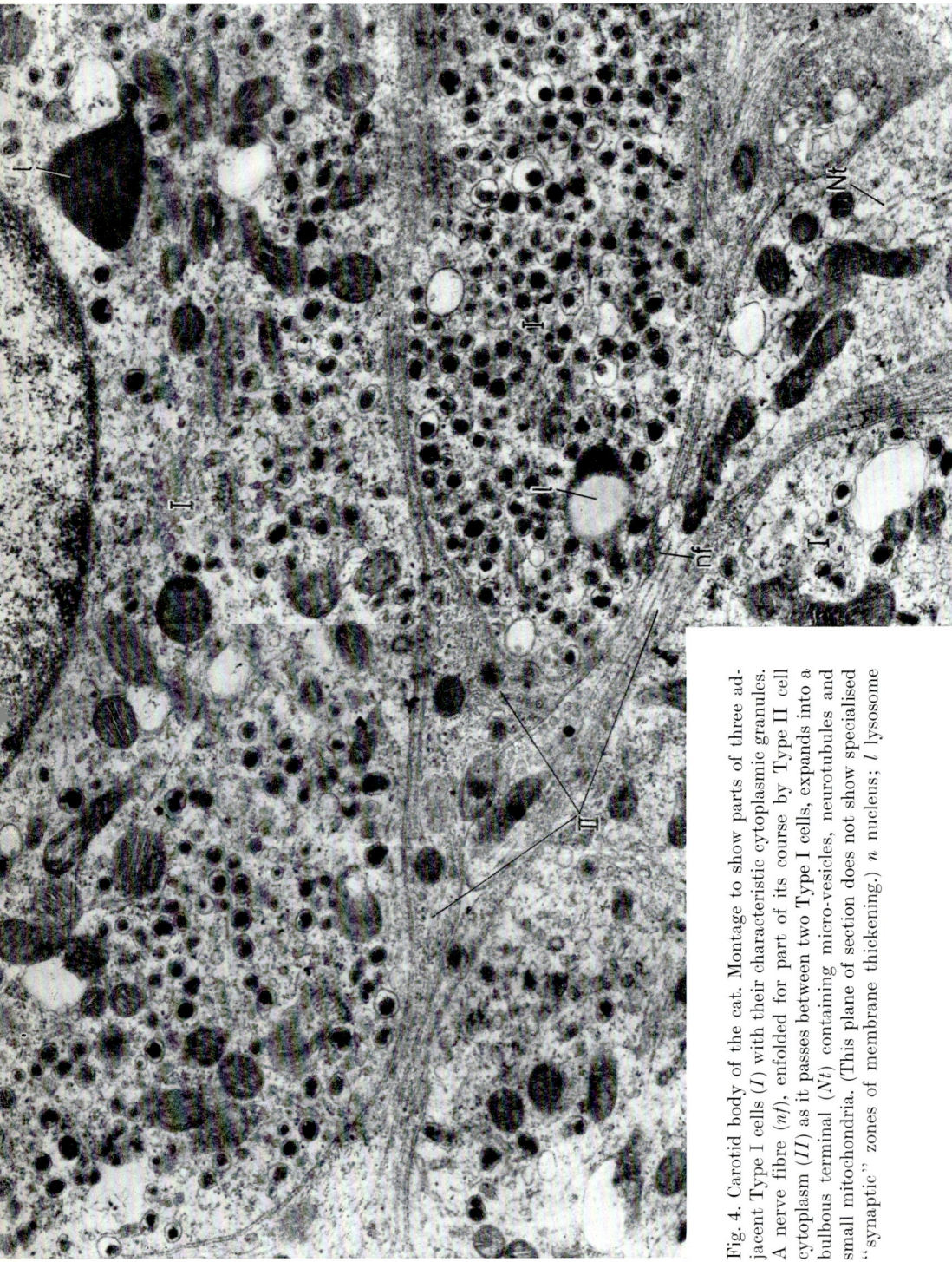

Fig. 4. Carotid body of the cat. Montage to show parts of three adjacent Type I cells (*I*) with their characteristic cytoplasmic granules. A nerve fibre (*nf*), enfolded for part of its course by Type II cell cytoplasm (*II*) as it passes between two Type I cells, expands into a bulbous terminal (*Nt*) containing micro-vesicles, neurotubules and small mitochondria. (This plane of section does not show specialised "synaptic" zones of membrane thickening.) *n* nucleus; *l* lysosome

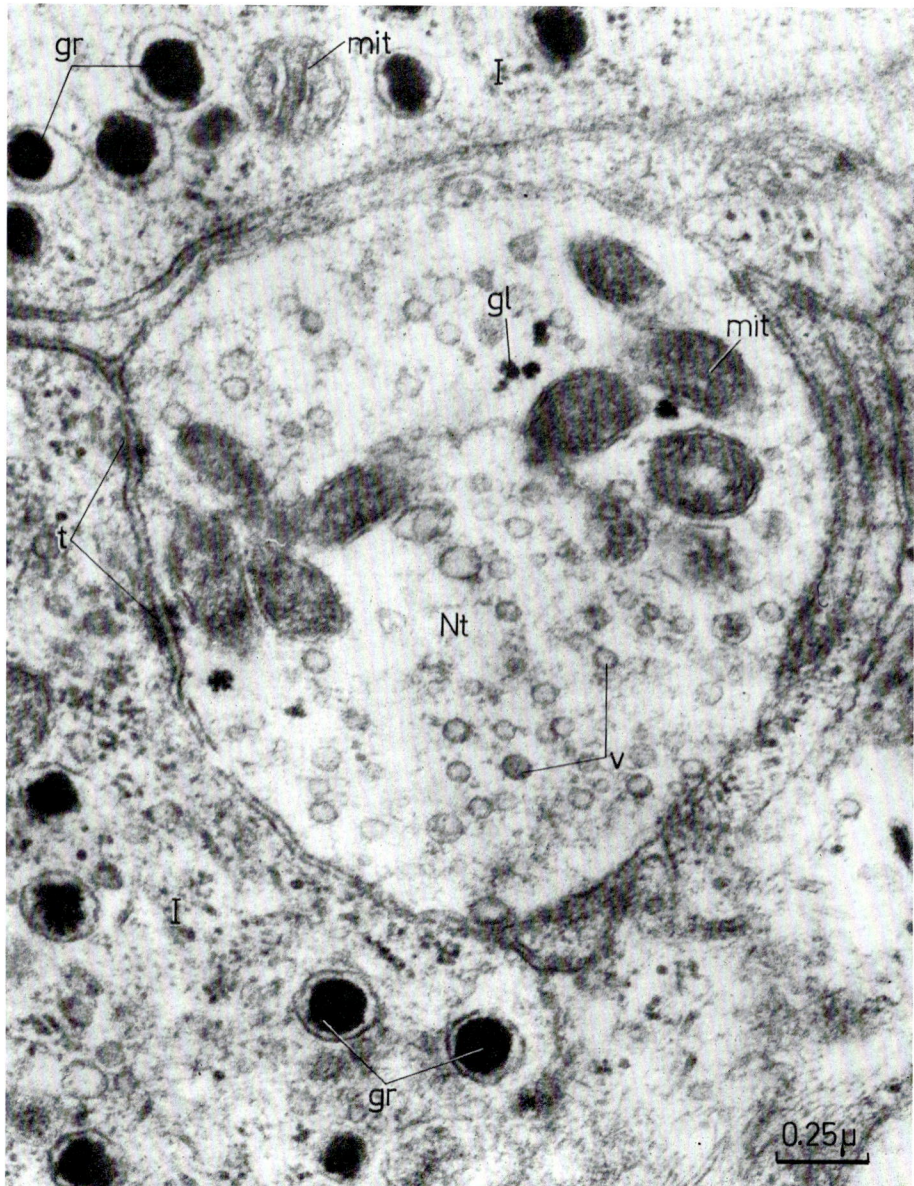

Fig. 5. Carotid body of the cat. High power view of parts of two Type I cells (*I*) with their characteristic membrane-bound cytoplasmic granules (*gr*). A nerve terminal (*Nt*) is in "synaptic" association with one of them, the apposed plasma membranes being free from investing Type II cytoplasm at this point and exhibiting the typical thickened zones (*t*). The terminal contains the characteristic abundant microvesicles (*v*), small mitochondria (*mit*), and aggregations of glycogen granules (*gl*)

or more electron-dense zones or thickenings (up to $0.5\,\mu$ long) of the apposed plasma membranes of the nerve terminal and the Type I cell cytoplasm. The

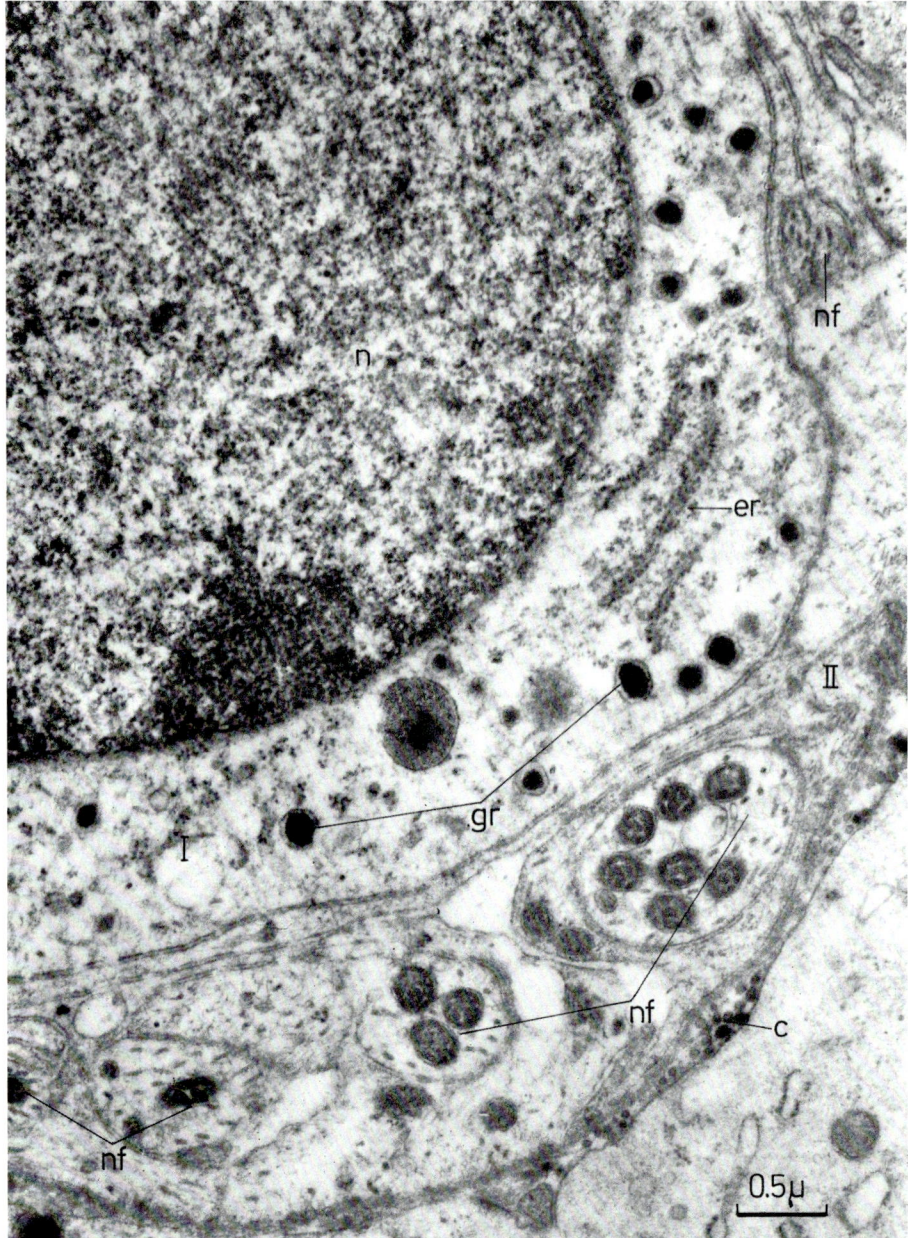

Fig. 6. Carotid body of the cat. Small unmyelinated nerve fibres (*nf*), containing neurotubules and small mitochondria and apparently invested in Type II cell cytoplasm, on the periphery of a Type I cell (*I*). Whether these fibres are pre-terminal or en passage is not known. *n* nucleus; *gr* membrane-bound granules; *er* endoplasmic reticulum; *c* collagen bundles

membranes may be separated by a "synaptic" intercellular cleft of some 80 to 300 Å. These nerve endings are characterised by their high content of small vesicles ("synaptic vesicles") 400–700 Å in diameter, and by the presence of numerous small mitochondria and aggregations of glycogen granules (Fig. 5).

Examination of large numbers of random sections suggests that usually only one nerve ending innervates each Type I cell, although occasionally 2 or 3 are seen (BISCOE and STEHBENS, 1965, 1966). These may conceivably be the endings of different fibres impinging on a common glomus cell. Alternatively, they may derive from the same nerve fibre coursing over the Type I cell surface, either cut more than once in section or actually making multiple contacts.

Two varieties of nerve ending on Type I cells have been reported by both AL-LAMI and MURRAY (1968) and HOGLUND (1967), viz: (a) compact bulbous endings crowded with micro-vesicles, mitochondria and glycogen granules and having thickenings of their plasma membrane; (b) broad flat endings having relatively "empty" cytoplasm with few vesicles; these have extensive contact with Type II cells also.

C. Nature of the Terminals on the Type I Glomus Cells—Afferent or Efferent?

The appearance of the nerve endings on Type I cells is suggestive of their subserving an efferent function (BISCOE and STEHBENS, 1965, 1966, 1967), with the nerve terminal as the pre-junctional and the glomus cell the post-junctional element. However, morphological criteria alone are insufficient for the characterisation of the function of a nerve fibre and its terminals. Micro-("synaptic") vesicles have been described in many sensory structures: rods and cones presynaptic element (DE ROBERTIS and FRANCHI, 1956); Pacinian corpuscles (PEASE and QUILLIAM, 1957); Organ of Corti (SMITH and DEMPSEY, 1957); Taste buds (DE LORENZO, 1958); Cristae ampullares (ENGSTRÖM and WERSÄLL, 1958); Meissner's corpuscles (CAUNA and ROSS, 1960); dendrites (GRAY, 1963); dendrodendritic synapses (RALL, SHEPHERD, REESE, and BRIGHTMAN, 1966).

Electron-dense thickenings (desmosomes) are similarly a feature of many non-nervous cell contacts.

Degeneration studies have been performed to elucidate the matter. Since the cell bodies of the chemoreceptor fibres are situated in the petrous ganglion, section above (central) to the ganglion should not cause degeneration if they are efferent with cell bodies lying in the CNS. Section of the sinus nerve, or of the glossopharyngeal nerve itself below the petrous ganglion must cause degeneration of all endings of IX nerve origin on Type I cells, whether they be endings of afferent or efferent fibres.

DE CASTRO (1928) reported that no change could be detected in the nerve endings on Type I cells twelve days after intracranial section of IX (and sometimes X and XI). His observations were confined to the study of silver stained material with the light microscope.

DE CASTRO and RUBIO (1968) and BISCOE, LALL, and SAMPSON (1970) have repeated such experiments using the electron microscope. Their conclusions differ completely from each other. DE CASTRO and RUBIO find that whereas degeneration develops in the Type I endings within 48 hours of infra-petrosal (extracranial) section of IX, no degeneration could be detected 30 days after cutting the roots of IX and X intracranially (i.e. central to the ganglion). They conclude, as DE CASTRO did before, that the fibres innervating Type I are afferent in nature.

BISCOE et al. reported that 60% of the microvesicle—containing nerve endings on Type I cells degenerated after cutting IX intracranially. The time course of the degeneration was prolonged. BISCOE et al. made electroneurographic recordings of chemoreceptor activity from the sinus nerve with the same ease on the "operated" side as on the non-operated side. They concluded that all endings on Type I cells are *efferent* and that their cell bodies are probably located in the brain stem.

Judgment between these diametrically opposed findings must depend on the quality of the evidence offered. The electronmicrographs of BISCOE et al. are far superior to those of DE CASTRO and RUBIO and leave little room for doubt that intracranial section of IX and X has indeed caused degeneration of *many* of the terminals on Type I cells. However, their own counts show that 40% of these endings survived, which makes their claim that *all* nerve endings on Type I cells are efferent somewhat paradoxical. It seems to us likely that the chemoreceptor impulse activity which they recorded came at least in part from surviving fibres whose terminals ended on Type I cells.

If all the Type I nerve endings are efferent, as BISCOE et al. assert, then the chemoreceptor fibres must end elsewhere. It is possible that the small unmyelinated nerves enveloped by Type II cells or by folded membrane systems may subserve this function. DE KOCK and DUNN (1968) have suggested that such nerve fibres may end without showing any morphological specialization. Whether Type II cells themselves form a functional part of the chemoreceptor apparatus is quite unknown.

III. Function

A. The Influence of Hypoxia on Chemoreceptor Activity, Glomus Blood Flow and Oxygen Usage

There are four types of hypoxia:

(i) Hypoxic—a fall in arterial pO_2;

(ii) Anaemic—a reduction in (HbO_2) in the arterial blood;

(iii) Stagnant—a reduction in the blood flow;

(iv) Histotoxic—paralysis of the enzymes of the respiratory transfer chain which renders the mitochondria of the cell incapable of using oxygen.

Of these, histotoxic hypoxia (iv) of course does not occur naturally. Sodium cyanide causes histotoxic hypoxia by blocking the enzyme cytochrome a_3 (see p. 66). The chemoreceptive reflex role of the carotid body was shown first by the injection of sodium cyanide and the drug provides a useful tool for proving that the chemoreceptors and their nerve fibres are functional and for examining the response of chemoreceptor units to this disruption of their metabolism (see p. 66). Most of this section however will be devoted to the chemoreceptor responses to more "physiological" types of hypoxia.

(i) Hypoxic hypoxia: The chemoreceptor discharge recorded from a single or few-fibre preparation of the sinus nerve cut centrally in an animal spontaneously

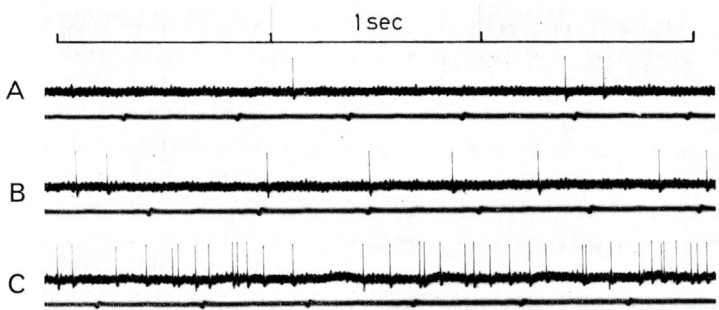

Fig. 7 A–C. Chemoreceptor activity in a few fibre preparation made from the carotid sinus nerve of a cat breathing spontaneously. Each slip shows electroneurogram (above) and electrocardiogram (below). A Breathing air—B.P. = 110 mm Hg. B Breathing 5% CO_2 in air—B.P. = 105 mm Hg. C Breathing 5% O_2 in N_2—B.P. = 105 mm Hg

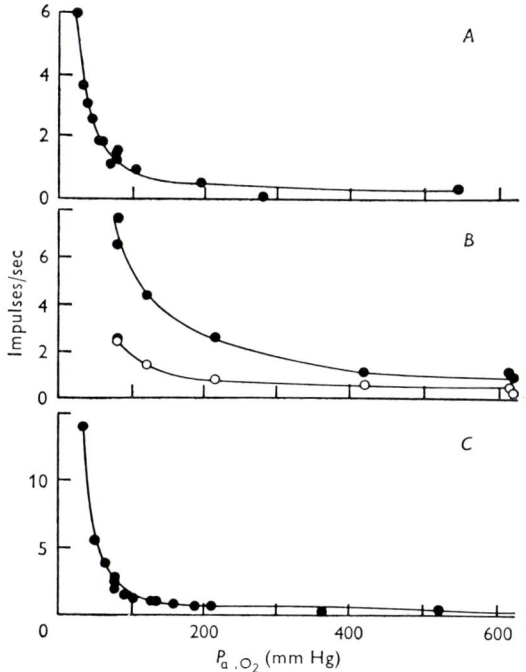

Fig. 8 A–C. The rate of carotid chemoreceptor discharge (impulses/second) in single fibres, plotted against the arterial O_2 tension (mm Hg) all from the same cat. B shows two fibres (●, ○) from the same strand, A and C are from single fibres. Arterial pCO_2 was 28–31 mm Hg for A and B and 31–34 mm Hg for C. Mean arterial pressure was 95 ± 7 mm Hg in A and B and was 123 ± 6 mm Hg in C. (From BISCOE, PURVES, and SAMPSON, 1970)

breathing air (arterial pO_2 85–100 mm Hg) is sparse and irregular (Fig. 7A). If oxygen is substituted as the inspired gas, chemoreceptor discharge almost vanishes but an occasional impulse can still be seen on the record. Some of this hyperoxic residual discharge is due to the influence of CO_2 on chemoreceptor discharge. Thus the substitution of oxygen for air in the spontaneously breathing anaesthetized animal depresses the respiration because the higher arterial pO_2 reduces the afferent chemoreceptor activity in the intact innervated chemoreflex zones. Consequently, the arterial pCO_2 rises slightly and this influence itself is sufficient to provoke some chemoreceptor discharge. Only if the animal is artificially hyperventilated with room air does the chemoreceptor discharge virtually disappear.

Hypoxic hypoxia is an important stimulant of the chemoreceptors (Fig. 7C). The effect of hypoxia alone on the vascularly isolated carotid body perfused by blood at known pO_2, pCO_2 and pH has been reported by BISCOE, PURVES and SAMPSON (1970). The cats used were paralyzed with gallamine triethiodide and were artificially ventilated so as to provide a source of arterial blood at a constant pCO_2. Single chemoreceptor units were peeled from the carotid sinus nerve (cut centrally). The responses of three of these units to arterial pO_2 values between 30 and 450 mm Hg are shown in Fig. 8.

The response of each unit is hyperbolic. There is little change in impulse discharge when the arterial pO_2 is raised from 100 to 450 mm Hg. Discharge increases rapidly as the arterial pO_2 falls below 100 mm Hg.

(ii) Anaemic hypoxia: Earlier studies of the effects of anaemic hypoxia (as produced by progressive carboxyhaemoglobinaemia) on chemoreceptor excitation were reported by COMROE and SCHMIDT (1938) and by DUKE, GREEN, and NEIL (1952). DUKE et al. were unable to find any evidence of chemoreceptor discharge in carboxyhaemoglobinaemia, but they used multifibre preparations of the sinus nerve. However, PAINTAL (1967) reported that single chemoreceptor units of the aortic nerve showed an increased activity when the animals were ventilated with 2% CO in air. As no measurements of arterial pO_2 were made, it was uncertain whether any hypoxic hypoxia occurred concurrently in these experiments of PAINTAL. MILLS and EDWARDS (1968) reinvestigated the matter using cats breathing or ventilated by 0.3% or 1% CO in air. They found an obvious increase in chemoreceptor activity in few-fibre units of the carotid sinus nerve when the blood (COHb) exceeded 20%. When the percentage saturation of the blood by CO reached 40% the impulse frequency increased five-fold.

These results of MILLS and EDWARDS are important because they show clearly that the oxygen requirements of the chemoreceptors are not satisfied simply by oxygen in simple solution in the blood (TORRANCE, 1969). As the arterial pO_2 in their experiments remained almost constant, the increased discharge in carboxyhaemoglobinaemia indicates that the A-V O_2 difference across the glomus is significant under normal circumstances in animals breathing air.

(iii) Stagnant hypoxia: LANDGREN and NEIL (1951) reported a striking increase in chemoreceptor discharge recorded from multifibre carotid chemoreceptor preparations in artificially ventilated cats subjected to arterial hypotension by bleeding. No measurements of arterial pO_2, pCO_2 or pH were made but it was assumed that these values were reasonably constant following the haemorrhage because of the constant artificial ventilation which the animals received. This assumption is not above suspicion because the perfusion of different parts of the lungs may alter in circumstances of haemorrhagic hypotension.

LEE, MAYOU, and TORRANCE (1964) extended these findings using single or few-fibre preparations of aortic chemoreceptor nerves. They showed an obvious increase of impulse activity as the arterial pressure fell below 100 mm Hg (Fig. 9A). When the cats were ventilated with hypoxic mixtures (Fig. 9B), impulse activity was of course greater at normal arterial pressure levels than on air, but still showed a considerable increase when the systemic pressure was lowered by bleeding.

FLOYD and NEIL (1952) showed that increased activity in the sympathetic fibres supplying the carotid body was implicated in the carotid chemoreceptor

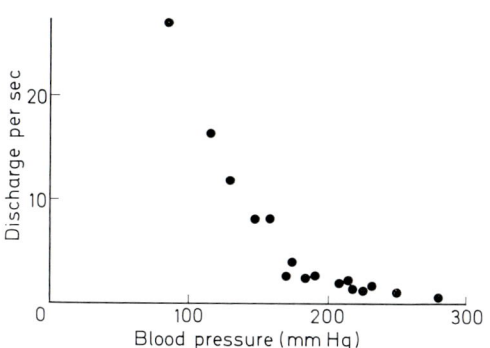
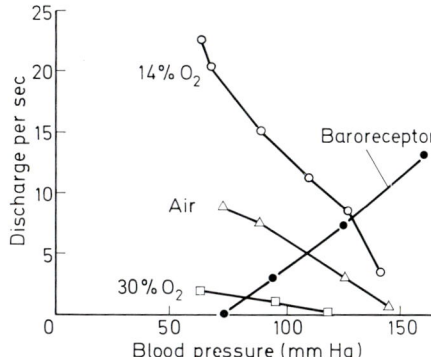

Fig. 9A. Cat. Closed chest. Vagi cut. Artificial respiration with air. Few fibre chemoreceptor preparation of the right aortic nerve. Blood pressure varied by repeated occlusions of the abdominal aorta. Steady state discharge plotted against blood pressure

Fig. 9B. Cat. Closed chest. Vagi cut. Artificial respiration on 14% O_2 in N_2, air and 30% O_2 in N_2. As for Fig. 9A. The discharge of an aortic baroreceptor fibre (●) plotted as impulses per heart beat is included for comparison. (From LEE, MAYOU, and TORRANCE, 1964)

response to haemorrhagic hypotension. Thus, haemorrhagic hypotension strikingly increased impulse discharge in the post-ganglionic fibres supplying the carotid body. Correspondingly, when the post-ganglionic sympathetic branch to the glomus was cut, chemoreceptor impulse activity recorded during haemorrhagic hypotension was reduced. FLOYD and NEIL (1952) attributed the effect of sympathetic discharge on chemoreceptor activity to a vasoconstrictor influence on the glomeral vessels but made no measurements of carotid body blood flow.

B. Blood Flow through the Carotid Body and the Role of the Sympathetic Innervation

DALY, LAMBERTSEN, and SCHWEITZER (1954) recorded blood flow through the vascularly isolated carotid body by collecting the venous effluent from the glomus and reported a value of some 40 μl/min. Blood flow was found to vary almost linearly with mean arterial blood pressure. Stimulation of the local sympathetic supply to the glomus reduced the blood flow and section of the sympathetic caused a slight increase in flow. Measurements of the A-V O_2 difference were made and these, coupled with those of the blood flow, enabled the calculation of the oxygen usage of the carotid body. The A-V O_2 difference proved almost impossible to determine in normal circumstances with the method (Roughton-Scholander microsyringe technique) available at that time. However, by reducing the arterial pressure and thereby lowering the flow, they were able to obtain repeatable results for the A-V O_2 difference and to calculate (from flow and A-V O_2) an oxygen usage of approximately 0.18 µl/min. Taking a flow of 40 µl/min then, an oxygen usage of 0.18 µl/min presupposes an A-V O_2 difference of less than 0.5 ml/100 ml. It is thus not surprising that the veins of the glomus appear bright red (which aids their identification during the dissection).

These figures of 40 µl/min blood flow and 0.18 µl/min O_2 usage seem quite trivial but have of course to be considered in terms of the mass of tissue concerned. The carotid body of the cat weighs not more than 2 mg, so the blood flow is of the order of 2000 ml/100 g/min (twenty times that through 100 g of liver) and the oxygen usage is some 9 ml/100 g/min (nearly three times that of 100 g of brain tissue). The high oxygen consumption of the carotid body is disguised by the bright red appearance of its venous blood, in a manner analogous to that seen in the kidney.

The findings of DALY et al. have been widely confirmed and have been recently extended by studies of BISCOE, BRADLEY, and PURVES (1970) and PURVES (1970a and b). BISCOE et al. reported that the blood flow of the vascularly isolated carotid body changed from some 10 µl/min at a mean perfusion pressure of 60 mm Hg to one of 60 µl/min at 160 mm Hg pressure and PURVES (1970a) showed a roughly linear relationship between carotid body blood flow and mean pressure when pO_2 and pCO_2 of the blood were held constant. He reported mean values for flow, A-V O_2 difference and O_2 usage of the glomus in cats as follows:

Average blood flow = 41.5 µl/min (range 33–68)
Average A-V O_2 difference = 0.34 ml/100 ml (range 0.21–0.46)
Average O_2 usage = 0.147 µl/min (range 0.115–0.195)

BISCOE et al. (1970) found that changes of perfusion pressure in the range 70–160 mm Hg had virtually no effect on *steady state* chemoreceptor activity (unit discharge) if pO_2, pCO_2 and pH of the arterial blood were kept constant. They did note that some two-thirds of the single chemoreceptor fibres tested responded within 5 sec to an abrupt change of perfusion pressure with an increased frequency (if the pressure fell) and with a decreased discharge (if the pressure rose). One third of their preparations were irresponsive to changes of pressure in the range of 70–160 mm Hg, even with the initial rise or fall of pressure. The *steady* discharge of each unit was not affected by the perfusion pressure, providing this did not fall below 70 mm Hg or exceed 160 mm Hg. These results are quite different from what might have been forecast from the results on multifibre preparations of the carotid chemoreceptors obtained by LANDGREN and NEIL (1951) and from those on single and very few-fibre preparations of the aortic nerve by LEE et al. (1964).

Both LANDGREN and NEIL (1951) and LEE et al. (1964), however, studied chemoreceptor responses to changes of pressure in the whole animal and in each case the sympathetic supply to the chemoreceptor tissue under study was intact. BISCOE et al. (1970) also preserved the local sympathetic supply to the glomus but, as they had cut the carotid sinus nerve from the glomus under investigation, such results as they obtained on the perfused carotid body could not have been influenced by any reflex alteration of local sympathetic discharge because the afferent arm of the reflex had been destroyed. When the whole animal is subjected to hypotension the reduction in baroreceptor activity alone causes powerful reflex stimulation of the sympathetic nervous system (for references see HEYMANS and NEIL, 1958) and the local sympathetic vasoconstrictor discharge exacerbates the chemoreceptor activity.

BISCOE et al. nevertheless did record some chemoreceptor unit responses to changes of systemic pressure in the whole animal, maintaining pO_2, pCO_2 and pH constant and compared these results with their findings obtained on the perfused preparation at the same mean pressures and blood gas tensions. Chemoreceptor

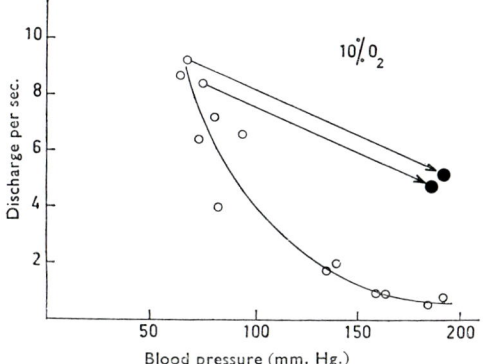

Fig. 10. Cat. Closed chest. Vagi cut. Artificial respiration with 10% O_2 in N_2. As in Fig. 9 A graph is plotted (○) of steady state discharge of a few fibre preparation of the right aortic nerve at different blood pressures obtained by repeated occlusions of the abdominal aorta. The effects of two carotid occlusions (●) are compared with the control curve (○). (From LEE, MAYOU, and TORRANCE, 1964)

activity differed by only 1.2 impulses/sec (over the pressure range of 60–160 mm Hg) in the two methods of "natural" and "artificial" perfusion. Thus the explanation offered above for the discrepancy between the results of BISCOE et al. and the earlier findings of LANDGREN and NEIL and of LEE et al. is not fully satisfactory.

LEE et al. (1964) provided persuasive evidence of the role of the local sympathetic supply to the aortic body in exacerbating chemoreceptor discharge at any given systemic pressure. Fig. 10 shows that carotid occlusion, which causes reflex hypertension due to cardiac and vasomotor sympathetic stimulation, provokes a chemoreceptor discharge well in excess of that induced by raising the blood pressure by transfusion or by occlusion of the abdominal aorta. LEE et al. proved that this effect could not be attributed to any difference in arterial pO_2. MILLS (1968) has shown that the artificial stimulation of the sympathetic supply to the aortic body increases chemoreceptor discharge. BISCOE and PURVES (1967) have reported that cervical sympathetic activity aroused by passive movements of the hind limb induces a rapid chemoreceptor response. Lastly, MILLS and SAMPSON (1969) have shown that cervical sympathetic stimulation induces hyperpnoea in decerebrate unanaesthetized cats, if and only if the corresponding carotid sinus nerve is intact.

There is thus ample confirmation that sympathetic discharge to the glomus increases the chemoreceptor activity for a given pO_2 and arterial pressure. Hitherto, this effect has been ascribed to sympathetic vasoconstriction, but further details of the action of the sympathetic on the glomus have been provided lately by Purves (1970a and b).

C. The Influence of the Sympathetic Nerve Supply upon Glomeral Oxygen Usage

Purves examined the effects of hypoxia (arterial $pO_2 = 30$–40 mm Hg) or hypercapnia (arterial $pCO_2 = 50$ mm Hg) on carotid blood flow, A-V O_2 difference and oxygen usage in the vascularly isolated carotid body supplied by blood. The local sympathetic innervation and the carotid sinus nerve were both intact. Each of these chemical stimuli slightly increased blood flow through the glomus but halved the A-V O_2 difference and caused an average fall of glomus oxygen usage of the order of 36%.

When the sympathetic supply was cut in animals ventilated with air, Purves (1970b) found the carotid body flow rose by 9.2 µl/min and the A-V O_2 difference also increased by an average of 0.09 ml/100 ml. Hence, the carotid body oxygen usage increased by 0.075 µl/min. Stimulation of the sympathetic supply caused the opposite changes. Vasoconstriction induced by sympathetic stimulation was enhanced if the arterial pCO_2 were high.

After sympathectomy, hypercapnia or hypoxia still caused a rise in glomeral blood flow but induced a fall in A-V O_2 difference, so that the O_2 usage remained constant during hypoxia or hypercapnia. Lastly, after sympathectomy, a deliberate reduction of perfusion pressure lowered flow but increased the A-V difference so that the glomeral oxygen usage remained constant. When the sympathetic supply was intact, a lowered perfusion pressure lowered glomeral flow but A-V O_2 difference remained constant so that glomeral O_2 usage decreased *pari passu* with the reduction in pressure.

Thus, whether the sympathetic supply to the glomus is intact or not, hypoxia or hypercapnia increase the flow. As the sympathetic itself is vasoconstrictor, then hypoxia or hypercapnia must themselves cause non-sympathetic effects which override this vasoconstrictor action. These effects may either be direct on the calibre of the glomus vessels or, more probably (see p. 70), via reflex excitation of parasympathetic efferent nerve fibres coursing in the sinus nerves (which were intact in Purves' experiments). Perhaps of more significance is the effect of sympathetic discharge in lowering the oxygen usage of the carotid body. In discussion of his results, Purves discarded the proposition that catecholamines liberated by such sympathetic stimulation exert a direct effect on the glomus cells, citing the results of Eyzaguirre and Lewin (1961), who showed that chemoreceptor discharge recorded from the carotid body superperfused *in vitro* (see p. 72) is not affected by sympathetic stimulation. He suggested that sympathetic activity might affect O_2 usage by diverting blood within the glomus, from areas of high metabolism to areas of low metabolism. The resultant hypoxia of such areas of high metabolic activity would favour anaerobic metabolism and a fall in oxygen consumption of the glomus as a whole, with a resultant increase in the oxygen content of the venous effluent.

D. Possible Significance of A-V Anastomoses

At this point it is appropriate to consider the possible role of A-V anastomoses repeatedly described as an invariable feature of chemoreceptor tissue by DE BOIS-SEZON (1943), CELESTINO DA COSTA (1944), DE CASTRO (1940, 1951), and DE CASTRO and RUBIO (1968) and already referred to (see p. 50, 53).

DE CASTRO (1951) described in great detail the alteration in the appearance of the glomus and its venous drainage during hypertension (provoked by adrenaline or by occlusion of the abdominal aorta) and during hypoxia, hypercapnia, or hypocapnic hyperoxia. Hypertension following an adrenaline injection (i.v.) caused an increase in size of the glomus and a slowing of the venous outflow. However, neither the hypertension caused by occluding the abdominal aorta nor the local fall of pressure in the carotid bifurcation caused by clamping the common carotid induced any change in the appearance of the glomus or its venous outflow. Hypoxia caused a shrinkage in the size in the glomus and an increased velocity of flow through the glomeral veins, and hypercapnia produced similar changes. Hypocapnic hyperoxia led to swelling of the glomus and marked slowing of venous flow. DE CASTRO considered that in circumstances associated with chemoreceptor stimulation, the A-V anastomoses opened, although he was unable to define the mechanism of this effect, because of our total ignorance of the innervation of these anastomoses. He also considered that the carotid body was equipped with a reflex pathway which helped to secure a stabilization of glomeral flow in the face of even striking changes of blood pressure, thus allowing the sensory endings to act to the best advantage as chemoreceptors. The afferent limb of this reflex arc he concluded to be the baroreceptors situated in the walls of the occipital artery and the vessels supplying the glomus. He suggested that the efferent limb consisted of preganglionic glossopharyngeal fibres which relayed impulses to microganglia situated in the connective tissue surrounding the glomus and whose fibres were destined for the small arteries and arterioles of the glomus.

There is now evidence that the efferent stimulation of the carotid sinus nerve does cause hyperaemia (NEIL and O'REGAN, 1969; BISCOE, BRADLEY, and PURVES, 1970) and that this increase in flow is accompanied by an increase in A-V O_2 difference and hence an increase in glomeral oxygen usage. This would suggest that stimulation of sinus nerve efferents directs more blood into the metabolic areas of the carotid body, and is hardly in keeping with the proposition of DE CASTRO. The matter remains *sub judice*. Nevertheless, the proposition that some of the blood collected by the venous drainage cannula has not traversed the glomus capillaries (and has not suffered any loss of oxygen) because it has been "shunted" requires serious consideration (JOELS and NEIL, 1963). FORSTER (1968) has calculated that the pO_2 of the glomus cells cannot be more than 10 mm Hg below that of the arterial blood, so that with an arterial pO_2 of 100 mg Hg tissue pO_2 would be 90 mm Hg. Elsewhere in the organism cellular pO_2 must drop to a critical value of 1–3 mm Hg before aerobic metabolism is disturbed. Yet the carotid and aortic chemoreceptors are discharging at an arterial pO_2 of 100 mm Hg and this discharge increases steadily as the arterial pO_2 falls.

E. The Role of Energy Rich Phosphates in the Chemoreceptors

In their monograph, ANICHKOV and BELEN'KII (1963) summarized a large volume of work on the effects of metabolic inhibitors on chemoreceptor discharge (see also KRYLOV and ANICHKOV, 1968; JOELS and NEIL, 1968). ANICHKOV and BELEN'KII advanced the proposition that chemoreceptor discharge occurred as a result of a reduction of the ATP concentration of the chemoreceptor cell.

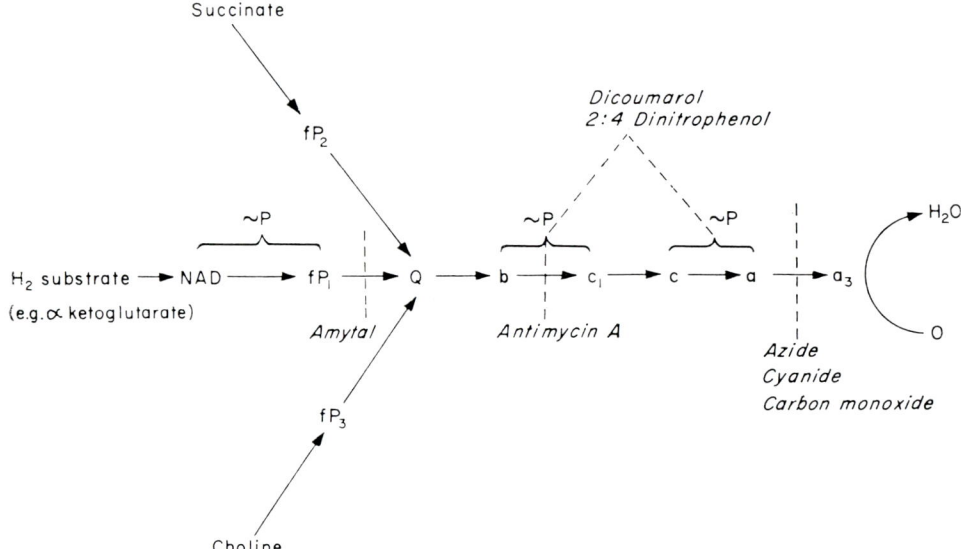

Fig. 11. The respiratory phosphorylating chain. Substances in italics disrupt the normal metabolic processes at the points indicated. (From JOELS and NEIL, 1968)

The role of the cellular respiratory transfer chain in removing hydrogen from a hydrogenated substrate such as α-Ketoglutarate and then passing it by a combination of hydrogen and electron transfers, finally to provide the hydrogen which uses oxygen to form water is depicted in Fig. 11.

During this process three ATP molecules are formed and this is known as oxidative phosphorylation. Different inhibitors can block the respiratory chain and each of these stimulates chemoreceptor discharge, just as does hypoxic hypoxia (JOELS and NEIL, 1964).

ANICHKOV and BELEN'KII showed that the prolonged perfusion of hypoxic blood through the carotid body led to a progressive "exhaustion" of the chemoreceptor response. The introduction of ATP into the perfusing blood restored the responsiveness of the preparation to hypoxia and to other chemoreceptor agents.

TORRANCE (1968) quotes LONGMUIR (1966) in giving evidence that a falling pO_2 reduces the ATP concentration of a tissue well before it reduces the oxygen usage of the tissue. Thus if the tissue pO_2 is adequate, aerobic metabolism may raise the ATP:ADP ratio to 1000:1 and glycolysis starts only when the ATP:ADP ratio falls to 1:1. Glycolysis becomes significant only when the tissue pO_2 falls to a level low enough to reduce the oxygen usage of that tissue. Hence, it is quite possible that a fall of ATP concentration initiates chemoreceptor discharge before any reduction of oxygen usage occurs. It is still not clear why the ATP concentration should fall at relatively high tissue oxygen tensions. Cytochrome a_3 has a K_m below 1 mm Hg and this is one of the reasons for the low critical pO_2 of most cells of the organism.

JOELS and NEIL (1962) showed that solutions equilibrated with carbon monoxide at tensions above 500 mm Hg would excite the carotid chemoreceptors when perfused through the carotid body. This excitation was reversible by strong light which is highly suggestive of a cytochrome oxidase inhibition by CO. LEE and MATTENHEIMER (1964) suggested that some inhibitor was present in the glomus cells which raised the K_m of cytochrome oxidase.

More biochemical investigations are required before ANICHKOV and BELEN'KII's views can be substantiated.

F. The Effects of CO_2 and pH on Chemoreceptor Activity, Carotid Body Blood Flow and Oxygen Usage

Reflex studies by HEYMANS and his colleagues (for references see HEYMANS, BOUCKAERT, and REGNIERS, 1933; HEYMANS and NEIL, 1958) showed that raised pCO_2 or [H⁺] of the blood perfused through the vascularly isolated innervated carotid body caused powerful hyperpnoea and vasoconstriction. BISCOE, PURVES, and SAMPSON [1970] have provided evidence of the response of single chemoreceptor units to changes of pCO_2 in blood perfused at steady pressure and constant pO_2 through the glomus. As noted by EULER et al. (1939), the discharge increased linearly over a limited range of CO_2 (Fig. 12). Even if the arterial pH were kept constant, chemoreceptor discharge still increased in response to a rise of arterial pCO_2.

LEE, MCCLOSKEY, and TORRANCE (1964) showed that the chemoreceptor response to CO_2 was independent of blood pressure and MCCLOSKEY and TORRANCE (1965) found (using the denervated carotid body of the rabbit) that over a range of 80–150 mm Hg, carotid body blood flow was independent of blood pressure in eight out of thirty cases, provided that the pCO_2 of the blood supplying the glomus was normal or above normal. Hyperventilation abolished this "myogenic" autoregulation. As MCCLOSKEY (1968) stated, "a carotid body which is sensitive to the level of hypoxia and to the blood pressure is functionally a receptor which reports inadequacy of oxygen *transport*; a carotid body which is sensitive only to the level of hypoxia, however, is more truly a 'chemoreceptor' being functionally a receptor for oxygen tension".

MCCLOSKEY points out that any results which show a dependence of hypoxic discharge on blood pressure above a pressure of 80–90 mm Hg must argue against the presence of autoregulation. Both LANDGREN and NEIL (1951) and LEE et al. (1964), who reported that hypoxic chemoreceptor discharge was affected by systemic pressure, artificially ventilated their animals and it is quite likely that the arterial CO_2 tension in their experimental preparations was correspondingly too low for autoregulation to occur.

McCloskey (1970) has extended his findings, showing that the phenomenon of autoregulation can be demonstrated in the carotid body of cats, provided that barbiturate anaesthesia is avoided. However, even under chloralose anaesthesia only three cats out of six showed autoregulation. These findings were obtained on the denervated carotid body; both the local sympathetic activity and efferent traffic in the sinus nerve may normally modify the blood flow response in circumstances of hypoxic or asphyxial chemoreceptor stimulation, or when marked changes of systemic blood pressure occur. The possibility of autoregulation of carotid body blood flow is attractive but further evidence is required.

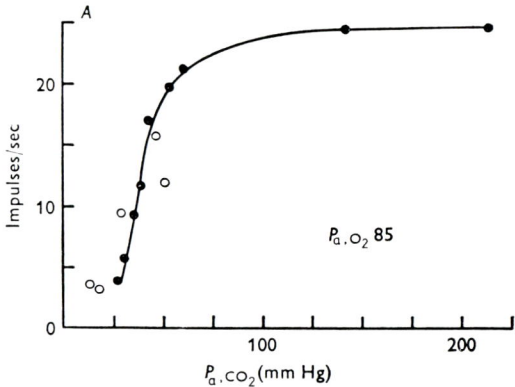

Fig. 12. Single carotid chemoreceptor fibre discharge plotted against arterial CO_2 tension. Rate and volume of artificial ventilation were varied to change arterial CO_2 tension and so also the pH. Arterial oxygen tension was 85 mm Hg throughout. ● results after the CO_2 tension was increased; ○ results after the CO_2 tension was decreased. (From Biscoe, Purves, and Sampson, 1970)

Purves (1970a) found that an increase of arterial CO_2 tension (and $[H^+]$) caused an increase in carotid blood flow, a fall in A-V O_2 difference and a small fall in oxygen consumption of the *innervated* carotid glomus. Whether the fall in oxygen consumption is due to "shunting" of the blood via arteriovenous anastomoses remains undecided. The sympathetic innervation was intact; after sympathectomy, although a rise in arterial CO_2 tension still increased flow, it now caused a decrease in the A-V O_2 difference, so that the oxygen usage remained almost unchanged.

The combination of the effects of hypoxia and hypercapnia has not been studied using single chemoreceptor units. Hornbein, Griffo, and Roos (1961) and Neil and Joels (1961, 1963) have shown evidence of such interaction in multifibre units, but without the results of an analysis of the response of single chemoreceptor units the problem remains unsolved. It is known that CO_2 and O_2 lack exert a more than additive effect on the breathing (Nielsen and Smith, 1951; Lloyd, Jukes, and Cunningham, 1958), and it seems likely, though not proven, that this effect is exerted via the chemoreceptors.

G. The Influence of Efferent Components in the Sinus and Aortic Nerves on Chemoreceptor Activity

JOELS (1960) recorded efferent impulse activity in the sinus nerve cut near its junction with the carotid body. The discharge recorded was sparse and irregular and was little affected by moderately severe hypoxia, asphyxia, or hypotension.

EYZAGUIRRE and LEWIN (1961) showed that some of the centrifugal activity thus recorded was due to the presence of post-ganglionic sympathetic fibres derived from the adjacent superior cervical ganglion. Such fibres joined the glossopharyngeal trunk and coursed downwards into the sinus nerve towards the glomus.

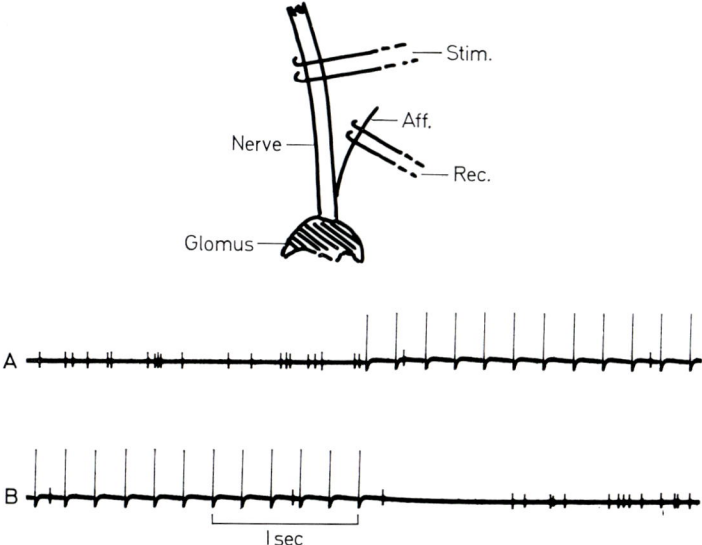

Fig. 13. The effects of efferent stimulation of the cut aortic nerve on the chemoreceptor afferent activity recorded from a slip of this nerve. The arrangement of the preparation is shown schematically in the upper section. *Stim* stimulating electrodes; *Rec* recording electrodes; *Aff* afferent slip. Lower section: A = control and early period of a stimulation 25 V, 5 c/s (1 msec) which lasted for a total period of 30 sec. B = end of stimulatory and early recovery periods. Large deflection = stimulus artifact. (From R. G. O'REGAN, 1970)

BISCOE and SAMPSON (1967) analyzed the situation more fully and reported two types of discharge—rhythmic and non-rhythmic discharge. Rhythmic discharge was abolished by cutting the cervical sympathetic trunk and thus probably featured the impulse activity of post-ganglionic sympathetic nerves. Non-rhythmic discharge was unaffected by sympathetic section but was markedly increased by the intravenous injection of small doses of adrenaline—in amounts which barely raised the blood pressure.

The influence of such efferent discharge upon chemoreceptor afferent activity was shown to be inhibitory by NEIL and O'REGAN (1969a, 1971a). Stimulating electrodes were placed on the central part of the intact sinus nerve and a thin slip containing afferent fibres was peeled from the sinus nerve trunk near its

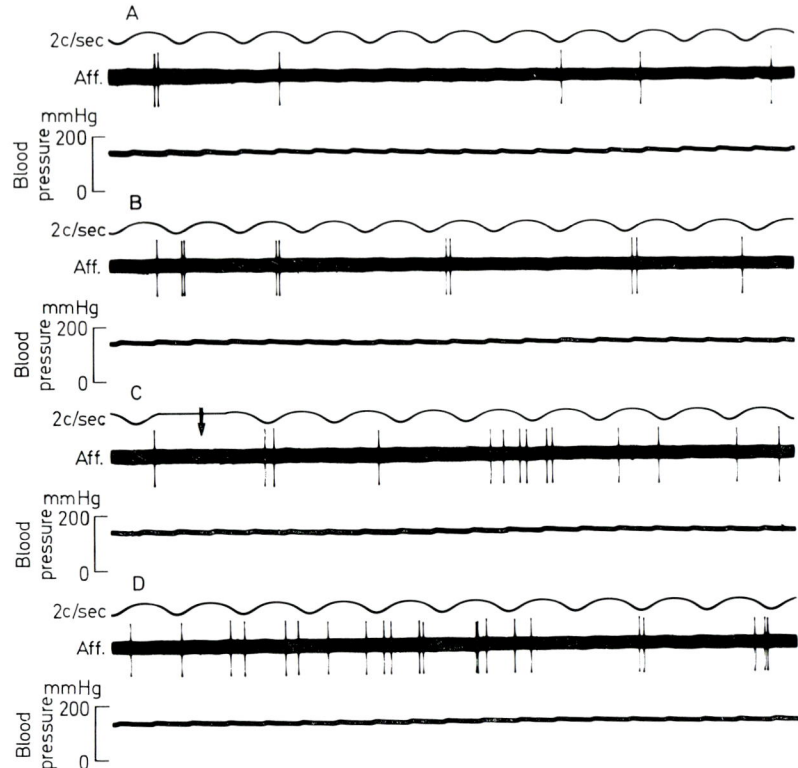

Fig. 14. Cat. The effects of central section of the left sinus nerve on the chemoreceptor afferent traffic of a slip of the same nerve. Track A = recording taken during artificial ventilation with air; sinus nerve "intact". Track B = recording taken 75 sec subsequent to the onset of artificial ventilation with 15% O_2 in N_2; nerve "intact". Track C = recording taken 90 sec subsequent to the onset of artificial ventilation with 15% O_2. At the arrow the sinus nerve was cut central to the afferent slip. Track D = recording taken 20 sec after the central section of the sinus nerve while the animal was still artificially ventilated with 15% O_2. (From R. G. O'REGAN, 1970)

entrance to the carotid body. Electrical stimulation of the central part of the nerve reduced or even silenced afferent impulse activity. The same results were obtained using the aortic nerve (which is technically much more easy), and a typical record of the inhibitory effect of efferent stimulation on aortic chemoreceptor discharge is shown in Fig. 13.

Efferent stimulation of the sinus nerve also caused an increase in carotid body blood flow, but the hyperaemia thus aroused showed a different time course from that of the depression of impulse activity. Moreover, a local intraarterial injection of atropine abolished the hyperaemia but did not abolish the depression of afferent activity caused by electrical stimulation.

NEIL and O'REGAN (1969b, 1971b) showed that the impulse activity recorded in a few-fibre preparation of chemoreceptors peeled off the otherwise intact sinus nerve was always increased by cutting the sinus nerve trunk centrally (Fig. 14).

SAMPSON and BISCOE (1970) have confirmed this. NEIL and O'REGAN (1969b) also showed that efferent impulse activity in the otherwise intact sinus nerve recorded during systemic hypoxia was greater than if the sinus nerve was cut near its junction with the carotid body. They found that such efferent activity could be reflexly promoted by the local injection or perfusion of chemoreceptor stimulants such as cyanide or acetaldehyde into the vascularly isolated ipsilateral carotid body. No reflex response of carotid body efferents occurred when such stimulants were injected into the contralateral carotid body.

Thus there exists a feed back mechanism between the chemoreceptors and the efferent fibres which innervate the glomus complex. Increased chemoreceptor discharge promotes increased efferent impulse activity which serves to damp down the afferent response to the original chemical stimulus. The complication of the situation resulting from this effect of efferent fibre activity must cast some doubt on the quantitative applicability of results on chemoreceptor response to chemical stimuli plotted graphically, as all such experimental findings have hitherto been obtained on preparations made from the sinus nerve *cut centrally*.

BISCOE, BRADLEY, and PURVES (1970) proved that the efferent stimulation of the carotid sinus nerve increased the oxygen consumption of the glomus. As the sinus nerve contains also post-ganglionic sympathetic fibres which when stimulated *lower* glomeral oxygen usage, the influence of these (inferentially) parasympathetic efferents in promoting glomus oxygen consumption must be marked.

At present the site of the cell bodies of these glossopharyngeal efferent fibres is not known; presumably it is in the brain stem in the dorsal motor nucleus of IX and/or X. We have no data about the effects of cortico-hypothalamic or other supra-medullary influences on these cell bodies. Further work is required on the role of these efferents, which seem to exercise a function similar to that of the olivo-cochlear bundle of Rasmussen in inhibiting or reducing the impulse activity of auditory sensory units (FEX, 1965).

As detailed on p. 57, 58, BISCOE and his coworkers have shown that many if not all of the fibre endings which act upon the Type I cells are efferent.

YATES et al. (1970) have claimed that efferent stimulation of the sinus nerve discharges the electron dense cored vesicles of the Type I cells which are known to contain catecholamines. Whether this release is involved in the inhibitory effect of the efferents on chemoreceptor discharge requires further investigation.

H. The Sensory Unit and its Excitation

DE CASTRO's work in 1928 firmly entrenched the view that hypoxia, hypercapnia or acidaemia affected the metabolism of the glomus (Type I) cell and thereby excited its adjacent sensory ending. WINDER (1937) proposed that the common stimulant of the sensory endings was the hydrogen ion. Thus hypoxia causes the liberation of lactic acid from cells and hypercapnia and acidaemia themselves provide an excess $[H^+]$ in the ECF. WINDER based his views on his findings that sodium iodoacetate would abolish the reflex hyperpnoea provoked by perfusing hypoxic blood through the carotid bodies, without affecting that elicited by the perfusion of hypercapnic or acidaemic blood. Thus the idea was born of "anaerobic metabolites" which, when released from the glomus cells, stimulated the sensory endings by virtue of their acidic properties.

Unfortunately, WINDER's claims could not be substantiated by experiments in which chemoreceptor discharge was recorded—if sodium iodoacetate were perfused at a concentration sufficient to abolish the hypoxic response it also abolished that to acidaemia or hypercapnia.

At this time acetyl choline (ACh) had been proved to be a transmitter at many synapses and at the neuromuscular junction. ANICHKOV et al. (1936) and HEYMANS, BOUCKAERT, FARBER and HSU (1936) showed that ACh induced carotid chemoreflexes. SCHWEITZER and WRIGHT (1938), who confirmed this, proposed that ACh might be a transmitter between the glomus cell and the afferent nerve ending. There has been a formidable amount of work on this problem since 1938 and HEYMANS and NEIL (1958), ANICHKOV and BELEN'KII (1963), and TORRANCE (1968) summarize the evidence. It is fair to say that most workers nowadays do not support the proposition that ACh is a transmitter causing chemoreceptor discharge in natural circumstances. EYZAGUIRRE however has been a notable supporter of the proposal. EYZAGUIRRE developed an ingenious technique for investigating chemoreceptor discharge by *superfusing* the carotid body itself using a stream of oxygenated Ringer solution. Whereas this technique is simple, it means forfeiting many of the natural features of the glomus—most important that of its overall natural blood flow and possible "intraglomeral" adjustments of that flow. Nevertheless, although a proportion of the superfused glomus—at its centre—must be severely asphyxiated or dead, those cells and nerve endings at its surface respond to stimulants in the superfusing fluid and are rendered quiet by hyperoxia, suggesting that they are adequately functional.

A long series of papers (EYZAGUIRRE and KOYANO, 1965a and b; EYZAGUIRRE, KOYANO, and TAYLOR, 1965; EYZAGUIRRE and ZAPATA, 1968a–c; ZAPATA, HESS, BLISS, and EYZAGUIRRE, 1969) examine this and cognate problems.

One of EYZAGUIRRE's most striking findings is obtained on a "sensory-type LOEWI" experiment. Two carotid bodies are placed in series in the superfusing stream. The "upstream" carotid body can be asphyxiated by lifting it out of the tube through which flows the oxygenated Ringer solution, into a side tube where it gets no oxygen supply. On replacing this asphyxiated glomus in the stream, the "downstream" carotid body chemoreceptors discharge vigorously, as recorded from multifibre preparations of the sinus nerve. The nature of the substance(s) liberated is not known, but eserine enhances and mecamylamine blocks the response of the second carotid body; both of these drug actions are compatible with the conclusion that acetylcholine, released from the asphyxiated glomus, stimulates the second.

The carotid body contains a large amount of ACh—25 µg/g weight (EYZAGUIRRE et al., 1965) and, as TORRANCE (1968) has stated, such a large concentration would imply that the glomus is generously supplied with the requisite synthetic enzyme choline acetyl-transferase. However, HEBB (1968) reported that the glomeral concentration of this enzyme is only 5% of that in the superior cervical ganglion. As the ACh concentration in the two structures is similar she advanced two suggestions: (a) that the glomus may contain some inhibitory substance so that the capacity of the tissue for ACh synthesis is underestimated; (b) that the glomeral content of ACh is high because destruction is low. However, this second possibility would imply a very slow turnover of ACh which would argue against a role of ACh as transmitter (TORRANCE, 1968).

A serious criticism of the ACh hypothesis was provided by DOUGLAS (1952). He showed that hexamethonium would abolish the chemoreflex response to acetylcholine but did not affect that provoked by hypoxia. Fuller details of the ACh controversy may be found in the symposium volume edited by TORRANCE (1968).

Catecholamines have also been implicated as transmitter substances. However, ZAPATA et al. (1969) have shown that the catecholamines exert no direct effect on the glomus and there is general agreement that, although the Type I cells contain catecholamine granules and that the content of these is depleted by efferent sinus nerve stimulation (YATES et al., 1970), there is no evidence that the catecholamines cause direct chemoreceptor stimulation. Indeed, stimulation of the sinus nerve efferents actually depresses chemoreceptor discharge. NEIL and JOELS (1963), who did describe the stimulation of chemoreceptor discharge caused by adding nor-adrenaline to the fluid perfusing the carotid body could never dissociate this excitation from the effects of reduced flow which the catecholamine caused by inducing vasoconstriction.

Other candidates for transmitter substances such as 5-hydroxytryptamine and histamine have been proposed. 5-HT is a chemoreceptor stimulant and its action is not blocked by hexamethonium. There is little further evidence in its favour.

According to PAINTAL (1967), NEIL (1951) and LANDGREN and NEIL (1951) postulated that chemoreceptor excitation was mediated by metabolites because it was believed at that time that the metabolic requirements of the carotid body were low. This is not strictly the case. COMROE and SCHMIDT (1938) had indeed claimed that the carotid body had a low usage of oxygen, which could be satisfied by the amount of oxygen dissolved in the plasma at normal arterial pO_2.

KENNEY and NEIL (1950), however, showed that section of the sinus nerves in bled animals caused a fall of systemic pressure which they attributed to an interruption of carotid chemoreflex excitation of the vasomotor centre.

Discussing these results at the Stockholm Symposium on Chemoreceptors, NEIL (1951) pointed out that "it is extremely difficult to understand the activity of the chemoreceptors after haemorrhage if the SCHMIDT-COMROE theory be correct. In the experiments of KENNEY and NEIL (1950) the oxygen tension of the blood was not reduced, as all animals showed hyperpnoea—moreover evidence of increased chemoreceptor activity following haemorrhage could still be obtained if the animals were subjected to constant artificial respiration. If the metabolic usage of oxygen of the chemoreceptors is as low as SCHMIDT and COMROE suggest, then the reduction of blood flow after haemorrhage should hardly prejudice the absolute supply of oxygen in unit time. However, it is possible that reduction of the circulation following haemorrhage may influence chemoreceptor discharge by allowing the accumulation of metabolites in the vicinity of the chemoreceptor cells. Such an accumulation of metabolites might itself be the stimulating agent ... Increased chemoreceptor discharge in hypoxic conditions may be elicited in two ways: (1) By reduction of the blood flow through the glomus tissue (e.g. haemorrhage). Here the rate of formation of anaerobic metabolites may be the same as in normal conditions but, owing to their low rate of removal, accumulation occurs and chemoreceptor discharge ensues. (2) By reduction of the oxygen tension of the arterial blood supplying the glomus—the rate of formation of anaerobic metabolites outstrips their rate of removal by the blood flow and the accumulation of these substances causes chemoreceptor firing.

Thus, the argument proposed that the metabolism of the glomus was appreciably higher than that suggested by COMROE and SCHMIDT and that the blood flow through the glomus was an important factor in influencing chemoreceptor discharge. The direct measurements of DALY et al. (1954) showed that although the glomus oxygen usage was indeed high the blood flow exceeded that of any other tissue weight for weight. Such was the reasoning behind the support given by NEIL (1951) and LANDGREN and NEIL (1951) to the "anaerobic metabolite" hypothesis. HEYMANS and NEIL (1958) later referred to the supporting evidence of BOGUE and STELLA (1935) that intense discharge of the chemoreceptors persisted for as long as 30 min after death.

PAINTAL rejected the postulate that chemoreceptor excitation is mediated through metabolites on the basis of experimental results in which he showed (a) that carboxyhaemoglobinaemia did cause the stimulation of single aortic chemoreceptor units — the first time this was shown; (b) that the activity of a single aortic unit — already increased by the ventilation of the animal by pure nitrogen for two or three minutes — showed only a transient further increase on circulatory arrest (produced by injecting 30 ml air into the right ventricle) and then subsided to a level somewhat below that recorded during ventilation with nitrogen. These results are puzzling in that the rise of [H^+] and pCO_2 lead to an increased chemoreceptor discharge themselves.

However, when the "anaerobic metabolite" hypothesis was favoured, one of the main reasons for its support lay in the fact that it was thought that carboxyhaemoglobinaemia did not stimulate the chemoreceptors, which made it difficult to understand how a simple drop in tissue pO_2 could be the essential cause of chemoreceptor activity. The resolution of this difficulty by PAINTAL (1967), MILLS and EDWARDS (1968) and EDWARDS and MILLS (1969) does indeed make the "metabolite" hypothesis less likely. PAINTAL (1967) has proposed that two kinds of cells exist in glomus tissue—one specifically sensitive to changes in the local pO_2—the pO_2 sensor, and the other responsible for the high rate of O_2 usage—the metabolic cell. He suggests that the glomus cell Type I fulfills the latter function and that the sustentacular (Type II) cell serves as the O_2 sensor. Whether the O_2 sensor or the metabolic cell is primarily affected by CO_2 and H^+ is not defined.

BISCOE (1971) has suggested that the high oxygen usage of the glomus is due to the metabolism of the thousands of bare sensory endings themselves. This high metabolic rate, it is argued, is necessary for the nerve endings to maintain their state of polarization. A fall in pO_2 or a disruption in their metabolism (e.g. a disruption of oxidative phosphorylation as caused by dinitrophenol) inactivates the pumping mechanisms of the cell membranes of these endings. It is not clear why the adjacent baroreceptors should be so utterly irresponsive to hypoxia and to these metabolic inhibitors. BISCOE and TAYLOR (1963), discussing the random discharge of single carotid chemoreceptor fibres, considered whether this might be due to thermal agitation but discarded the proposition on calculating that, for this to be the case, the nerve endings concerned must be less than 0.1 μ in diameter. At that time no such small nerve endings had been found. With the development of electronmicroscopy there is now evidence of terminals of this size in the glomus.

If the multiplicity of nerve endings is the major cause of the high oxygen usage of the glomus cells one is left in doubt as to the contribution made by the Type I cells. Their EM appearance suggests an active metabolism—hypoxia must presumably cause them to "leak" their intracellular cations—notably K^+. BISCOE et al. (1970) consider that their entire innervation is efferent and it is known that stimulation of these efferents reduces chemoreceptor discharge. The Type I cells, having occupied the starring role for so long, seem to have been relegated to the wings. It is even possible that they may subserve some other function—possibly an endocrine function once investigated by CHRISTIE (1933).

I. The Role of the Chemoreceptors in Eupnoeic Respiration

The initial claims of HEYMANS et al. (1933) stressed that the chemoreceptors were more sensitive to carbon dioxide than was the central respiratory complex. COMROE and SCHMIDT (1940) vigorously contested this view and suggested that the chemoreceptors played little if any part in eupnoeic breathing, being important only in acute hypoxia and similarly adverse conditions. With recent developments it seems that the carotid bodies, and inferentially the aortic bodies, are indeed concerned with monitoring eupnoeic breathing, at least in the anaesthetized cat.

Multifibre studies of chemoreceptor discharge show this to have a rhythm with the same period as respiration (BISCOE and PURVES, 1967). The authors conclude that this chemoreceptor rhythm represents the moment to moment changes in blood gas tensions and not a response to the arterial pressure changes with respiration.

PURVES (1966) showed that the carotid blood arterial pO_2 oscillated with the same frequency as respiration and BAND, CAMERON, and SEMPLE (1969a and b) found a respiratory oscillation of 0.0–0.02 pH which they ascribed to fluctuations in arterial pCO_2.

BLACK and TORRANCE (1967) injected saline, equilibrated with 100% CO_2, retrogradely down the external carotid artery to reach the carotid body and were able to increase the tidal volume of the breath occurring at the time, provided that the solution reached the chemoreceptor during inspiration. If the injection arrived at the glomus during expiration the expiratory effect was prolonged. The effect of this acknowledged stimulus thus depended on its timing in relation to the respiratory cycle. As BAND, CAMERON, and SEMPLE (1970) observe, such a finding would mean that a continuously rising arterial pCO_2 would affect both inspiration and expiration and thus might not be regarded as a stimulus similar to that provided by repeated rises of pCO_2 which coincided with inspiratory movements only. BAND et al. confirmed BLACK and TORRANCES' findings and showed that the effect of CO_2 saline injections was abolished if the appropriate sinus nerve was temporarily blocked by procaine. They showed also that injections of lactic acid, which provoked changes of pH comparable with those caused by the CO_2 saline injections, had no effect on respiration. The disparity here was ascribed to the rapid diffusion of CO_2 into the glomeral ECF compared with that of H^+ or HCO_3^-.

References

Abbott, C. P., Daly, M. de B., Howe, A.: Early ultrastructural changes in the carotid body after degenerative section of the carotid sinus nerve in the cat. Acta anat. (Basel) (in the press) (1971).

Al-Lami, F., Murray, R. G.: Fine structure of the carotid body of normal and anoxic cats. Anat. Rec. **160**, 697–718 (1968).

Anichkov, S. V., Belen'Kii, M. L.: Pharmacology of the carotid body chemoreceptors. Trans. R. Crawford. London: Pergamon 1963.

— Zakusov, V. V., Kuznetzov, A. I., Polyakov, N. G.: Fiziol. Zh. S.S.S.R. **21**, 809 (1936). Cited by S. V. Anichkov and A. I. Belen'Kii. In: Pharmacology of the carotid body chemoreceptors, p. 64. London: Pergamon 1963.

Band, D. M., Cameron, I. R., Semple, S. J. G.: Oscillations in arterial pH with breathing in the cat. J. appl. Physiol. **26**, 261–267 (1969a).

— — — Effect of different methods of CO_2 administration on oscillations of arterial pH in the cat. J. appl. Physiol. **26**, 268–273 (1969b).

— — — The effect on respiration of abrupt changes in carotid artery pH and pCO_2 in the cat. J. Physiol. (Lond.) **211**, 479–494 (1970).

Biscoe, T. J.: Carotid body: Structure and function. Physiol. Rev. **51**, 437–495 (1971).

— Bradley, G. W., Purves, M. J.: The relation between carotid body chemoreceptor discharge, carotid sinus pressure and carotid body venous flow. J. Physiol. (Lond.) **208**, 99–120 (1970).

— Lall, A., Sampson, S. R.: Electron microscopic and electrophysiological studies on the carotid body following intracranial section of the glossopharyngeal nerve. J. Physiol. (Lond.) **208**, 133–152 (1970).

— Purves, M. J.: Observations on the rhythmic variation in the cat carotid body chemoreceptor activity which has the same period as respiration. J. Physiol. (Lond.) **190**, 389–412 (1967).

— — Sampson, S. R.: The frequency of nerve impulses in single carotid body chemoreceptor afferent fibres recorded *in vivo* with intact circulation. J. Physiol. (Lond.) **208**, 121–131 (1970).

— Sampson, S. R.: Spontaneous activity recorded from the central cut end of the carotid sinus nerve of the cat. Nature (Lond.) **216**, 294–295 (1967).

— Stehbens, W. E.: Electron microscopic observations on the carotid body. Nature (Lond.) **208**, 708–709 (1965).

— — Ultrastructure of the carotid body. J. Cell Biol. **30**, 563–568 (1966).

— — Ultrastructure of the denervated carotid body. Quart J. exp. Physiol. **52**, 31–36 (1967).

— Taylor, A.: The discharge pattern recorded in chemoreceptor afferent fibres from the cat carotid body with normal circulation and during perfusion. J. Physiol. (Lond.) **168**, 332–344 (1963).

Black, A. M. S., Torrance, R. W.: Chemoreceptor effects in the respiratory cycle. J. Physiol. (Lond.) **189**, 59–61 P (1966).

Bogue, J. Y., Stella, G.: Afferent impulses in the carotid sinus nerve (nerve of Hering) during asphyxia and anoxaemia. J. Physiol. (Lond.) **83**, 459–465 (1935).

Cauna, N. C., Ross, L. L.: The fine structure of Meissners touch corpuscles of human fingers. J. biophys. biochem. Cytol. **8**, 467–482 (1960).

Celestino da Costa: Sur les dispositifs glomiques du corpuscle carotidien. Arch. port. Sci. biol. **8**, 129–132 (1944).

Chen, I-Li, Yates, R. D.: Electron microscopic radioautographic studies of the carotid body following injections of labelled biogenic amine precursors. J. Cell Biol. **42**, 794–803 (1969).

— — Duncan, D.: The effects of reserpine and hypoxia on the amine-storing granules of the hamster carotid body. J. Cell Biol. **42**, 804–816 (1969).

Christie, R. V.: The function of the carotid gland (glomus caroticum). I. The action of extracts of a carotid gland tumour in man. Endocrinology **17**, 421–432 (1933).

Coleridge, H., Coleridge, J. C. G., Howe, A.: Aortico-pulmonary glomus tissue in the cat. Nature (Lond.) **211**, 1187 (1966).

COLERIDGE, H., COLERIDGE, J. C. G., HOWE, A.: A search for pulmonary arterial chemoreceptors in the cat, with a comparison of the blood supply of the aortic bodies in the new-born and adult animal. J. Physiol. (Lond.) **191**, 353–374 (1967).
— — — Thoracic chemoreceptors in the dog: A histological and electrophysiological study of the location, innervation, and blood supply of the aortic bodies. Circulat. Res. **26**, 235–247 (1970).
COMROE, J. H., Jr.: The location and function of the chemoreceptors of the aorta. Amer. J. Physiol. **127**, 176–191 (1939).
— SCHMIDT, C. F.: The part played by reflexes from the carotid body in the chemical regulation of respiration in the dog. Amer. J. Physiol. **121**, 75–97 (1938).
DALY, M. DE B., LAMBERTSEN, C. J., SCHWEITZER, A.: Observations on the volume of blood flow and oxygen utilization of the carotid body in the cat. J. Physiol. (Lond.) **125**, 67–89 (1954).
DEARNALEY, D. P., FILLENZ, M., WOODS, R. I.: The identification of dopamine in the rabbit's carotid body. Proc. roy. Soc. B **170**, 195–203 (1968).
DE BOISSEZON, P.: Les appareils de réglage de la circulation du corpuscule carotidien. Bull. d'histologie appl. **20**, 136–150 (1943).
DE CASTRO, F.: Sur la structure et l'innervation de la glande intercarotidienne (Glomus caroticum) de l'homme et des mammifères, et sur un nouveau système d'innervation autonome du nerf glossopharyngien. Études anatomiques et expérimentales. Trab. Lab. Invest. biol. Univ. Madr. **24**, 365–432 (1926).
— Sur la structure et l'innervation du sinus carotidien de l'homme et des mammifères. Nouveaux faits sur l'innervation et la fonction du glomus caroticum. Études anatomiques et physiologiques. Trab. Lab. Invest. biol. Univ. Madr. **25**, 331–380 (1928).
— Nuevas observaciones sobre la inervación de la región carotidea. Los quimio- y pressoreceptores. Trab. Inst. Cajal Invest. biol. **32**, 297–384 (1940).
— Sur la structure de la synapse dans les chemorecepteurs: leur mécanisme d'excitation et role dans la circulation sanguine locale. Acta physiol. scand. **22**, 14–43 (1951).
— Sur la vascularisation et l'innervation des corpuscles carotidiens aberrants. Arch. int. Pharmacodyn. **139**, 212–224 (1962).
— RUBIO, M.: The anatomy and innervation of the blood vessels of the carotid body and the role of chemoreceptive reactions in the autoregulation of the blood flow. In: Arterial chemoreceptors, ed. by R. W. TORRANCE, pp. 267–270. Oxford: Blackwell 1968.
DE KOCK, L. L.: Histology of the carotid body. Nature (Lond.) **167**, 611–612 (1951).
— Intraglomerular tissues of the carotid body. Acta anat. (Basel) **21**, 101–116.
— DUNN, A. E. G.: Electron-microscopic investigation of the nerve endings in carotid body. In: Arterial chemoreceptors, ed. by R. W. TORRANCE, pp. 179–188. Oxford: Blackwell 1968.
DE LORENZO, A. J.: Electron-microscopic observations on the taste buds of the rabbit. J. biophys. biochem. Cytol. **4**, 143–150 (1958).
DE ROBERTIS, E., FRANCHI, C. M.: Electron microscope observations on synaptic vesicles in synapses of the retinal rods and cones. J. biophys. biochem. Cytol. **2**, 307–318 (1956).
DOUGLAS, W. W.: The effect of a ganglion blocking drug hexamethonium on the response of the cat's carotid body to various stimuli. J. Physiol. (Lond.) **118**, 373–383 (1952).
DUKE, H. N., GREEN, J. H., NEIL, E.: Carotid chemoreceptor impulse activity during inhalation of carbon monoxide mixtures. J. Physiol. (Lond.) **118**, 520–527 (1952).
EDWARDS, M. W., MILLS, E.: Arterial chemoreceptor oxygen utilization and oxygen tension. J. appl. Physiol. **27**, 291–294 (1969).
ENGSTRÖM, H., WERSÄLL, J.: The ultra structural organisation of the organ of Corti and of the vestibular sensory epithelia. Exp. Cell Res., Suppl. **5**, 460–492 (1958).
EULER, U. S. VON, LILJESTRAND, G., ZOTTERMAN, Y.: The excitation mechanism of the chemoreceptors of the carotid body. Skand. Arch. Physiol. **83**, 132–152 (1939).
EYZAGUIRRE, C., KOYANO, H.: Effects of hypoxia, hypercapnia and pH on the chemoreceptor activity of the carotid body *in vitro*. J. Physiol. (Lond.) **178**, 385–409 (1965a).
— — Effects of some pharmacological agents on chemoreceptor discharges. J. Physiol. (Lond.) **178**, 410–431 (1965b).
— — TAYLOR, J. R.: Presence of acetylcholine and transmitter release from carotid body chemoreceptors. J. Physiol. (Lond.) **178**, 473–476 (1965).

Eyzaguirre, C., Lewin, J.: The effect of sympathetic stimulation on carotid nerve activity. J. Physiol. (Lond.) **159**, 251–267 (1961).
— Zapata, P.: Pharmacology of pH effects on carotid body chemoreceptors *in vitro*. J. Physiol. (Lond.) **195**, 557–588 (1968a).
— — The release of acetylcholine from carotid body tissues. Further study of the effects of acetylcholine and cholinergic blocking agents on the chemosensory discharge. J. Physiol. (Lond.) **195**, 589–607 (1968b).
— — Discussion of possible transmitter on generator substances in carotid body chemoreceptors. In: Arterial chemoreceptors, ed. by R. W. Torrance, pp. 213–247. Oxford: Blackwell 1968c.
Fex, J.: Auditory activity in centrifugal and centripetal cochlear fibers in cat. A study of a feed back system. Acta physiol. scand., Suppl. **189**, 1 (1965).
Fillenz, M.: In: Arterial chemoreceptors, ed. by R. W. Torrance, p. 266. Oxford: Blackwell 1968.
— Woods, R. I.: Some observations on the rabbit carotid body. J. Physiol. (Lond.) **186**, 39–40 P (1966).
Floyd, W. W., Neil, E.: The influence of the sympathetic innervation of the carotid bifurcation on chemoreceptor and baroreceptor activity in the cat. Arch. int. Pharmacodyn. **91**, 230–239 (1952).
Forster, R. E.: The diffusion of gases in the carotid body. In: Arterial chemoreceptors, ed. by R. W. Torrance, pp. 115–128. Oxford: Blackwell 1968.
Goormaghtigh, N., Pannier, R.: Les paraganglions du coeur et des zones vaso-sensibles carotidienne et cardio-aortique chez le chat adulte. Arch. Biol. (Liège) **50**, 455–533 (1939).
Gray, E. G.: Electron microscopy of presynaptic organelles of the spinal cord. J. Anat. (Lond.) **97**, 101–106 (1963).
Hebb, C. O.: In: Arterial chemoreceptors, ed. by R. W. Torrance, pp. 138–139. Oxford: Blackwell 1968.
Hess, A.: Electron microscopic observations of normal and experimental cat carotid bodies. In: Arterial chemoreceptors, ed. by R. W. Torrance, pp. 51–56. Oxford: Blackwell 1968.
Heymans, C., Bouckaert, J. J.: Sinus caroticus and respiratory reflexes. J. Physiol. (Lond.) **69**, 254–266 (1930).
— — Farber, S., Hsu, F. J.: Influence réflexogène de l'acétylcholine sur les terminaisons nerveuses chimiosensitives du sinus carotidien. Arch. int. Pharmacodyn. **54**, 129–135 (1936).
— — Regniers, P.: Le sinus carotidien et la zone homologue cardio-aortique. Paris: Doin 1933.
— Neil, E.: Reflexogenic areas of the cardiovascular system. London: Churchill 1958.
— Rijlant, P.: Le courant d'action du nerf du sinus carotidien intact. C. R. Soc. Biol. (Paris) **113**, 69–73 (1933).
Heymans, J. F., Heymans, C.: Sur les modifications directes et sur la régulation réflexe de l'activité du centre respiratoire de la tête isolée du chien. Arch. int. Pharmacodyn. **33**, 273–372 (1927).
Hoglund, R.: An ultrastructural study of the carotid body of horse and dog. Z. Zellforsch. **76**, 568–576 (1967).
Hollinshead, W. H.: The origin of the nerve fibres to the glomus aorticum of the cat. J. comp. Neurol. **71**, 417–426 (1939).
— The innervation of the supracardial bodies in the kitten. J. comp. Neurol. **73**, 37–47 (1940).
— A cytological study of the carotid body of the cat. Amer. J. Anat. **73**, 185–211 (1943).
Hornbein, T. F., Griffo, Z. J., Roos, A.: Quantitation of chemoreceptor activity: interrelation of hypoxia and hypercapnia. J. Neurophysiol. **24**, 561–568 (1961).
Howe, A.: The vasculature of the aortic bodies in the cat. J. Physiol. (Lond.) **134**, 311–318 (1956).
— Morphological and functional studies of thoracic chemoreceptors—the aortic bodies—in the cat. Ph.D. Thesis University of London (1957).
Joels, N.: Chemoreceptor excitation and chemoreceptor respiratory reflexes. Ph.D. Thesis University of London (1960).

References

Joels, N., Neil, E.: The influence of anoxia and hypercapnia separately and in combination on the chemoreceptor impulse discharge. J. Physiol. (Lond.) **155**, 45–46 P (1961).
— — The action of high tensions of carbon monoxide on the carotid chemoreceptors. Arch. int. Pharmacodyn. **138**, 528–534 (1962).
— — The excitation mechanism of the carotid body. Brit. med. Bull. **19**, 21–24 (1963).
— — The action of some metabolic inhibitors on the chemoreceptor activity of the carotid body. J. Physiol. (Lond.) **171**, 43—44 P (1964).
— — The idea of a sensory transmitter. In: Arterial chemoreceptors, ed. by R. W. Torrance, pp. 153–176. Oxford: Blackwell 1968.
Kenney, R. A., Neil, E.: The contribution of aortic chemoreceptor mechanisms to the maintenance of arterial blood pressure of cats and dogs after haemorrhage. J. Physiol. (Lond.) **112**, 223–228 (1951).
Krylov, S. S., Anichkov, S. V.: The effect of metabolic inhibitors on carotid chemoreceptors. In: Arterial chemoreceptors, ed. by R. W. Torrance, pp. 103–109. Oxford: Blackwell 1968.
Landgren, S., Neil, E.: Chemoreceptor impulse activity following haemorrhage. Acta physiol. scand. **23**, 158–167 (1951).
Lee, K. D., Mattenheimer, H.: The biochemistry of the carotid body. Enzym. biol. clin. **4**, 199–216 (1964).
— Mayou, R. A., Torrance, R. W.: The effect of blood pressure upon chemoreceptor discharge to hypoxia and the modification of this effect by the sympathetic adrenal system. Quart. J. exp. Physiol. **49**, 171–183 (1964).
Lever, J. D., Lewis, P. R., Boyd, J. D.: Observations on the fine structure and histochemistry of the carotid body in the cat and rabbit. J. Anat. (Lond.) **93**, 478–490 (1959).
Lloyd, B. B., Jukes, M. G. M., Cunningham, D. J. C.: The relation between alveolar oxygen pressure and the respiratory response to carbon dioxide in man. Quart. J. exp. Physiol. **43**, 214–227 (1958).
Longmuir, I. S.: In: Advances in respiratory physiology, ed. by C. G. Caro, p. 219. London: Arnold 1966.
McCloskey, D. I.: Carbon dioxide and the carotid body. In: Arterial chemoreceptors, ed. by R. W. Torrance, pp. 279–293. Oxford: Blackwell 1968.
— Autoregulation of blood flow in the denervated carotid body of the cat. J. Physiol. (Lond.) **211**, 40 P (1970).
— Torrance, R. W.: Autoregulation of blood flow in the carotid body. J. Physiol. (Lond.) **179**, 37–39 P (1965).
Mills, E.: Activity of aortic chemoreceptors during electrical stimulation of the stellate ganglion of the cat. J. Physiol. (Lond.) **199**, 103–114 (1968).
— Edwards, McI. W.: Stimulation of aortic and carotid chemoreceptors during carbon monoxide enhalation. J. appl. Physiol. **25**, 494–502 (1968).
— Sampson, S. R.: Respiratory responses to electrical stimulation of the cervical sympathetic nerves in decerebrate unanaesthetized cats. J. Physiol. (Lond.) **202**, 271—282 (1969).
Morita, E., Chiocchio, S. R., Tramezzani, J. H.: Four types of main cells in the carotid body of the cat. J. Ultrastruct. Res. **28**, 399–410 (1969).
Muratori, G.: Contributo istologico allo studio dei riflessi aortici della carotide. Boll. Soc. ital. Biol. sper. **8**, 387–391 (1933).
— Connessioni tra tessuto paragangliare e zone recettrici aortiche in vari mammiferi. Monit. Zool. Ital. **45**, 300–310 (1935).
Neil, E.: Chemoreceptor areas and chemoreceptor circulatory reflexes. Acta physiol. scand. **21**, 54–65 (1951).
— Joels, N.: The carotid glomus sensory mechanism. In: The regulation of human respiration, ed. by D. J. C. Cunningham and B. B. Lloyd, pp. 163–171. Oxford: Blackwell.
— O'Regan, R. G.: Effects of sinus and aortic nerve efferents on arterial chemoreceptor function. J. Physiol. (Lond.) **200**, 69–71 P (1969a).
— — Efferent and afferent impulse activity in the "intact" sinus nerve. J. Physiol. (Lond.) **205**, 20–21 P (1969b).
— — The effects of efferent electrical stimulation of the cut sinus and aortic nerves on peripheral arterial chemoreceptor activity in the cat. J. Physiol. (Lond.) **215**, 15–32 1971a.
— — Efferent and afferent impulse activity recorded from few-fibre preparations of otherwise intact sinus and aortic nerves. J. Physiol. (Lond.) **215**, 33–47 1971b.

Nielsen, M., Smith, H.: Studies on the regulation of the breathing in acute hypoxia. Acta physiol. scand. **24**, 293–313 (1951).

O'Regan, R. G.: Studies on oxygen usage of the carotid body and the effects of efferent nerves on arterial chemoreceptor function in the cat. Ph.D Thesis University of London (1970).

Paintal, A. S.: Mechanism of stimulation of aortic chemoreceptors by natural stimuli and chemical substances. J. Physiol. (Lond.) **189**, 63–84 (1967).

Pease, D. C., Quilliam, T. A.: Electron microscopy of the Pacinian corpuscle. J. biophys. biochem. Cytol. **3**, 331–342 (1957).

Purves, M. J.: Fluctuations of arterial oxygen tension which have the same period as respiration. Resp. Physiol. **1**, 281–296 (1966).

— The effect of hypoxia, hypercapnia and hypotension upon carotid body blood flow and oxygen consumption in the cat. J. Physiol. (Lond.) **209**, 395–416 (1970a).

— The role of the cervical sympathetic nerve in the regulation of oxygen consumption of the carotid body of the cat. J. Physiol. (Lond.) **209**, 417–431 (1970b).

Rall, W., Shepherd, E. M., Reese, T. S., Brightman, M. W.: Dendrodendritic synaptic pathway for inhibition in the olfactory bulb. Exp. Neurol. **14**, 44–56 (1966).

Sampson, S. R., Biscoe, T. J.: Efferent control of the carotid body chemoreceptor. Experientia (Basel) **26**, 261–262 (1970).

Schweitzer, A., Wright, S.: Action of prostigmin and acetylcholine on respiration. Quart. J. exp. Physiol. **28**, 33–37 (1938).

Serafini-Fracassini, A., Volpin, D.: Some features of the vascularization of the carotid body in the dog. Acta anat. (Basel) **63**, 571–579 (1966).

Smith, C. A., Dempsey, E. W.: Electron microscopy of the organ of Corti. Amer. J. Anat. **100**, 337–367 (1957).

Torrance, R. W.: Prolegomena to arterial chemoreceptors, ed. by R. W. Torrance. Oxford: Blackwell 1968.

— The idea of a chemoreceptor. In: The pulmonary circulation and interstitial space, ed. by A. P. Fishman and H. H. Hecht, p. 223. Chicago: Chicago University Press 1969.

Winder, C. V.: On the mechanism of stimulation of carotid gland chemoreceptors. Amer. J. Physiol. **118**, 389–398 (1937).

Yates, R. D., Chen, I-Li., Duncan, D.: Effects of sinus nerve stimulation on carotid body glomus cell. J. Cell Biol. **46**, 544–552 (1970).

Zapata, P., Hess, A., Bliss, E. L., Eyzaguirre, C.: Chemical, electron microscopic and physiological observations on the role of catecholamines in the carotid body. Brain Res. **14**, 473–496 (1969).

Zotterman, Y.: Action potentials in glossopharyngeal nerve and in chorda tympani. Skand. Arch. Physiol. **72**, 73–77 (1935).

Chapter 3

Receptors of the Lungs and Airways

By

MARIANNE FILLENZ and J. G. WIDDICOMBE, Oxford (Great Britain)

With 23 Figures

Contents

Introduction	81
Methods of Study	82
Histology	82
Receptor Physiology	82
Reflex Physiology	83
Nasopharyngeal Receptors	84
The "Aspiration Reflex"	84
Epithelial Receptors	88
Cough and Irritant Receptors	88
Smooth Muscle Receptors	97
Pulmonary Stretch Receptors	97
Type J Receptors	101
Sensation Mediated by Respiratory Receptors	106
Other Receptors	107
Conclusions	108
References	109

Introduction

This review deals with the histology and receptor physiology of sensory endings in the lungs and respiratory tract. Reflex actions are mentioned briefly, and more detailed accounts of the motor responses can be obtained from the references cited. No description of the sensory receptors in the nose is attempted, partly because this subject is intrinsically linked with the special sense of smell, and also because the functions of receptors in the nose, other than those mediating the sensation of smell, have been rather surprisingly neglected; this is possibly because a more difficult surgical approach is needed for physiological studies on the nose, compared with lungs and larynx. The review will also omit discussion of receptors of the pharynx; although these have been shown to influence breathing, their primary action, swallowing, is related more to the alimentary than to the respiratory system.

Methods of Study

Histology

The histology of the innervation of the airways and lungs has received relatively little attention. GAYLOR (1934) gave an excellent account of the early studies using methylene blue and various silver staining methods. Results obtained with both methods have to be examined with care since the staining of reticulin and elastin fibres by these methods has been a source of much confusion; only under carefully controlled conditions is there selective staining of nerve fibres. Furthermore, although with these methods one can demonstrate both nerve fibres and ganglion cells in the walls of the airways, they do not enable one to distinguish sensory from motor neurones or fibres from terminals.

The airways receive a parasympathetic motor innervation from nerve fibres whose cell bodies are scattered in the walls of the trachea and large bronchi; selective motor denervation, leaving only sensory innervation, is therefore not possible. Whereas there are now histochemical criteria for catecholaminergic and cholinergic motor fibres and both light and electron microscope criteria for motor nerve terminals (FILLENZ and WOODS, 1970), no such criteria exist for sensory fibres or terminals. Their identification must therefore rest on indirect evidence. The presence of sensory fibres is inferred from the difference between the innervation pattern obtained with the use of empirical methods such as methylene blue and silver, and that obtained with the use of the histochemical methods for monoaminergic and cholinergic motor fibres. This is possible in areas such as the epithelium of the airways, where the histochemical methods fail completely to show any motor innervation, but both methylene blue and silver methods show terminal ramifications of nerve fibres. But it is much more difficult in areas such as the smooth muscle of the airways, where histochemical methods show a rich motor supply and one has to determine whether there is in addition a sensory innervation. The identification of sensory fibres in this situation at present rests on the appearance of terminal arborizations stained with methylene blue and silver which is different from that of the motor fibres. There is another method which has provided some information on the sensory innervation of the lungs. Supranodose section of the vagus, although leaving the postganglionic motor fibres intact, causes degeneration of the preganglionic fibres, which are myelinated. The sensory neurones, whose cell bodies are in the nodose ganglion, do not degenerate and any surviving myelinated fibres can therefore be presumed to be sensory.

Receptor Physiology

This has been studied entirely by recording activity from the nerve fibres connected to the receptors. Those from the lungs are found mainly in the vagus nerves, with a few in the sympathetic nerves; from the trachea in the recurrent laryngeal nerves; from the larynx in the superior laryngeal nerves; and from the epipharynx in the glossopharyngeal nerves. In the case of lung receptors it is rarely possible to identify the receptor attached to the fibre giving recorded action potentials; some degree of localization of mechanoreceptors may be determined by probing the lung, but isolation or precise localization has not proved possible owing to the vascularity of the lung. In the case of receptors in the trachea, larynx and upper respiratory tract, fairly precise localization can be made by probing the luminal wall of the opened respiratory passage. However the depth of any receptor in the airway wall cannot be confidently assessed.

For this reason the conclusion that a certain pattern of fibre activity corresponds to a certain histological receptor structure is always based on an accumulation of indirect evidence and the elimination of other possibilities. For example: the tracheal bifurcation is an exceptionally sensitive region for eliciting coughing; recording from vagal afferent fibres from rapidly adapting tracheal endings sensitive to mechanical and chemical irritation of the trachea shows that they are concentrated at the tracheal bifurcation; histological studies show that at this site there is a concentration of "epithelial receptors", and of no other type of ending. This indirect evidence seems convincing that these particular vagal fibres do really come from the tracheal epithelial receptors, even in the absence of more direct proof.

In the light of these considerations it is not surprising that there have been no studies of the mode of initiation of an action potential in any of the receptors in the lungs and respiratory tract. Receptor potentials have not been recorded, and the ways in which the nerve terminals of a receptor are influenced by the surrounding tissue elements have not been studied. Mechanical and chemical stimuli are applied and fibre discharge patterns recorded, but what happens in the receptors remains a mystery. Possibly studies on the receptors of the trachea or larynx, isolated *in situ*, would be of value.

Another problem in the case of lung receptors is that it is difficult to assess precisely a mechanical stimulus. Lung inflation or deflation, changes in airway resistance or lung compliance, or in pulmonary capillary pressure can all be measured for the lungs as a whole, but the measurement may not represent the change at a particular receptor site. Tissue stress relaxation, contraction of smooth muscle and distributional inequalities in ventilation and perfusion may all cause a particular receptor to give an unrepresentative response. To some extent this problem may be overcome by studying large numbers of each type of receptor, but even then quantitative conclusions about properties such as adaptation and threshold must be cautious. With chemical stimuli it is often difficult to say if they are acting directly on the receptor, or exciting it mechanically by inducing changes in the surrounding tissue.

Conduction velocities of fibres from respiratory receptors are valuable for categorization. These are fairly easily obtained for lung receptors with vagal fibres, but there have been few values for tracheal and upper respiratory tract fibres, because of the shortness of the dissected nerves. Recording from C-fibres from the lungs has been carried out mainly by PAINTAL (e.g. 1970); this is clearly more difficult than the study of myelinated A-fibres, and does not seem to have been done for tracheal and upper respiratory tract endings.

Reflex Physiology

In studying reflex responses from the lungs it is usually difficult to devise a stimulus which excites only one group of receptors. For example, large inflations of the lungs stimulate at least three types of receptor (pulmonary stretch, lung irritant and type J), and deflations of the lungs stimulate two types of receptor (lung irritant and type J) and inhibit another (pulmonary stretch). For this reason the reflex action of each group of receptors is based mainly on the correlation of receptor properties (from single-fibre recording) with the patterns of reflex response. In the case of the three types of lung receptor described in this review the correlations now appear sufficiently extensive to justify, when taken with supplementary evidence, clear conclusions about many of the reflex responses from the receptors. With receptors in the trachea and upper respiratory tract it is possible to localize and control the stimulus sufficiently accurately to allow reasonable certainty about the type of receptor being stimulated and causing the reflex.

The use of unphysiological chemical stimuli can be valuable; although none of these stimuli is specific for one group of endings, in some instances the concentration threshold for one modality of receptors is much lower than the thresholds for others. In such a case the chemical, provided it has been used carefully and tested on all types of receptor, can tentatively be accepted as a "specific" stimulus and used to study the reflex concerned. Differential block of conduction in nerves can also be useful in separating different pathways of reflexes, provided it is correlated with single-fibre recordings. The recent use of anodal block of vagal conduction (GUZ and TRENCHARD, 1970) indicates that it may be more reliable than the conventional block of conduction by cooling.

Conclusions drawn from electrical stimulation of afferent fibres in the vagus nerve have proved unreliable, and have been criticized frequently and severely. The technique may be valuable as a means of influencing the respiratory centre, but can say little about the function of particular modalities of sensory fibre in the vagus nerve.

Nasopharyngeal Receptors

The "Aspiration Reflex"

The properties of afferent end-organs in the nasopharynx of the cat have been studied by NAIL et al. (1969). (Strictly speaking this region should be called the "epipharynx", since the soft palate separates the nasal cavity from the pharynx and structures cranial to this level should not be considered part of the pharynx. However nasopharynx is common usage.) Gentle mechanical stimulation of the epithelium of the nasopharynx with a nylon fibre elicits an irregular discharge in afferent single fibre preparations of the glossopharyngeal nerve with instantaneous frequencies up to 330 impulses/sec (Fig. 1). The same endings are stimulated by a fine jet of air sufficient to produce a dimple in the mucosa, and by airflow through the nasopharynx provided the rate is sufficient to distend the lumen. In all instances the discharge of action potentials is irregular, and also transient for a single maintained application of the stimulus. An "off-response" is sometimes seen, especially with airflow. The sensitivity of the receptors to chemical stimuli, airborne or blood-borne, has not been adequately tested, but most of them do not respond to strong concentrations of ammonia in air (Fig. 1). A. M. S. WHITE (personal communication), in frozen sections of the cat's epipharyngeal region stained with SCHOFIELD's silver method, has found myelinated fibres ranging from 1.5–4 μ in diameter ramifying beneath the surface epithelium and sending fine branches between the epithelial cells (Fig. 2).

The evidence that these fibres are attached to receptors like that of Fig. 2 is indirect: the action potential studies are consistent with a submucosal site for the receptors; no other submucosal endings in the nasopharynx have been described by histologists; and there have been recordings from no other fibres in the glossopharyngeal nerve with properties appropriate to a submucosal receptor site, although other fibres have shown behaviour characteristics of deeper endings, e.g. muscle spindles (NAIL et al., 1969). Conduction velocities of the fibres from nasopharyngeal endings have not been measured, but the action potentials are typical of myelinated fibres.

The same stimuli that produce a discharge in glossopharyngeal single fibre preparations elicit the "aspiration reflex". This was so-named and first studied

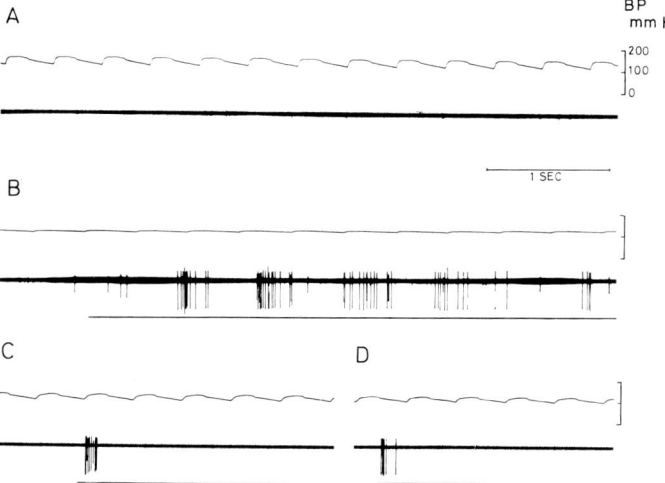

Fig. 1 A–D. Blood pressure and action potentials in a strand of the pharyngeal branch of the glossopharyngeal nerve during stimulation of the epipharynx. A 3 ml ammonia vapour at signal; no response. B Mechanical stimulation with nylon fibre during signal; transient discharges with each movement of the thread. C and D Air flow at 6 l/min through a catheter in the left nostril; rapidly adapting discharges at the start of air flow only. Long horizontal bars mark the duration of each stimulus. (From NAIL et al., 1969)

by IVANCO and KORPAS (1954) and IVANCO et al. (1956), who have made extensive studies of its physiology. Gentle mechanical stimulation of the nasopharyngeal epithelium causes repeated contractions of the diaphragm. In the cat these occur at a frequency of 8–10 Hz, and correspond to bursts of action potentials in the diaphragm and phrenic nerve (Fig. 3) (TOMORI and WIDDICOMBE, 1969) and appropriate changes in intrapleural pressure and airflow (TOMORI, 1965). Recordings from the phrenic nerve show a surprisingly high impulse frequency in each burst, up to 400 impulses/sec transiently, and the reflex seems to activate preferentially a group of diaphragmatic motoneurones with high thresholds and large diameter fibres, possibly analogous to "twitch" motoneurones in other muscles. Electrical stimulation of afferent fibres in the branch of the glossopharyngeal nerve that enter the wall of the pharynx causes similar diaphragmatic and phrenic

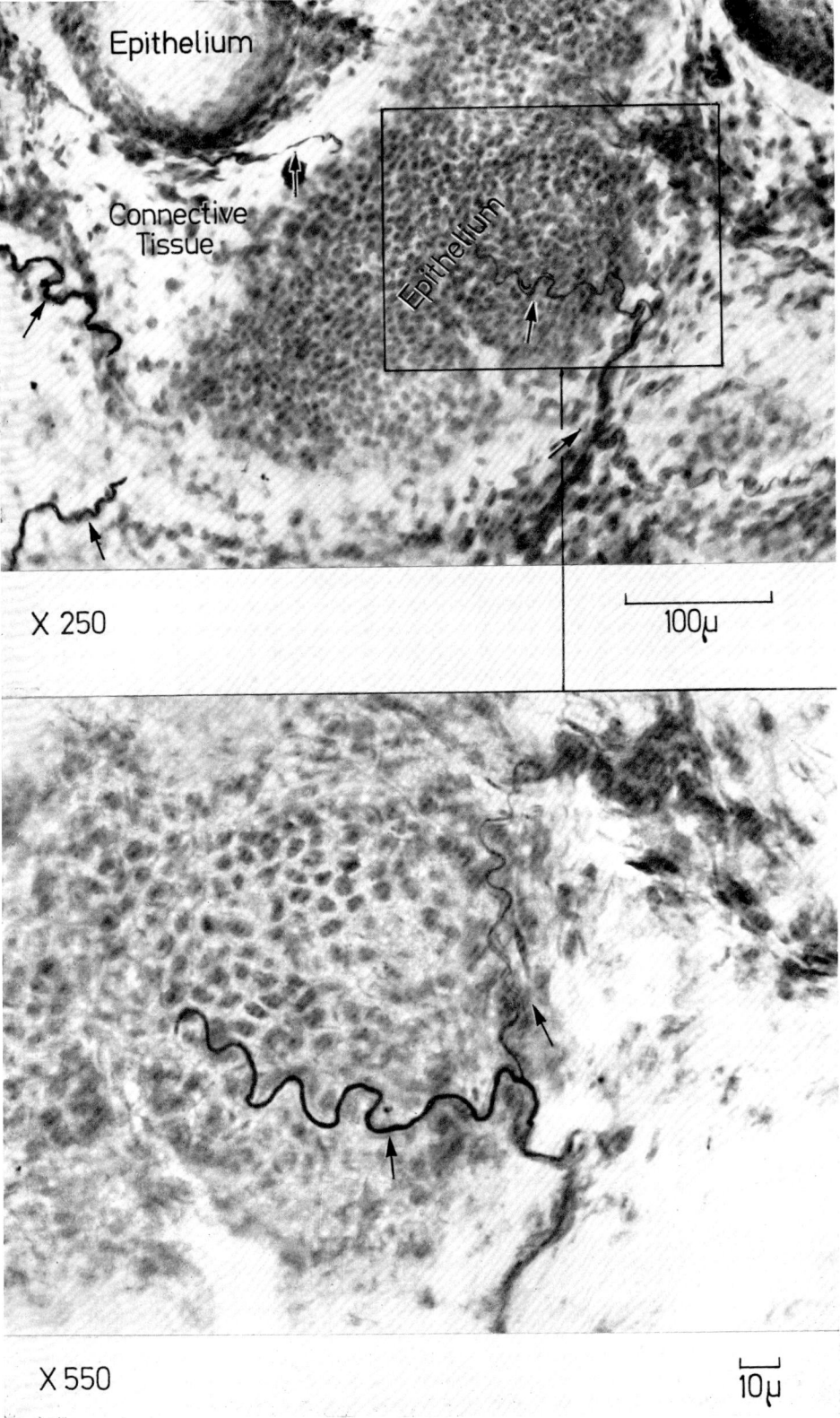

Fig. 2

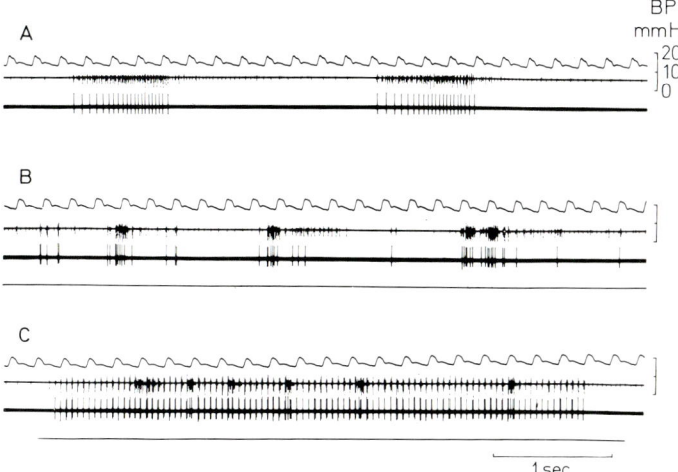

Fig. 3 A–C. Response of phrenic motoneuron discharge to epipharyngeal stimulation. Each record shows from above down: systemic arterial blood pressure (*B.P.*); multi-fibre phrenic discharge; single phrenic motoneuron discharge. Horizontal bars below records B and C represent duration of stimulus. A Control ventilation. B Mechanical stimulation of epipharyngeal mucosa. C Electrical stimulation of glossopharyngeal nerve at 20 shocks per sec. (From NAIL et al., 1971)

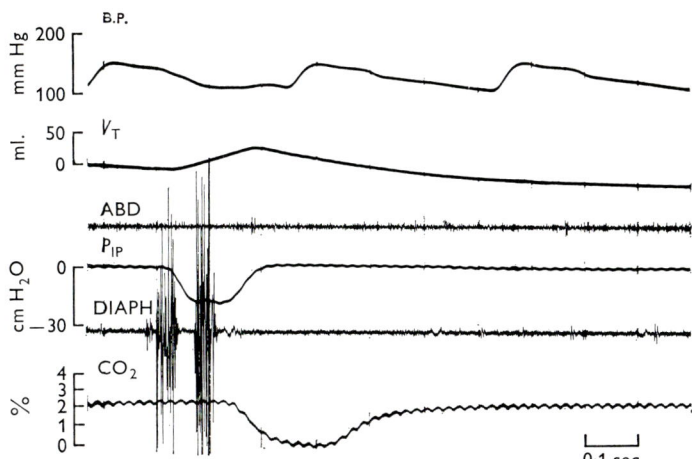

Fig. 4. Changes in abdominal (*ABD*) and diaphragmatic (*DIAPH*) electromyograms, blood pressure (*B.P.*), tidal volume (V_T), intrapleural pressure (P_{IP}) and tidal CO_2% during a single inspiratory-expiratory cycle of the aspiration reflex in an anaesthetized spontaneously breathing cat. (From TOMORI and WIDDICOMBE, 1969)

Fig. 2. Frozen sections of cat epipharyngeal region showing nerve fibres ramifying among epithelial cells. Schofield's silver stain. Photograph by Mrs. A. M. S. WHITE

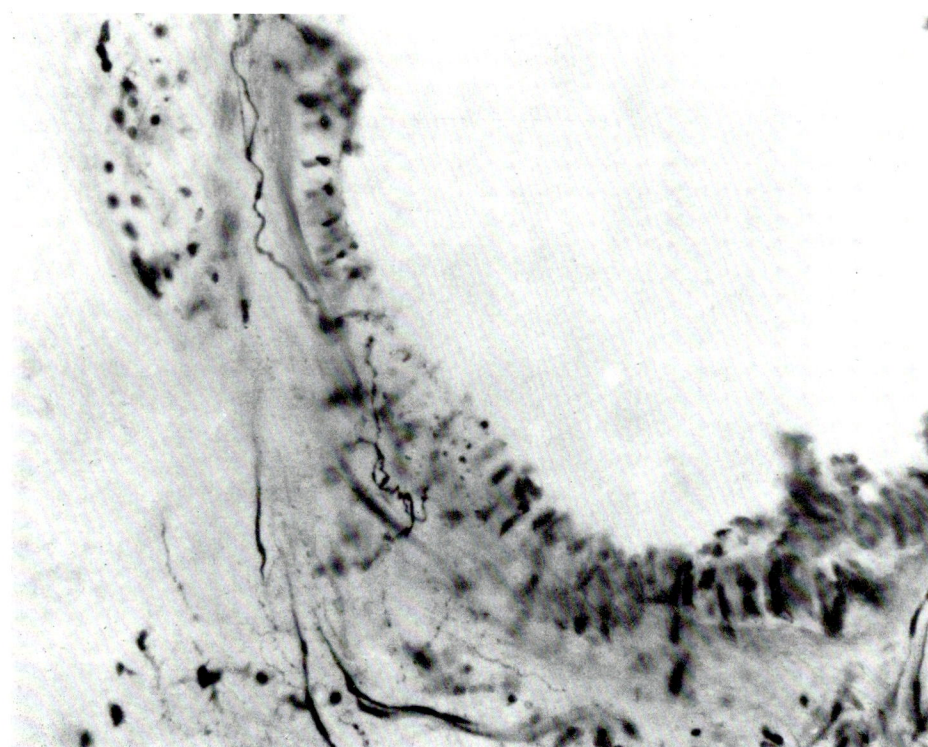

Fig. 6. Rabbit bronchiole stained with methylene blue. Fine varicose motor fibres run between smooth muscle cells in bronchiolar wall. A thicker nerve fibre ramifies under the epithelium and sends varicose terminals between the epithelial cells. Photograph by Dr. R. I. Woods

sensory cells. Similar cells occur in the trachea of the guinea-pig, but their association with nerve fibres has not been confirmed in this species (FILLENZ, unpublished). FEYRTER (1938) has described a system of pale cells in the epithelium of the airways and BENSCH et al. (1965), in an electron-microscopic study of the bronchial epithelium of man, have described cells containing dense-cored vesicles. Both these sets of authors regard these cells as counterparts of the argentaffin cells in the gut. However the latter show green fluorescence with FALCK and HILLARP's method (SCHOFIELD et al., 1967), but no fluorescent cells have been found in the airway epithelium. FRÖHLICH (1949) has suggested that the pale cells resembe the α cells of the taste buds, and may be chemoreceptors. Non-fluorescent cells with dense cored vesicles are found in IGGO's corpuscles which are mechanoreceptors (IGGO and MUIR, 1963).

Mechanical stimulation of the epithelium of the respiratory tract from the larynx to the bronchi sets up discharges in vagal afferent fibres thought to come from epithelial receptors like those of Figs. 5–7. These endings in the lungs have come to be called irritant receptors because of their response to inhaled irritant aerosols and gases, and those in the larynx and trachea are sometimes called the cough receptors because of the reflex response to their stimulation.

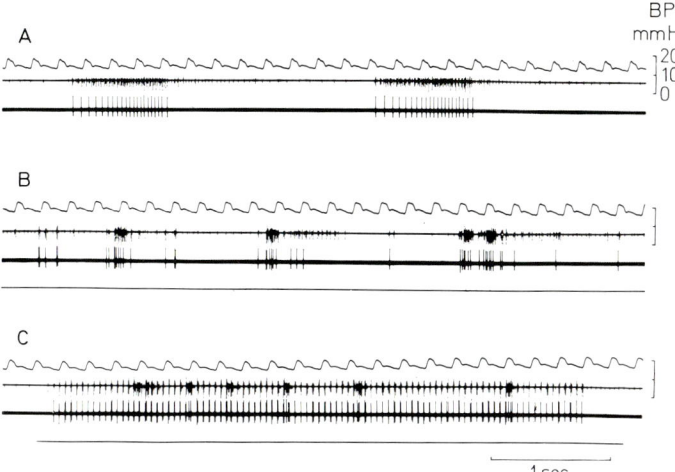

Fig. 3 A–C. Response of phrenic motoneuron discharge to epipharyngeal stimulation. Each record shows from above down: systemic arterial blood pressure (B.P.); multi-fibre phrenic discharge; single phrenic motoneuron discharge. Horizontal bars below records B and C represent duration of stimulus. A Control ventilation. B Mechanical stimulation of epipharyngeal mucosa. C Electrical stimulation of glossopharyngeal nerve at 20 shocks per sec. (From NAIL et al., 1971)

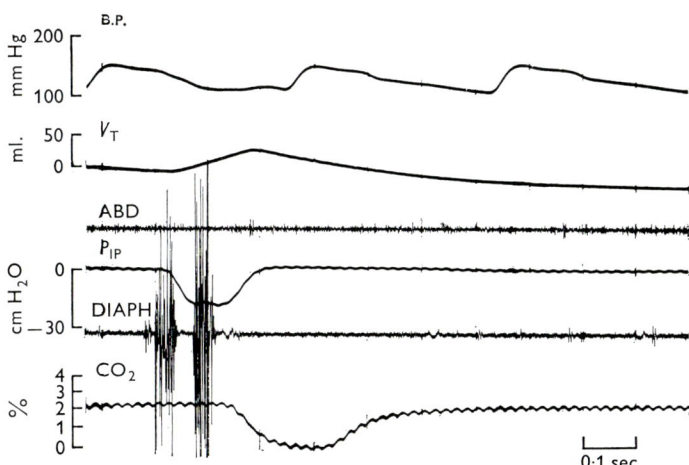

Fig. 4. Changes in abdominal (ABD) and diaphragmatic (DIAPH) electromyograms, blood pressure (B.P.), tidal volume (V_T), intrapleural pressure (P_{IP}) and tidal CO_2% during a single inspiratory-expiratory cycle of the aspiration reflex in an anaesthetized spontaneously breathing cat. (From TOMORI and WIDDICOMBE, 1969)

Fig. 2. Frozen sections of cat epipharyngeal region showing nerve fibres ramifying among epithelial cells. Schofield's silver stain. Photograph by Mrs. A. M. S. WHITE

motoneurone responses (NAIL et al., 1971). The inspiratory intercostal muscles contract at the same time as the diaphragm and, quite unlike sneezing and coughing, there is no contraction of expiratory muscles in the pauses between inspiratory efforts (Fig. 4). Mechanical stimulation of the nasopharyngeal epithelium also causes reflex broncho-dilatation and hypertension, the latter being the most prominent blood pressure response in the cat to mechanical stimulation of any mucosal site in the respiratory tract (TOMORI and WIDDICOMBE, 1969). These vaso- and bronchomotor responses are only seen in paralysed animals, since in spontaneously breathing ones they are masked by changes secondary to the inspiratory efforts.

The function of the aspiration reflex is not very clear, but it could be to lessen blockage at the back of the nasopharynx. The strong and repeated inspiratory efforts would pull an obstruction into the pharynx or larynx, whence it would be swallowed or coughed up. However this is a speculation, and it is not known whether mucus or similar obstructions are a sufficient stimulus for the reflex. There are also odd species differences: the dog and the rabbit lack an aspiration reflex (TEITELBAUM and RIES, 1935), although it is present in the newborn human and in the pig (TAKAGI et al., 1966). The hypertensive reflex response from the nasopharynx has been observed in paralysed human patients (CORBETT et al., 1969).

Epithelial Receptors

Cough and Irritant Receptors

Nerve fibres entering and ramifying in the epithelium of the airways from the trachea to the respiratory bronchioles can be demonstrated with silver and methylene blue staining methods (LARSELL, 1921; GAYLOR, 1934; HAYASHI, 1937; HONJIN, 1956; FISHER, 1964; SPENCER and LEOF, 1964). In the rabbit, fine processes, staining for nonspecific cholinesterase, continuous with nerve bundles, are also seen to enter the surface epithelium (FILLENZ and WOODS, 1970). There is reason to believe that these processes containing butyryl cholinesterase represent Schwann cells accompanying axons, rather than the axons themselves. Histochemical methods for acetylcholinesterase and catecholamines on the other hand show nerve fibres in the subepithelial layer, but none entering the epithelium. It would appear that all the fibres ramifying in the epithelium are sensory. In some cases one can see these epithelial terminations arising from a myelinated axon.

Histologically there are no differences in the appearances of the epithelial endings at different levels in the airways, e.g. the trachea and the bronchioles. Their different functional properties could be due to their different positions and their different reflex connections.

With methylene blue and silver staining extensive branching of myelinated fibres is seen in the subepithelial tissue and non-myelinated twigs then enter the epithelium. There is further branching giving rise to an extensive network at the base of the epithelial cells (Fig. 5), from which vertical twigs run towards the surface (Fig. 6). In methylene blue stained tissue the nerve fibres appear varicose.

Electronmicroscopy of tracheal epithelium shows the presence of varicose axons running towards the surface of the epithelium (Fig. 7). The varicosities

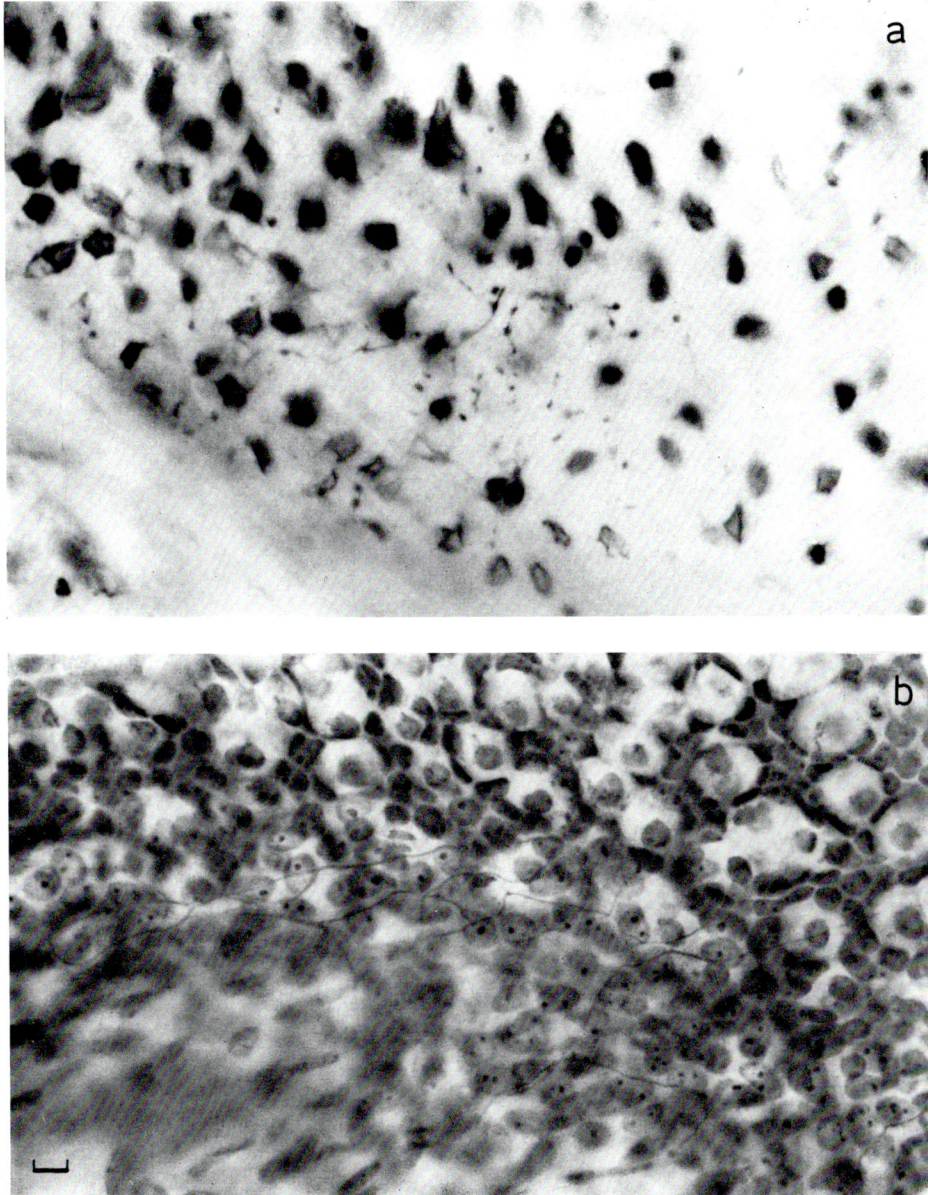

Fig. 5a and b. Terminal nerve fibre arborizations among epithelial cells lining airways. a Rabbit bronchus stained with methylene blue showing varicose appearance of fibres. b Dog bronchus stained with Richardson's silver method. Calibration 5 μm

lack the synaptic vesicles characteristic of autonomic motor fibres (FILLENZ and WOODS, 1970). COOK and KING (1969) have described nerve fibres in the avian airway associated with cells containing dense cored vesicles which they regard as

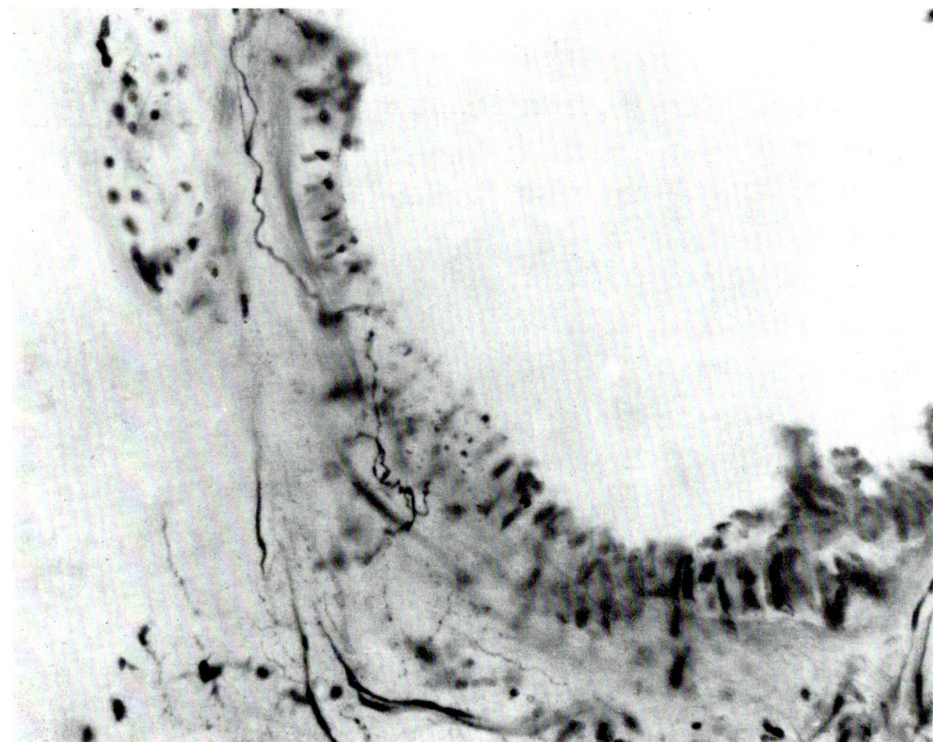

Fig. 6. Rabbit bronchiole stained with methylene blue. Fine varicose motor fibres run between smooth muscle cells in bronchiolar wall. A thicker nerve fibre ramifies under the epithelium and sends varicose terminals between the epithelial cells. Photograph by Dr. R. I. Woods

sensory cells. Similar cells occur in the trachea of the guinea-pig, but their association with nerve fibres has not been confirmed in this species (Fillenz, unpublished). Feyrter (1938) has described a system of pale cells in the epithelium of the airways and Bensch et al. (1965), in an electron-microscopic study of the bronchial epithelium of man, have described cells containing dense-cored vesicles. Both these sets of authors regard these cells as counterparts of the argentaffin cells in the gut. However the latter show green fluorescence with Falck and Hillarp's method (Schofield et al., 1967), but no fluorescent cells have been found in the airway epithelium. Fröhlich (1949) has suggested that the pale cells resembe the α cells of the taste buds, and may be chemoreceptors. Non-fluorescent cells with dense cored vesicles are found in Iggo's corpuscles which are mechanoreceptors (Iggo and Muir, 1963).

Mechanical stimulation of the epithelium of the respiratory tract from the larynx to the bronchi sets up discharges in vagal afferent fibres thought to come from epithelial receptors like those of Figs. 5–7. These endings in the lungs have come to be called irritant receptors because of their response to inhaled irritant aerosols and gases, and those in the larynx and trachea are sometimes called the cough receptors because of the reflex response to their stimulation.

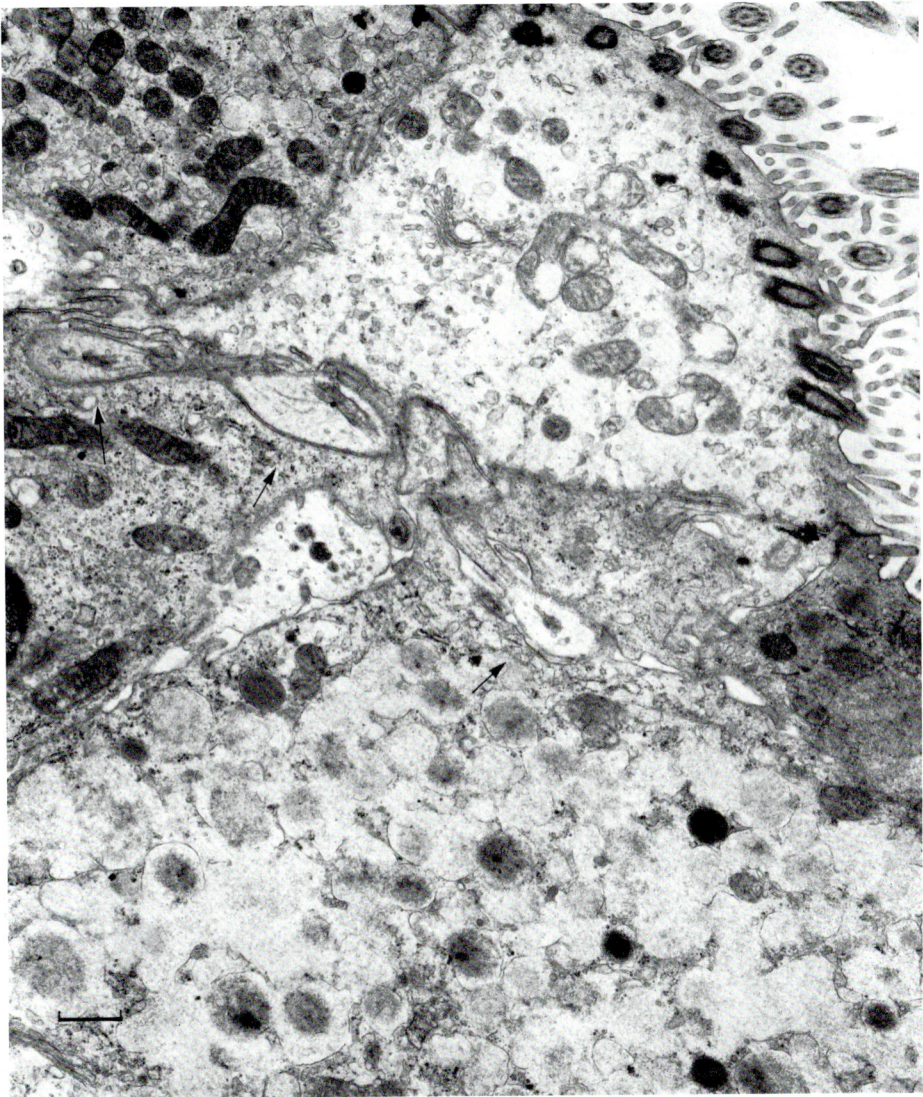

Fig. 7. Epithelial cells in guinea pig trachea with varicose axon. Arrows mark varicosities. Acrylic aldehyde and glutaraldehyde in sodium metavanadate fixation. Lead stain. Calibration 0.5 μm

The receptor properties have been studied mainly by recording action potentials from single fibre preparations of the cut vagus nerve. There are properties in common for receptors at each level: larynx (ANDREW, 1956), trachea (WIDDICOMBE, 1954b; WIDDICOMBE et al., 1962) and intrapulmonary bronchi (HOMBERGER, 1968; FERRER and KOLLER, 1969; MILLS et al., 1969, 1970; SELLICK and WIDDICOMBE, 1969, 1970). All are stimulated by an intraluminal catheter, and

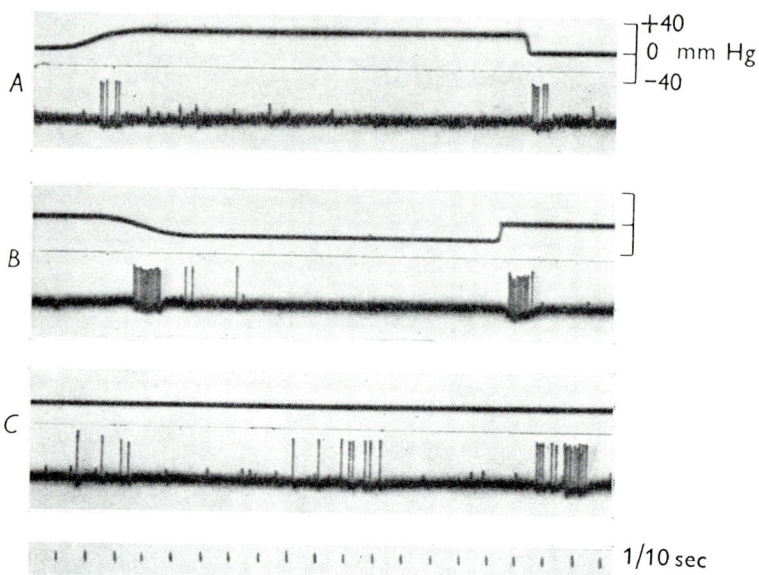

Fig. 8 A–C. Action potentials (lower traces) from a rapidly adapting receptor in the trachea of a cat. Upper traces, intratracheal pressure. Inflation (A) and deflation (B) of the trachea causes rapidly adapting discharges from the receptor. In C, the tracheal epithelium was gently touched with a catheter, causing further activity. (From WIDDICOMBE, 1954b)

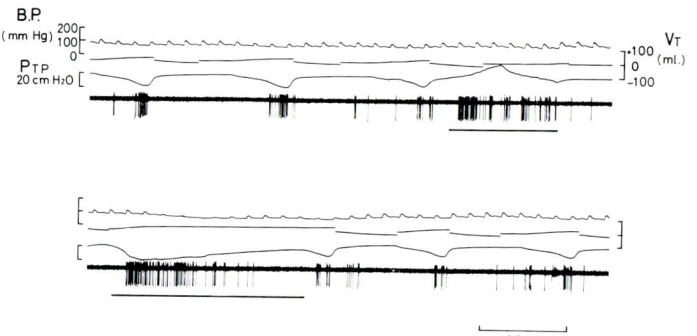

Fig. 9. Responses of a pulmonary irritant receptor to deflation (upper record) and inflation (lower record) of the lungs. From above down: systemic arterial blood pressure (B.P.), tidal volume changes (V_T, trace zeroing at points of zero airflow; inflation upwards), transpulmonary pressure (P_{TP}) and action potentials in a single vagal fibre. Deflation and inflation were during the horizontal signal bars. Note the rapidly adapting irregular discharges. (From MILLS et al., 1969)

those in the trachea and lungs by distension and collapse of the airway (Figs. 8, 9) and by intraluminal carbon dust (Fig. 10) (laryngeal endings not tested). The discharges to a maintained mechanical stimulus are irregular (Fig. 9), and also rapidly adapting especially for the larynx and trachea. However the receptors differ in other properties. Those in the trachea, but not the lungs, have an off-response to volume changes (Fig. 8). The tracheal and especially the pulmonary

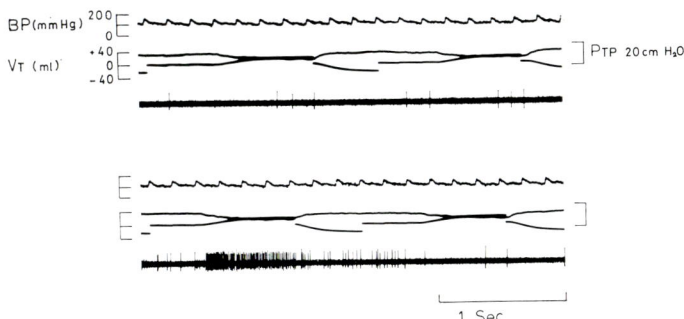

Fig. 10. Response of a lung irritant receptor to inhalation of carbon dust. Traces from above down: systemic arterial blood pressure (B.P.), trans-pulmonary pressure (P_{TP}), tidal volume (V_T) zeroing at points of zero airflow, and action potentials in a single vagal nerve fibre from a lung irritant receptor. Upper record, control showing slow spontaneous discharge; lower record, during inhalation of dust, showing maximum stimulation of the receptor. The rabbit was paralysed and artificially ventilated and vagotomised (SELLICK and WIDDICOMBE, unpublished)

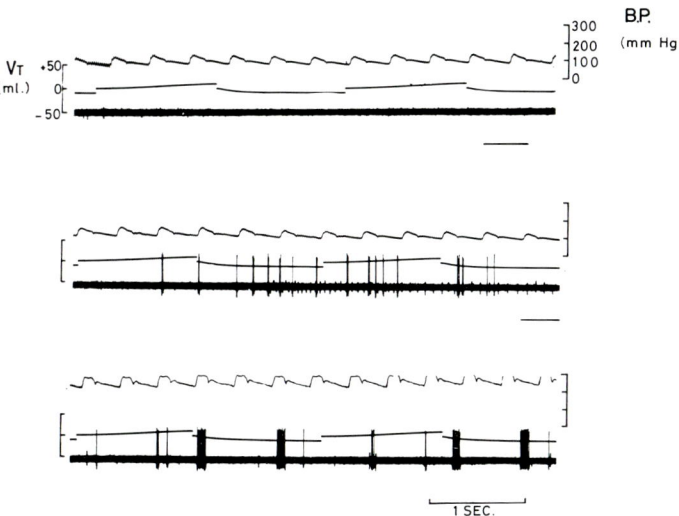

Fig. 11. Response of a pulmonary irritant receptor to intra-right atrial injection of histamine acid phosphate, 100 μg/kg (at signal in uppermost record) in a vagotomized, paralysed, artificially ventilated rabbit. 5.5 sec between upper two traces, 20.5 sec between lower two. Histamine caused an increase in blood pressure and a receptor discharge without clear relation to respiratory phase. (From MILLS et al., 1969)

receptors are stimulated by histamine injections (Fig. 11) or aerosols because of the contraction of underlying smooth muscle (MILLS et al., 1969; NADEL and WIDDICOMBE, unpublished), but this drug given intravenously does not usually affect the laryngeal endings (RICHARDSON and WIDDICOMBE, unpublished). The receptors in the lungs are very sensitive to chemical irritants such as ammonia, ether vapour and cigarette smoke, whereas those in the trachea and larynx are less sensitive.

The lung irritant receptors also have distinctive properties due to their intrapulmonary site. Thus they are stimulated by changes in the mechanical conditions of the lungs that increase the pull of the lung parenchyma on the airway walls: by pulmonary congestion, microembolism, anaphylaxis, atelectasis, and pneumothorax. In general these receptors seem to be sensitive to any mechanical change in their environment, and their threshold to such changes is low.

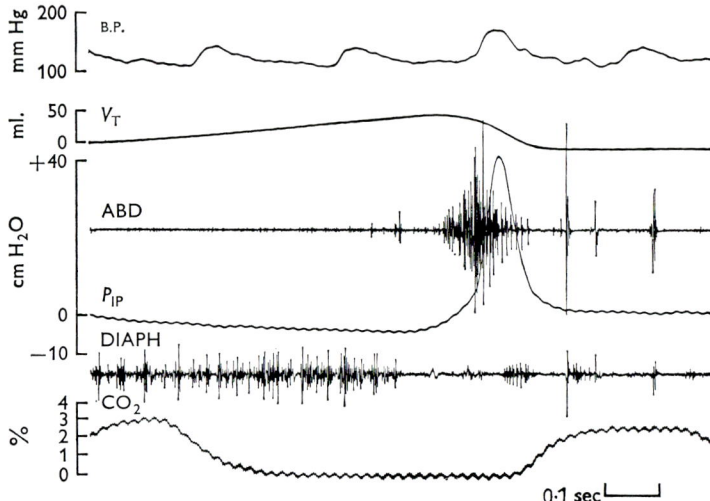

Fig. 12. Changes in abdominal (ABD) and diaphragmatic ($DIAPH$) electromyograms, blood pressure ($B.P.$), tidal volume (V_T), intrapleural pressure (P_{IP}) and tidal CO_2% during a single inspiratory cycle of laryngopharyngeal coughing in an anaesthetized spontaneously breathing cat. (From Tomori and Widdicombe, 1969)

The properties of the airway irritant receptors studied by single fibre recording are consistent with the histological appearance of receptors lying between the epithelial cells of the airways (Figs. 5–7). In particular their sensitivity to inert dust (particle diameter less than 16 μ) and irritant gases and aerosols, and their stimulation by gentle application of a fine intraluminal catheter, support this site. Furthermore the recordings are from myelinated fibres, although only the conduction velocities of those from lung irritant receptors have been measured (3.6–25.8 m/sec), and the epithelial endings have myelinated connections. Physiological studies have not shown whether irritant receptors occur in airways smaller than the cartilaginous bronchi, although there is histological evidence that they do.

In spite of the similarity of the histological properties of the epithelial receptors at different sites in the airways, and of some of their patterns of response to stimulation, the reflexes that they mediate show clear differences. Mechanical stimulations of the larynx and trachea cause coughing; in cats coughing from the larynx has more prominent expiratory efforts and also simultaneous contractions of the inspiratory and expiratory muscles somewhat like retching (Fig. 12), while those from the trachea and extrapulmonary bronchi have expiratory efforts alternating with strong inspirations. A similar difference is seen between coughing due to mechanical compared with chemical stimulation of the large airways; the latter is characterized by strong inspiratory efforts, presumably because the more distal receptors are more sensitive to chemical irritants. Stimulation of irritant receptors in the lungs does not usually cause coughing, but instead tachypnoea and hyperpnoea. This has been studied chiefly in the cat and rabbit, where the pulmonary conditions known to stimulate lung irritant receptors have been shown to cause a vagal reflex increase in breathing without coughing; this is true of

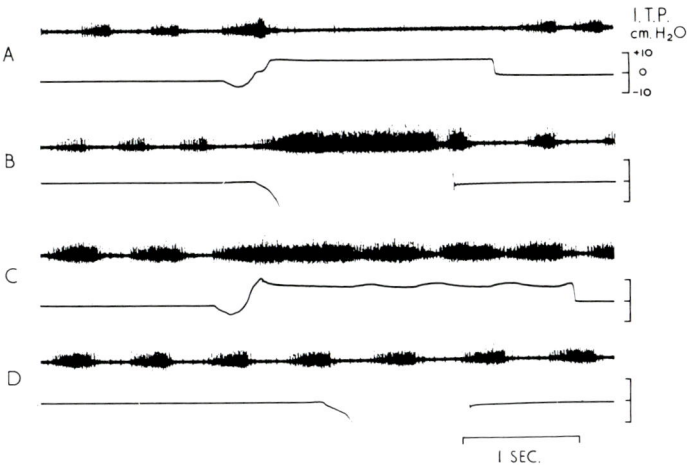

Fig. 13 A–D. Effect of inflation (A and C) and deflation (B and D) of the lungs of a rabbit on a diaphragmatic electromyogram (upper traces). In A and B the vagus nerves are not cooled; inflation causes a transient increase in diaphragmatic activity followed by inhibition, whereas deflation greatly increases activity. In C and D the vagus nerves are cooled to 5°C; inflation now increases diaphragmatic activity (Head's paradoxical reflex), while deflation has no effect. (From WIDDICOMBE, 1964)

pulmonary congestion (AVIADO et al., 1951; DOWNING, 1957), microembolism (WHITTERIDGE, 1950; GUZ and TRENCHARD, 1970), anaphylaxis (KOLLER, 1968; KARCZEWSKI and WIDDICOMBE, 1969b), histamine broncho-constriction (LETONA et al., 1961; DE KOCK et al., 1966; KARCZEWSKI and WIDDICOMBE, 1969a), atelectasis (CULVER and RAHN, 1952) and pneumothorax (SIMMONS and HEMINGWAY, 1957; SELLICK and WIDDICOMBE, 1969). In man coughing is absent or not prominent in these clinical conditions, although hyperventilation is characteristic.

Since lung irritant receptors are stimulated by deflation of the lungs and by pneumothorax, and since their reflex action is to cause hyperpnoea and hyperventilation, it has been concluded that they play a major role in the Hering-Breuer deflation reflex (SELLICK and WIDDICOMBE, 1970). The inhibition of pulmonary stretch receptor discharge by deflation of the lungs will also exert an excitatory action on breathing. Large inflations of the lungs can cause an inspiratory effort, or gasp, by a vagal reflex (Fig. 13) (HEAD, 1889; LARRABEE and KNOWLTON, 1946; CROSS, 1962; WIDDICOMBE, 1967). This is seen with intact vagi,

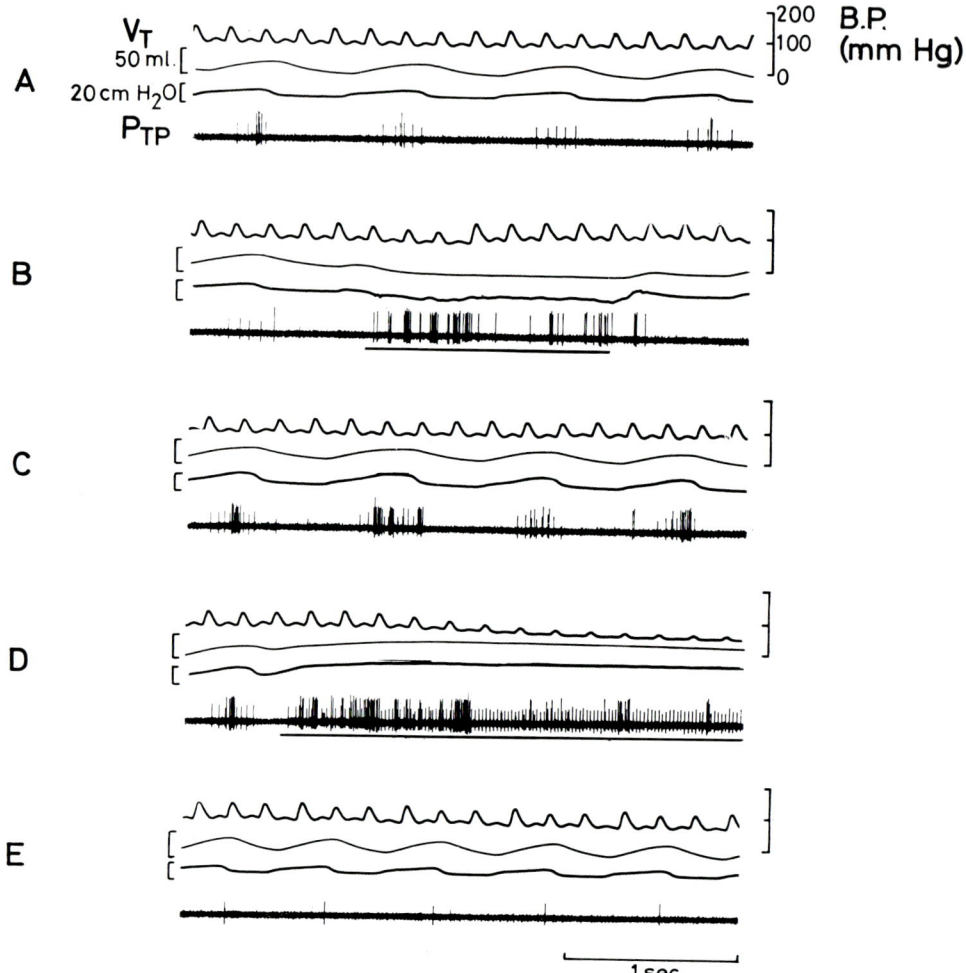

Fig. 14 A–E. Responses of a lung irritant receptor (larger action potentials) and a slowly adapting pulmonary stretch receptor (smaller action potentials) to deflation and inflation of the lungs. Traces from above down: systemic arterial blood pressure ($B.P.$), tidal volume (V_T), transpulmonary pressure (P_{TP}) and vagal action potentials. A Control showing spontaneous discharge form both receptors. C Deflation of the lungs during horizontal signal mark, showing cessation of discharge of the pulmonary stretch receptor and stimulation of the irritant receptor with an irregular discharge. C Control after deflation, showing increased firing of the irritant receptor and increased transpulmonary pressure swings. D Inflation of the lungs during signal, showing slowly adapting discharge of the pulmonary stretch receptor and rapidly adapting irregular discharge of the irritant receptor. E Control after inflation, showing cessation of discharge of the irritant receptor, inhibition of firing of the pulmonary stretch receptor, and decrease in transpulmonary pressure swings. (From SELLICK and WIDDICOMBE, 1970)

and when the vagus nerves are partially blocked by cold inflation, causes a maintained inspiratory effort, "HEAD's paradoxical reflex" (Fig. 13). Lung irritant receptors are a likely candidate for this reflex. They are stimulated by inflation of the lungs and their reflex action is to augment breathing. In addition the "gasp reflex" is enhanced by the collapse of the lungs and inhibited by over-inflation and increased compliance (REYNOLDS and HILGESON, 1965); exactly similar properties have been described for lung irritant receptors (Fig. 14) (SELLICK and WIDDICOMBE, 1970).

In addition to the reflex action of irritant receptors on breathing, their stimulation causes vascular and bronchomotor changes. In spontaneously breathing animals changes in blood pressure and bronchial calibre are mainly passive and secondary to the vigorous coughing, hyperpnoea and hyperventilation. However when the receptors are stimulated in paralysed and artificially ventilated animals, primary reflexes are revealed. Irritation of the larynx and trachea causes hypertension in cats and rabbits, and bronchoconstriction both from these sites and from lung irritant receptors (TOMORI and WIDDICOMBE, 1969). Cardiovascular reflex responses from lung irritant receptors have not been determined.

In the intact animal inhalation of irritant gases and aerosols will give rise to a complex reflex response, since both cough receptors and lung irritant receptors will be stimulated, and the chemicals may also have direct stimulant action on airway smooth muscle (BANISTER et al., 1950). Vagotomy greatly reduces the changes in breathing, and any residual respiratory effect is abolished by bilateral sympathectomy (WIDDICOMBE, 1954a). The bronchoconstriction is also abolished or greatly reduced by vagotomy or administration of atropine, indicating that the reflex component is large and the direct component small.

Smooth Muscle Receptors

Pulmonary Stretch Receptors

The smooth muscle cells in the walls of the airways have a motor innervation which consists of cholinergic and in some species also monoaminergic terminal fibres. These motor terminal fibres can be demonstrated by methylene blue staining and by certain of the silver methods (RICHARDSON's), as well as by the appropriate histochemical methods. The use of methylene blue and silver and gold methods stains some terminal nerve fibre ramifications among the smooth muscle cells which are quite different in appearance from the motor nerve terminals (HAYASHI, 1937; GAYLOR, 1934; HONJIN, 1956; TOUSSAINT and TOUSSAINT, 1959; FISHER, 1964; SPENCER and LOEF, 1964).

In some cases the terminal arborisations can be seen to come from myelinated fibres. These smooth muscle endings were first described by LARSELL (1922) and, although called "smooth muscle spindles", HAYASHI (1937) commented on their similarity to the sensory endings described by SUNDER-PLASSMANN (1933) in the carotid sinus, aortic arch and chambers of the heart. The sensory nature of these terminals is confirmed by their failure to stain with the histochemical methods for cholinergic and monoaminergic motor fibres. In rabbits with supranodose section of the vagus nerve (which would lead to degeneration of the myelinated preganglionic fibres) surviving myelinated fibres are found among smooth muscle cells (FILLENZ and WOODS, 1970); these are afferent fibres, and could be supplying terminals to either the smooth muscles or the epithelium.

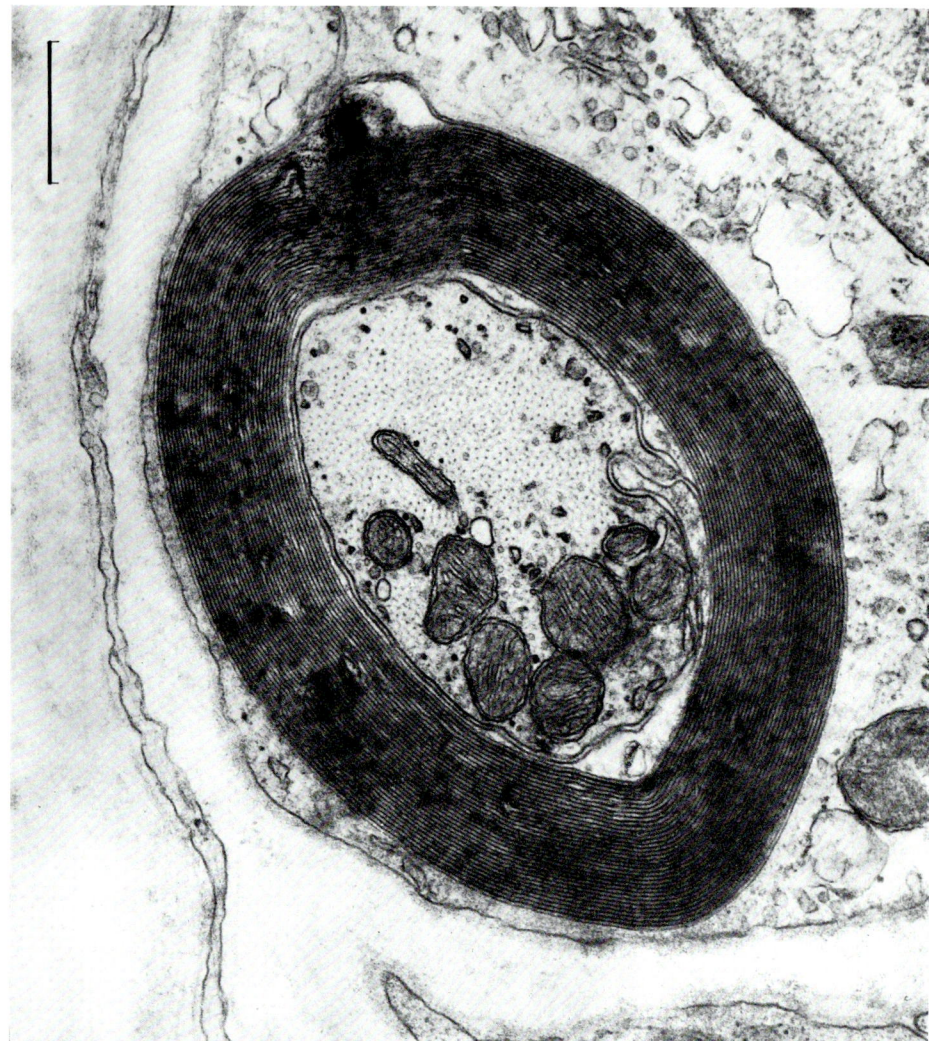

Fig. 15. Myelinated fibre containing mitochondria close to smooth muscle cells in guinea pig trachea. Fixation and staining as in Fig. 7. Calibration 0.5 μm

Electronmicroscopy shows myelinated fibres containing numerous mitochondria close to the smooth muscle cells (Fig. 15) and two types of terminal fibres among the smooth muscle cells (Fig. 16) (i) the motor nerve terminals with their characteristic synaptic vesicles and (ii) axons crowded with mitochondria, and containing glycogen and a few vesicles of a wide range of diameters. This appearance resembles that of some of the sensory fibres in the wall of the carotid sinus described by REES (1967).

There is now much indirect evidence, summarized below, that the smooth muscle receptors are the pulmonary stretch receptors studied by recording single fibre activity in the vagus nerves. They were the first lung receptors to be thus

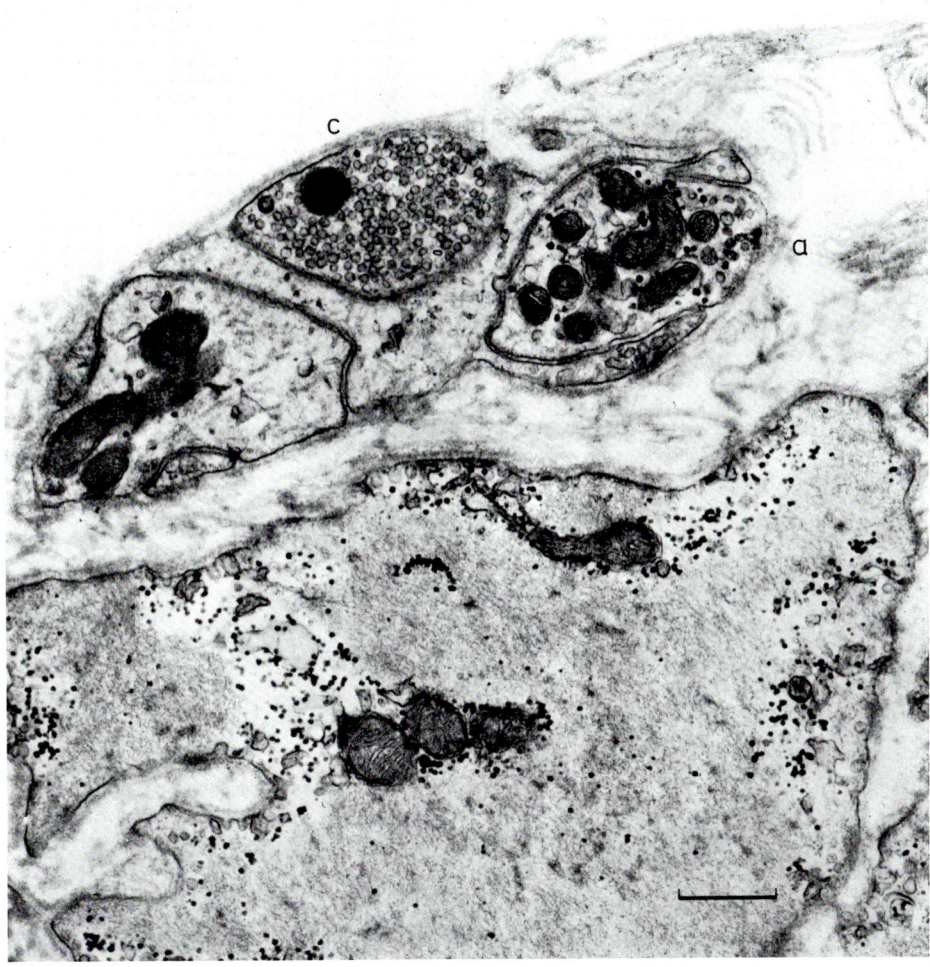

Fig. 16. Smooth muscle cell in guinea pig trachea with axon bundle. Axon (c) has the characteristic appearance of a cholinergic motor terminal. Axon (a) contains numerous mitochondria, glycogen, and a few vesicles of wide size range. Calibration 0.5 μm

studied (ADRIAN, 1933) and, although other receptors in the lungs are mechanosensitive and respond to "stretch", the term pulmonary stretch receptor has come to be restricted to slowly adapting endings stimulated by increases in lung volume (Figs. 14, 17) and thought to mediate the Hering-Breuer inflation reflex. What used to be called "rapidly adapting pulmonary stretch receptors" (KNOWLTON and LARRABEE, 1946) are now known to be cough receptors or irritant receptors in the airway epithelium. Most physiological studies have been with cats and rabbits, but human pulmonary stretch fibres have been recorded from (LANGREHR, 1964).

Pulmonary stretch receptors have low volume thresholds, usually within the eupnoeic tidal volume range or at a volume below functional residual capacity (i.e. some discharge continuously during the expiratory pause and are only inhibited by deflations of the lungs). Some receptors are stimulated by deflation as well as by inflation, and this is especially true of endings in the walls of the extrapulmonary airways with properties similar to those of pulmonary stretch receptors (WIDDICOMBE, 1954c); similar endings responding only to deflation are rare (PAINTAL, 1963). Maximum impulse frequencies are 100–300 impulses/sec in the cat for inflation pressures up to about 30 cm H_2O. Although called "slowly adapting", the degree of adaptation to a maintained inflation of the lungs is appreciable, and varies greatly between receptors. The variation may depend on

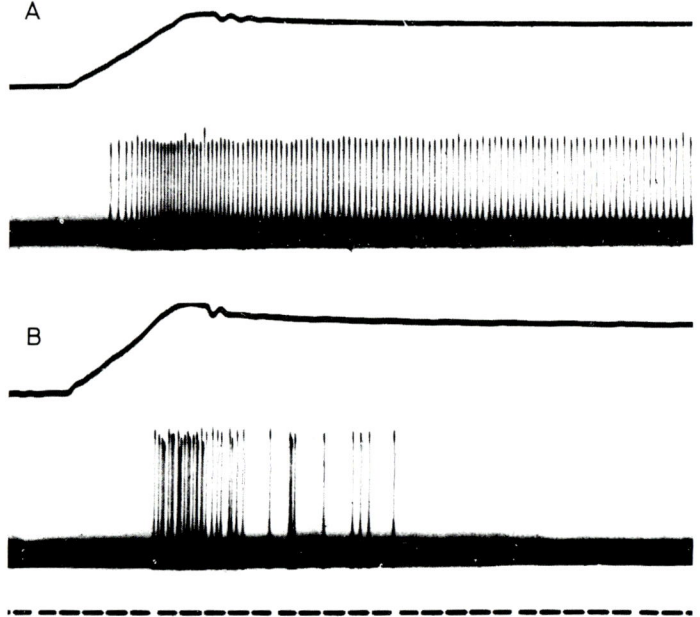

Fig. 17 A and B. Responses to inflation of the lungs of two kinds of afferent fibre in the vagus of a cat. The first receptor (A) adapts slowly, the second (B), very rapidly to maintained inflation of the lungs. Upper traces, intratracheal pressure; lower traces action potentials; time in 0.1 sec. (From KNOWLTON and LARRABEE, 1946)

the site of the endings, those in the larger airways having a slower adaptation. Part of the adaptation is because lungs show stress relaxation during inflation at constant volume, and the discharge of the receptors correlates better with transpulmonary pressure than with lung volume changes (DAVIS et al., 1956). In the cat, conduction velocities are in the range 14–59 m/sec (PAINTAL, 1953).

Considerable evidence has now accumulated that pulmonary stretch receptors are endings in the airway smooth muscle (Fig. 16). They are unlikely to be in the pleura or alveolar wall, since they are not destroyed by removing the visceral pleura and they lack a cardiac rhythm (WIDDICOMBE, 1954c). They are sensitized in pulmonary congestion, oedema and severe bronchoconstriction, and these responses are more consistent with localization in the walls of the airways than in the walls of the alveoli, since there is a greater mechanical pull on the airways with the decrease in lung compliance. They are sensitive to drugs injected into the bronchial arterial circulation (WIDDICOMBE, 1954c, 1961a). Smooth muscle receptors have large diameter myelinated fibres and account for over half of all lung endings (ELFTMAN, 1943); pulmonary stretch fibres are by far the commonest myelinated afferent nerves from the lungs (PAINTAL, 1953), and their receptors are localized chiefly at the points of bronchial branching where there is most smooth muscle, and smooth muscle receptors are most frequently seen (ELFTMAN, 1943; WIDDICOMBE, 1954b).

Since pulmonary stretch receptors lie in smooth muscle one would expect their activity to be influenced by the tone of the surrounding muscle. Bronchoconstriction caused by histamine usually increases the activity of the endings, but this could be secondary to increased distension of the airways associated with collapse of the lungs, i.e. a greater elastic pull of the lung parenchyma on the airway walls. In the absence of this indirect mechanical effect there is some evidence that smooth muscle contraction induced by drugs inhibits the discharge of adjacent receptors (WIDDICOMBE, 1954c). However this question has not been adequately studied.

Inflation of the lungs inhibits inspiratory activity by the Hering-Breuer inflation reflex (Fig. 13). This reflex has properties similar to those of pulmonary stretch receptors. Both have low volume thresholds, are slowly adapting and are conducted by vagal myelinated fibres. Vagotomy slows and deepens breathing in experimental animals by abolition of this reflex. The phasic action of the reflex seems to cut short inspiration and thus to control the rate and depth of breathing (ADRIAN, 1933; PAINTAL, 1963; WIDDICOMBE, 1964), and breathing may thereby be adjusted to an optimal pattern which minimizes the work or the inspiratory force of respiration (OTIS et al., 1950; MEAD, 1960). The optimal pattern depends on the mechanical properties of the lungs. Most of the characteristics of pulmonary stretch receptors are consistent with them sensing the mechanical properties of the lungs and adjusting breathing to its most economical pattern. This makes it surprising that the reflex is weak in conscious man (WIDDICOMBE, 1961b) and that in eupnoeic man blockade of vagal conduction does not change the pattern of breathing (GUZ et al., 1970a); however pulmonary stretch receptors are discharging in man in eupnoea (GUZ et al., 1970a). Possibly in man the reflex only becomes quantitatively important in conditions other than eupnoea. In addition to their action on breathing, pulmonary stretch receptors in experimental animals cause a reflex bronchodilatation (WIDDICOMBE and NADEL, 1963) and the same is probably true of man (NADEL and TIERNEY, 1961). They may also cause tachycardia (DALY and SCOTT, 1958).

Type J Receptors

In 1955 PAINTAL recorded action potentials from single vagal nonmyelinated fibres in the cat, which came from pulmonary endings which were then named "specific deflation receptors". However their response to deflation was transient and weak, and they have now been renamed J-receptors ("juxta-pulmonary capillary receptors") since they are thought to lie close to the pulmonary capillaries. As well as their occasional stimulation by deflation of the lungs and by pneumothorax, some are weakly stimulated by large lung inflations (PAINTAL, 1970;

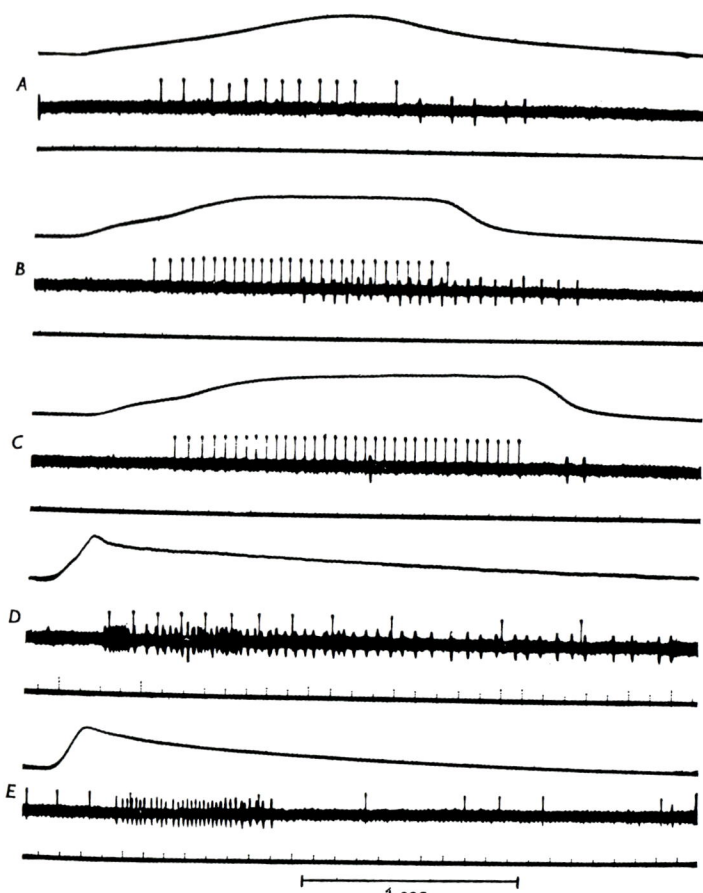

Fig. 18A–E. Responses of type J receptor (small diphasic spike) to inflation of the lung (A, B, and C) with open chest and insufflation with halothane (60 ml) in D and E. The large monophasic spikes in A, B, C, and D are those of a pulmonary stretch fibre and these show the consistent response to inflation of the lung with 60 ml in A and about 150 ml in B and C in contrast to the variable response in the type J fibre which is excited during the deflation phase in A, during inflation in B (but after a significant delay); there is comparatively little effect in C (again about 150 ml inflation). Insufflation of halothane had no excitatory effect on pulmonary stretch fibre in D (normal circulation) in contrast to the marked excitation of the type J receptor which is excited with a similar latency in E after cutting the great vessels and removing the ventricles. From above downwards in each record, intratracheal pressure, impulses in filament and 0.1 sec time marks. Gain of amplifier for intratracheal pressure in A, D, and E is twice that in B and C. (From PAINTAL, 1969)

SELLICK and WIDDICOMBE, 1970; COLERIDGE et al., 1965). They are also strongly excited by pulmonary congestion, pulmonary oedema, microembolism, and inhalation of strong irritants (Fig. 18) (PAINTAL, 1955, 1957, 1969, 1970; HOMBERGER, 1968). The evidence for their alveolar site is that they are stimulated within 2.5 sec by drugs such as phenyl diguanide injected into the right atrium (Fig. 19), and within 0.3 sec by inhalation of ether or halothane vapour; the

Fig. 19. Action potentials from afferent fibres that responded to deflation of the lungs, in the vagus of a cat. From above down, ECG and vagal action potentials, and time in 0.1 sec. At the horizontal bar 175 μg phenyl diguanide was injected intravenously. A vagal discharge follows. (From PAINTAL, 1957)

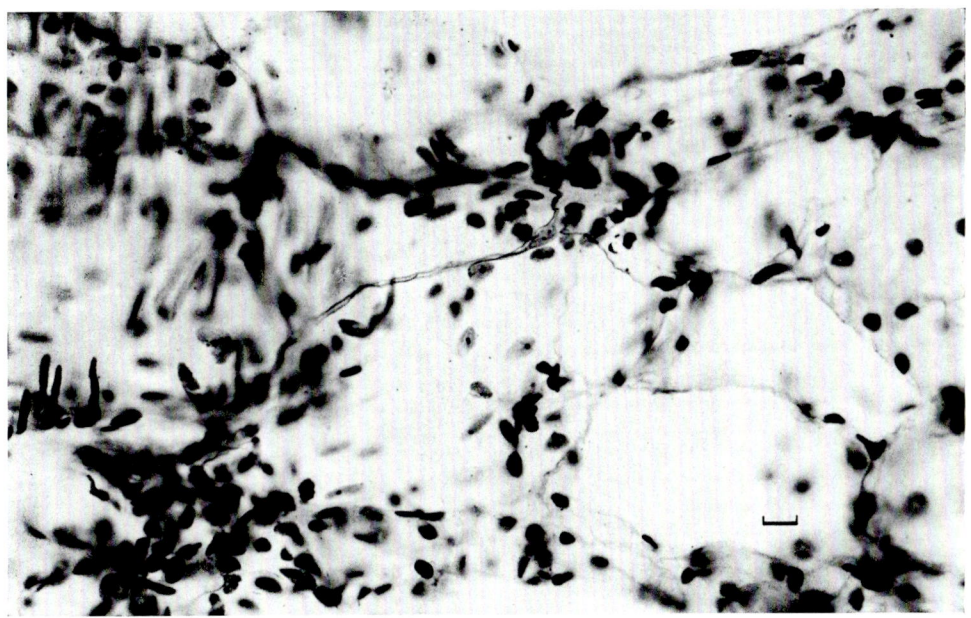

Fig. 20. Axon bundle with Schwann cell in lung parenchyma. 100 μm frozen section of dog lung stained with Richardson's silver method. Calibration 10 μm

response times are the same after circulatory arrest. The afferent vagal fibres have conduction velocities in the range 0.8 to 7.0 m/sec, so presumably some of the fibres are myelinated, but the majority conduct at less than 3.0 m/sec. Impulse frequencies during strong stimulation reach peak values of 20–50 impulses/sec, and average values of 7.5 impulses/sec.

Histological study of the innervation of the alveoli is complicated by the presence of scattered elastin and reticulin fibres, which tend to show up with most of the methods used for the demonstration of nerve fibres: they stain with silver, with methylene blue and show green autofluorescence with the method of HILLARP and FALCK, used for the demonstration of catecholaminergic fibres.

With the use of RICHARDSON's silver method, which stains nonmyelinated axons and their accompanying Schwann cell nuclei, occasional nerve fibres are seen in the lung parenchyma (Fig. 20). They are sparse and usually come from the network of nerve fibres surrounding the small arterioles (FILLENZ, 1970): this

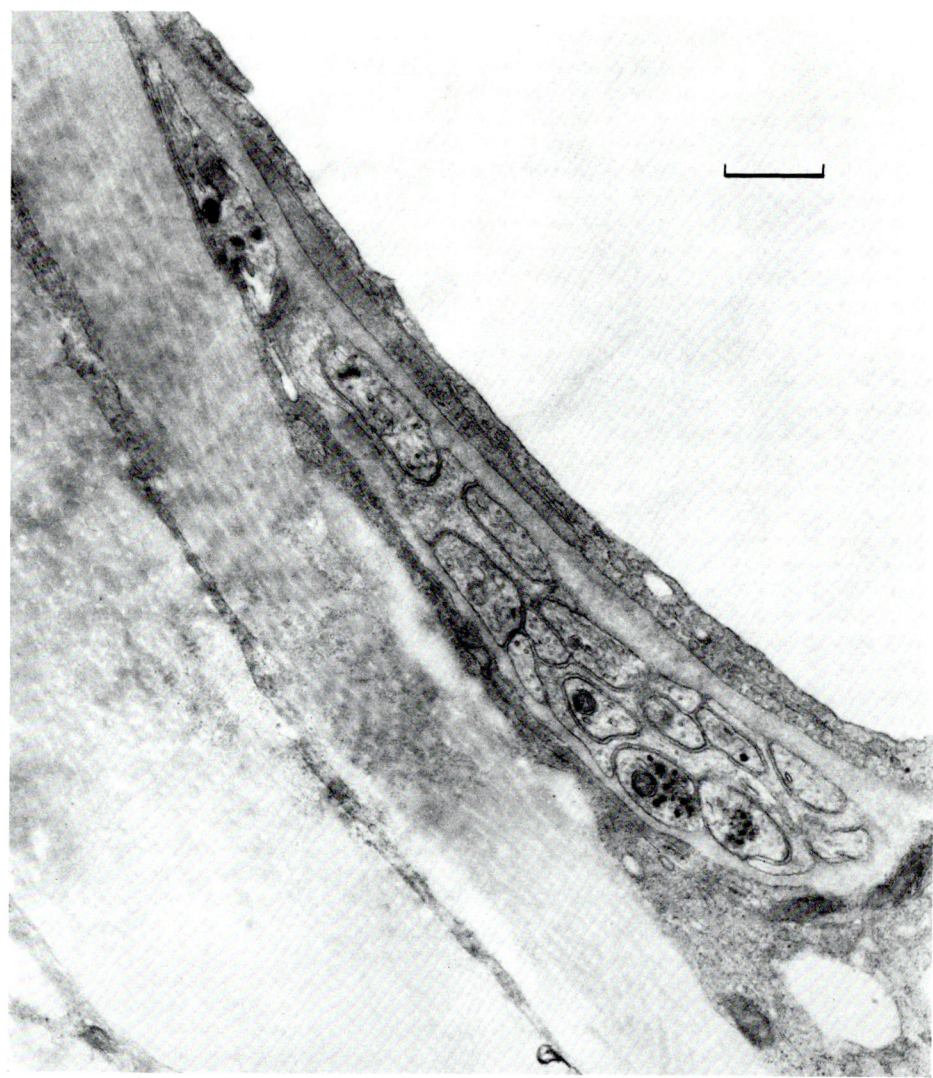

Fig. 21. Axon bundle in lung parenchyma running close to capillary endothelium. Four of the axons contain dense cored vesicles characteristic of sympathetic motor terminals; the others contain neurotubules and occasional mitochondria only, and some of these could be sensory. Calibration 0.5 μm

silver method does not enable one to distinguish between motor and sensory fibres, and between nerve endings and nerve fibres: hence one cannot say whether these nerve fibres are simply running through the lung parenchyma or innervating some structure in it. The extensive afferent innervation described by HIRSCH et al. (1968a, b) is based on the use of a silver staining method specific for reticulin.

DIJKSTRA (1967) has described cholinesterase-containing fibres in the lung parenchyma, but since it is not clear whether it is acetylcholinesterase or nonspecific cholinesterase, no conclusion can be drawn about the nature of the nerve fibres. Fig. 21 shows an electronmicrograph of an axon bundle in intimate relation to endothelial cells of pulmonary capillaries: some of the axons contain vesicles characteristic of cholinergic and noradrenergic autonomic fibres, but some of the axons do not contain such vesicles and could conceivably be sensory fibres. MEYRICK and REID (personal communication) have also observed fibres in the alveolar wall which appear more likely to be afferent. Whether these are connected to J-receptors has not been determined.

The reflex action of the J-receptors has been studied chiefly by the use of phenyl diguanide. This is not a specific stimulant for J-receptors, since it can also excite lung irritant receptors (MILLS et al., 1969) and peripheral chemoreceptors (DAWES et al., 1952), however the short latency of its response when the drug is

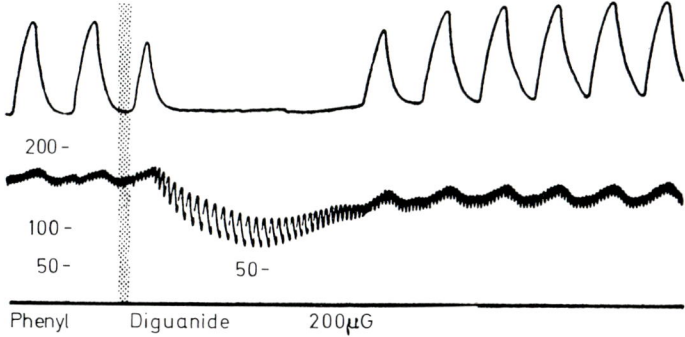

Fig. 22. Changes in respiration (upper trace) and carotid arterial blood pressure (lower trace) on injection of phenyl diguanide into the right auricle of a cat. The shaded vertical bar represents the time during which the injection was made. (From DAWES and COMROE, 1954)

injected into the right heart suggests that its main reflex action in the cat and rabbit is via the J-receptors. This has been confirmed by GUZ and TRENCHARD (1970) who found that phenyl diguanide produced its reflex response in the rabbit when all the vagal myelinated fibres had been blocked by anodal current. The drug causes an expiratory apnoea followed by rapid shallow breathing in the cat, and an inspiratory apnoea followed by rapid shallow breathing in the rabbit. In both species there is reflex hypotension and bradycardia (Fig. 22). Reflex actions on bronchomuscular tone have not been established. Injections of phenyl diguanide also inhibit the monosynaptic hind-limb reflex in the cat via a vagal afferent pathway (DESHPANDE and DEVANANDAN, 1970), but it is not certain whether this is a direct action of the J-receptor reflex or whether it is secondary to the reflex ventilatory and cardiovascular changes. PAINTAL (1969, 1970) has suggested that the true stimulus of the receptors is an increase in interstitial fluid volume in the alveolar wall secondary to an increase in pulmonary capillary pressure or permeability, and that the receptors may be stimulated in exercise as well as in the clinical conditions of pulmonary congestion and oedema. The endings could be part of a nociceptive system throughout the lungs.

Sensation Mediated by Respiratory Receptors

Human subjects are aware of the degree of distension of the lungs, but the sensation probably arises from the respiratory muscles and their associated joints (Campbell et al., 1970). There is no evidence that lung mechanoreceptors give rise to the sensation of lung distension or collapse, and subjects with conduction in both vagus nerves blocked by local anaesthesia have no obvious derangement of the sense of thoracic volume (Noble et al., 1970).

Similarly we can detect added elastic and viscous resistances to breathing but, since this sensation is unaffected by bilateral vagal blockade, it probably does not depend on the activity of receptors in the lungs (Campbell et al., 1970; Noble et al., 1970).

On the other hand there is now good evidence the vagal afferent fibres from the lungs can give rise to unpleasant respiratory sensation; this evidence is based chiefly on the work of Guz and his colleagues who have been performing bilateral vagal anaesthetization in healthy subjects and patients (Guz et al., 1966a and b; Noble et al., 1970). Many patients with dyspnoea due to lung disease have the symptom relieved by vagal blockade and their breath-holding time is prolonged (Guz et al., 1970b). Healthy subjects with vagal blockade no longer feel respiratory distress during hypercapnic hyperpnoea, although the fact that their ventilatory response to CO_2 is reduced may be a factor in the changed sensation: their breath-holding time is approximately doubled. Patients with functional block of conduction in the spinal cord at C 2–3, so that awareness of sensation from the diaphragm, intercostal, and abdominal muscles is abolished, can still feel unpleasant respiratory sensation on pulmonary congestion (Prys-Roberts, 1970), hypercapnia and breath-holding (Noble et al., 1970); the only likely mechanisms involve vagal afferent pathways which have unimpaired conduction. These and similar observations indicate that in some circumstances activity in vagal afferent fibres from the lungs can accentuate or give rise to unpleasant respiratory sensation. Direct evidence for the existence and nature of such a sensory mechanism is as follows:

1. If lung is first collapsed either by breathing oxygen or by centrifugation of subjects, on reinflation of the collapsed lung there is instantaneous pain or an unpleasant "tearing sensation" (Ernsting, 1960; Burger and Macklem, 1968). The nature and timing of this sensation can be explained only by mediation by lung receptors. In experimental animals the reinflation of collapsed lung causes a strong discharge from lung irritant (epithelial) receptors (Sellick and Widdicombe, 1970). Other receptors are not stimulated (J-receptors) or no more than on inflation of normal lung (pulmonary stretch receptors).

2. Inhalation of irritant gases and aerosols causes a painful or distressing sensation localized to the thoracic region, or a sensation of tightness in the chest. It has not been unequivocally established by vagal blockade that this sensation is due to stimulation of lung receptors, but the fact that the irritation or tissue damage is localized to the lungs and respiratory tract makes any other explanation implausible. J-receptors are stimulated by chlorine gas, halothane and ether (Paintal, 1970), and lung irritant receptors are stimulated by ether, ammonia vapour and cigarette smoke (Mills et al., 1969, 1970).

3. Passage of an endobronchial or endotracheal catheter in paralysed patients causes an immediate painful, burning or irritating sensation (PRYS-ROBERTS, 1970). This stimulus excites only cough and lung irritant receptors of those known to exist in the respiratory system (MILLS et al., 1970).

These observations establish that receptors in the lungs and lower respiratory tract can give rise to unpleasant respiratory sensation, and that they may contribute to dyspnoea. They certainly incriminate lung irritant (epithelial) and probably J-receptors. This conclusion is supported by the fact that there is a remarkably good correlation between the lung conditions which stimulate irritant and/or J-receptors in experimental animals and the occurance of dyspnoea in the corresponding condition in human patients. No such correlation is seen for other lung receptors or for proprioceptors in the respiratory motor system. Just as in the skin painful sensations are mediated primarily by nociceptive endings with nonmyelinated and δ-myelinated fibres, so it is possible that for the lungs J-receptors with nonmyelinated fibres and irritant receptors with δ-myelinated fibres can gave rise to a variety of unpleasant sensations such as pain and dyspnoea.

Other Receptors

The search for pulmonary chemoreceptors has led to much research, especially in the last decade. The term is restricted to receptors of which the physiological stimulus is pCO_2, pO_2, or pH in the blood, and most of the research has centred on the "pulmonary glomus", a small structure which might be supplied by blood from the pulmonary artery (KRAHL, 1962). The consensus of opinion seems to be that such pulmonary chemoreceptors are non-existent or unimportant. In any event they would be anatomically extrapulmonary, and there is no convincing physiological or histological evidence that there are chemoreceptors within the lungs of mammals (e.g. COLERIDGE et al., 1967). Of course many of the receptors already described respond to foreign chemicals injected into the blood stream or inhaled via the airways. Of considerable interest is the observation that in the lungs of birds there are receptors which are stimulated by a fall in airway pCO_2, and inhibited by a rise (Fig. 23) (FEDDE and PETERSON, 1970). Their reflex action is to inhibit inspiratory efforts. No equivalent receptor has been described for mammals, and their histological nature in birds has not been established.

There are histological studies of nerve endings in the visceral pleura of mammals (LARSELL, 1928; ELFTMAN, 1943). The endings are sparse, seem to have nonmyelinated fibres, and are concentrated at the hila of the lungs. Whether these are afferent, and if so what is their physiological stimulus and reflex action, is unknown.

Vagal single fibre recording can show activity from "mediastinal receptors" (ADRIAN, 1933; WIDDICOMBE, 1954b). These are slowly adapting endings with myelinated fibres, which can be localized to sites in the mediastinum, usually rostral to the heart or near the hila of the lungs. Their reflex function is unknown and they do not seem to have been studied histologically.

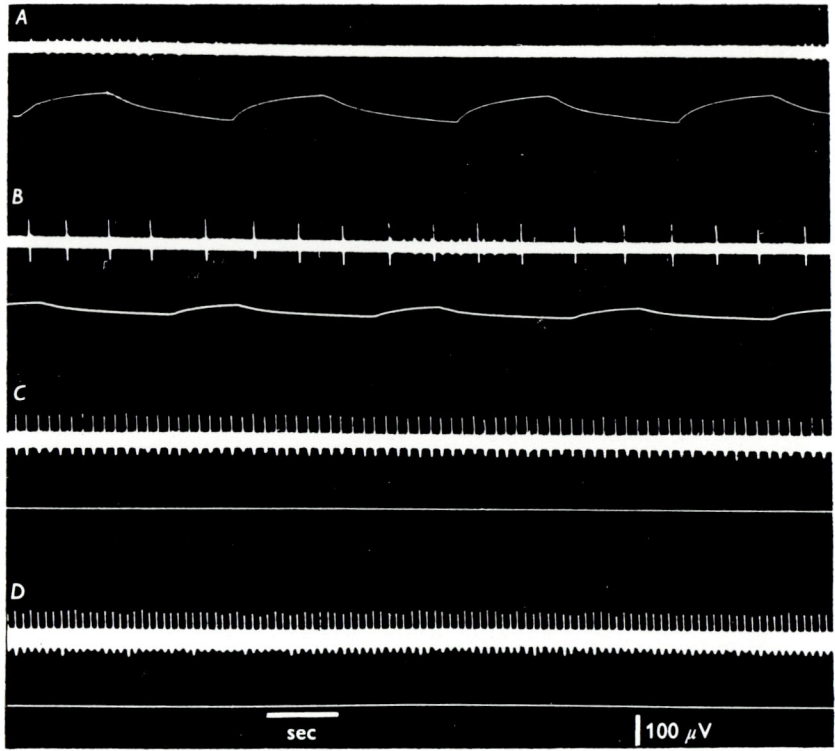

Fig. 23. Recordings from a vagal afferent fibre in a chicken whose impulse frequency markedly increased as the CO_2 content was decreased in the unidirectional ventilatory gas stream. A 9% CO_2; B 6% CO_2, C 3% CO_2; and D 0% CO_2. In each record the upper trace represents single-unit activity; the lower trace, sternal movement, inspiration upwards. (From FEDDE and PETERSON, 1970)

Conclusions

The last few years have seen notable advances in three aspects of the study of receptors in the lungs and respiratory tract: (1) histological studies which are more definitive because of the use of more sophisticated techniques; (2) a more comprehensive analysis of receptor properties by recording from single afferent fibres; and (3) the study of vagal reflexes from the lungs in healthy subjects and in patients with pulmonary disease.

Seven years ago the rash statement appeared that "at least nine respiratory reflexes originate in the thoracic viscera" (WIDDICOMBE, 1964). It now seems that all the reflexes at present known to arise from the lungs of mammals can be explained on the basis of three afferent pathways: from pulmonary stretch receptors, lung irritant receptors and type-J receptors. There may be other afferent fibres to be studied, and other lung receptors to be identified, but at present there is no need to postulate them. There have been histological descriptions of pulmonary endings other than those described above, but the absence of their confirmation in recent studies and the lack of a physiological role for them raises the suspicion that some at least may be artefacts.

Although in one respect the physiology of lung reflexes has been simplified, in another it is more complex. We now know that few if any physiological or pathological events in the lungs act on only one modality of receptor. As with the skin, so for the lungs there must be a varying pattern of activity from the different types of receptor in nearly all situations. This makes the analysis of lung reflexes more complicated. It may be that the analytical experiments with single fibre recording from vagal fibres which have been so prolific during the last forty years are due to be replaced by attempts to assess quantitatively the roles of the different reflexes in normal and abnormal physiology.

References

ADRIAN, E. D.: Afferent impulses in the vagus and their effect on respiration. J. Physiol. (Lond.) **79**, 332–358 (1933).

ANDREW, B. L.: A functional analysis of the myelinated fibres of the laryngeal nerve of the rat. J. Physiol. (Lond.) **133**, 420–432 (1956).

AVIADO, D. M., LI, T. H., KALOW, W., SCHMIDT, C. F., TURNBULL, G. L., PESKIN, G. W., HESS, M. E., WEISS, A. J.: Respiratory and circulatory reflexes from the perfused heart and pulmonary circulation of the dog. Amer. J. Physiol. **165**, 267–277 (1951).

BANISTER, J., FEGLER, G., HEBB, C.: Initial respiratory responses to the intratracheal inhalation of phosgene or ammonia. Quart. J. exp. Physiol. **35**, 233–250 (1950).

BENSCH, K. G., GORDON, G. B., MILLER, L. R.: Studies on the bronchial counterpart of the Kultschitzky (Argentaffin) Cell and innervation of bronchial glands. J. Ultrastruct. Res. **12**, 668–686 (1965).

BURGER, E. J., MACKLEM, P.: Airway closure: demonstration by breathing 100% O_2 at low lung volumes and by N_2 washout. J. appl. Physiol. **25**, 139–148 (1968).

CAMPBELL, E. J. M., AGOSTONI, E., DAVIS, J. N.: The respiratory muscles. London: Lloyd-Luke 1970.

COLERIDGE, H., COLERIDGE, J. C. G., HOWE, A.: Search for pulmonary arterial chemoreceptors in cat, with comparison of blood supply of aortic bodies in newborn and adult animal. J. Physiol. (Lond.) **191**, 353–374 (1967).

COLERIDGE, H. M., COLERIDGE, J. C. G., LUCK, J. G.: Pulmonary afferent fibres of small diameter stimulated by capsaicin and by hyperinflation of the lungs. J. Physiol. (Lond.) **179**, 248–263 (1965).

COOK, R. D., KING, A. S.: Nerves of the avian lung: electron microscopy. J. Anat. (Lond.) **105**, 202 (1969).

CORBETT, J. L., KERR, J. A., PRYS-ROBERTS, C., CRAMPTON-SMITH, A., SPALDING, J. M. K.: Cardiovascular disturbances in severe tetanus due to overactivity of the sympathetic nervous system. Anaesthesia **24**, 198–212 (1969).

CROSS, K. W.: Head's paradoxical reflex. Brain **84**, 529–534 (1962).

CULVER, G. A., RAHN, H.: Reflex respiratory stimulation by chest compression in the dog. Amer. J. Physiol. **168**, 686–693 (1952).

DALY, M. DE B., SCOTT, M. J.: The effects of stimulation of the carotid body chemoreceptors on heart rate in the dog. J. Physiol. (Lond.) **144**, 148–166 (1958).

DAVIS, H. L., FOWLER, W. S., LAMBERT, E. H.: Effect of volume and rate of inflation and deflation on transpulmonary pressure and response of pulmonary stretch receptors. Amer. J. Physiol. **187**, 558–566 (1956).

DAWES, G. S., COMROE, J. H., Jr.: Chemoreflexes from the heart and lungs. Physiol. Rev. **34**, 167–201 (1954).

— MOTT, J. C., WIDDICOMBE, J. G.: Chemoreceptor reflexes in the dog and the action of phenyldiguanide. Arch. int. Pharmacodyn. **90**, 203–222 (1952).

DE KOCK, M. A., NADEL, J. A., ZWI, S., COLEBATCH, H. J. H., OLSEN, C. R.: New method for perfusing bronchial arteries; histamine bronchoconstriction and apnoea. J. appl. Physiol. **21**, 185–194 (1966).

DESHPANDE, S. S., DEVANANDAN, M. S.: Reflex inhibition of monosynaptic reflexes by stimulation of type-J pulmonary endings. J. Physiol. (Lond.) **206**, 345–358 (1970).

DIJKSTRA, C.: Über die Lungeninnervation. Acta neuroveg. (Wien) **29**, 552–578 (1967).

DOWNING, S. E.: Reflex effects of acute hypertension in the pulmonary vascular bed of the dog. Yale J. Biol. Med. **30**, 43–56 (1957).

ELFTMAN, A. G.: The afferent and parasympathetic innervation of the lungs and trachea of the dog. Amer. J. Anat. **72**, 2–28 (1943).

ERNSTING, J.: Some effects of oxygen breathing on man. Proc. roy. Soc. Med. **53**, 96–98 (1960).

FEDDE, M. R., PETERSON, D. F.: Intrapulmonary receptor response to changes in airway-gas composition in *Gallus domesticus*. J. Physiol. (Lond.) **209**, 609–625 (1970).

FERRER, P., KOLLER, E. A.: Über die Vagusafferenzen des Meerschweinchens und ihre Bedeutung für die Spontanatmung. Helv. physiol. pharmacol. Acta **26**, 365–387 (1968).

FEYRTER, F.: Über diffuse endokrine epitheliale Organe. Leipzig 1938.

FILLENZ, M.: Innervation of pulmonary capillaries. Experientia (Basel) **25**, 842 (1969).

— WOODS, R. I.: Sensory innervation of the airways. Ciba Foundation Hering-Breuer Centenary Symposium: Breathing. London: Churchill 1970.

FISHER, A. W. F.: The intrinsic innervation of the trachea. J. Anat. (Lond.) **98**, 117–124 (1964).

FRÖHLICH, F.: Die „Helle Zelle" der Bronchialschleimhaut und ihre Beziehungen zum Problem der Chemorezeptoren. Frankfurt. Z. Path. **60**, 547–559 (1949).

GAYLOR, J. B.: The intrinsic nervous mechanisms of the human lung. Brain **57**, 143 (1934).

GUZ, A., NOBLE, M. I. M., EISELE, J. H., TRENCHARD, D.: The role of vagal inflation reflexes in man and other animals. In: Breathing: Hering-Breuer Centenary Symposium. ed. PORTER, R., pp. 17–40. London: Churchill 1970a.

— — — — Experimental results of vagal block in cardiopulmonary disease. In: Breathing: Hering-Breuer Centenary Symposium, ed. PORTER, R., pp. 315–328. London: Churchill 1970b.

— — WIDDICOMBE, J. G., TRENCHARD, D., MUSHIN, W. W.: The effect of bilateral block of vagus and glossopharyngeal nerves on the ventilatory response to CO_2 of conscious man. Resp. Physiol. **1**, 206–210 (1966b).

— — — — — MAKEY, A. R.: The role of vagal and glossopharyngeal afferent nerves in respiratory sensation, control of breathing and arterial pressure regulation in conscious man. Clin. Sci. **30**, 161–170 (1966a).

— TRENCHARD, D.: The role of non-myelinated vagal afferent fibres from the lungs in the genesis of tachypnoea in the rabbit. J. Physiol. (Lond.) **213**, 345–371 (1971).

HAYASHI, S.: Mikroskopische Studien zur Innervation der Lunge. J. orient. Med. **27**, 37–79 (1937).

HEAD, H.: On the regulation of respiration. J. Physiol. (Lond.) **10**, 1–70, 279–290 (1889).

HIRSCH, E. F., KAISER, G. C., BARNER, H. B., COOPER, T., RAUS, J. J.: Innervation of the mammalian lung. I. The afferent receptors. Arch. Path. **85**, 51–61 (1968a).

— — — NIGRO, S. L., HAMOUDA, F., COOPER, T., ADAMS, W, E.: The innervation of the mammalian lung. III. Regression of the intrinsic nerves and their afferent receptors following thoracic sympathectomy, cervical vagotomy or thoracic stripping of the vagus. Arch. Surg. **96**, 149–155 (1968b).

HOMBERGER, A. C.: Beitrag zum Nachweis von Kollapsafferenzen im Lungenvagus des Kaninchens. Helv. physiol. pharmacol. Acta **26**, 97–118 (1968).

HONJIN, R.: On the nerve supply of the lungs of the mouse, with special reference to the structure of the peripheral vegetative nervous system. J. comp. Neurol. **105**, 587–609 (1956).

IGGO, A., MUIR, A. R.: A cutaneous sense organ in the hairy skin of cats. J. Anat. (Lond.) **99**, 151 (1963).

IVANCO, I., KORPAS, J.: Kotázke laryngeálneho a tracheálneho kašla. Bratisl. lek. Listy **34**, 1391–1395 (1954).

Ivanco, L., Korpas, J., Tomori, Z.: Ein Beitrag zur Interoception der Luftwege. Physiol. bohemoslov. **5**, 84–90 (1956).
Karczewski, W., Widdicombe, J. G.: The rôle of the vagus nerves in the respiratory and circulatory responses to intravenous histamine and phenyl diguanide in rabbits. J. Physiol. (Lond.) **201**, 271–291 (1969a).
— — The role of the vagus nerves in the respiratory and circulatory reactions to anaphylaxis in rabbits. J. Physiol. (Lond.) **201**, 293–304 (1969b).
Knowlton, G. C., Larrabee, M. G.: A unitary analysis of pulmonary volume receptors. Amer. J. Physiol. **151**, 547–553 (1946).
Koller, E. A.: Breathing and circulation in anaphylactic bronchial asthma in guinea pigs. II. Importance of vagus nerve in respiratory and circulatory reaction. Helv. physiol. pharmacol. Acta **25**, 353–373 (1967).
Krahl, V. E.: The glomus pulmonale: its location and microscopic anatomy. In: Pulmonary structure and function, ed. Rueck, A. V. S. de, and O'Connor, M., pp. 53–69. Boston: Little, Brown 1962.
Langrehr, D.: Receptor-Afferenzen im Halsvagus des Menschen. Klin. Wschr. **42**, 239–244 (1964).
Larrabee, M. G., Knowlton, G. C.: Excitation and inhibition of phrenic motoneurones by inflation of the lungs. Amer. J. Physiol. **147**, 90–99 (1946).
Larsell, O.: Nerve terminations in the lung of the rabbit. J. comp. Neurol. **31**, 105–131 (1921).
— The ganglia, plexuses, and nerve-terminations of the mammalian lung and pleura pulmonalis. J. comp. Neurol. **35**, 97–132 (1922).
— The nerves and nerve-endings of the pleura pulmonaris histologically and experimentally. Phi Beta Pi Quart. 1–7 (1928).
Letona, J. M. L. de, Mata, R. C. de la, Aviado, D. M.: Local and reflex effects of bronchial arterial injection of drugs. J. Pharmacol. exp. Ther. **133**, 295–303 (1961).
Mead, J.: Control of respiratory frequency. J. appl. Physiol. **15**, 325–337 (1960).
Nadel, J. A., Tierney, D. F.: Effect of a previous deep inspiration on airway resistance in man. J. appl. Physiol. **16**, 717–719 (1961).
Mills, J., Sellick, H., Widdicombe, J. G.: The rôle of lung irritant receptors in respiratory responses to multiple pulmonary embolism, anaphylaxis, and histamine-induced bronchoconstriction. J. Physiol. (Lond.) **203**, 337–357 (1969).
Mills, J. E., Sellick, H., Widdicombe, J. G.: Epithelial irritant receptors in the lungs. Breathing: Hering-Breuer Centenary Symposium, ed. Porter, R., pp. 77–92. London: Churchill 1970.
Nail, B. S., Sterling, G. M., Widdicombe, J. G.: Epipharyngeal receptors responding to mechanical stimulation. J. Physiol. (Lond.) **204**, 91–98 (1969).
— — — Patterns of spontaneous and reflexly-induced activity in phrenic and intercostal motoneurones. J. Neurophysiol. in press (1971).
Noble, M. I. M., Eisele, J. H., Trenchard, D., Guz, A.: Effect of selective peripheral nerve blocks on respiratory sensations. In: Breathing: Hering-Breuer Centenary Symposium, ed. Porter, R., pp. 233–245. London: Churchill 1970.
Otis, A. B., Fenn, W. O., Rahn, H.: Mechanics of breathing in man. J. appl. Physiol. **2**, 592–607 (1950).
Paintal, A. S.: The conduction velocities of respiratory and cardiovascular afferent nerve fibres in the vagus nerves. J. Physiol. (Lond.) **121**, 341–359 (1953).
— Impulses in vagal afferent fibres from specific pulmonary deflation receptors. The response of these receptors to phenyl diguanide, potato starch, 5-Hydroxy-tryptamine and nicotine, and their rôle in respiratory and cardiovascular reflexes. Quart. J. exp. Physiol. **40**, 89–111 (1955).
— The location and excitation of pulmonary deflation receptors by chemical substances. Quart. J. exp. Physiol. **42**, 56–71 (1957).
— Vagal afferent fibres. Ergebn. Physiol. **52**, 74–156 (1963).
— Mechanism of stimulation of type J pulmonary receptors. J. Physiol. (Lond.) **203**, 511–512 (1969).
— The mechanism of excitation of type J receptors, and the J reflex. In: Breathing: Hering-Breuer Centenary Symposium, ed. Porter, R., pp. 59–70. London: Churchill 1970.

Prys-Roberts, C.: In: Breathing: Hering-Breuer Centenary Symposium, ed. Porter, R., p. 249. London: Churchill 1970.

Reynolds, L. D., Hilgeson, M. D.: Increase in breathing frequency following the reflex deep breath in anaesthetized cats. J. appl. Physiol. **20**, 491–499 (1965).

Schofield, G. C., Ho, A. K. S., Southwell, J. M.: Enterochromaffin cells and 5-hydroxytryptamine content of the colon of mice. J. Anat. (Lond.) **101**, 711–721 (1967).

Sellick, H., Widdicombe, J. G.: The activity of lung irritant receptors during pneumothorax, hyperpnoea and pulmonary vascular congestion. J. Physiol. (Lond.) **203**, 359–382 (1969).

— — Vagal deflation and inflation reflexes mediated by lung irritant receptors. Quart. J. exp. Physiol. **55**, 153–163 (1970).

Simmons, D. H., Hemingway, A.: Acute respiratory effects of pneumothorax in normal and vagotomized dogs. Amer. Rev. Tuberc. **76**, 195–214 (1957).

Spencer, H., Leof, D.: The innervation of the human lung. J. Anat. (Lond.) **98**, 599–610 (1964).

Sunder-Plassmann, P.: Über nervöse Receptorenfelder in der Wand der intrapulmonalen Bronchien des Menschen und ihre klinische Bedeutung, insbesondere ihre Schockwirkung bei Lungenoferatron. Dtsch. Z. Chir. **240**, 249–268 (1933).

Takagi, Y., Irwin, J. V., Bosma, J. F.: Effect of electrical stimulation of the pharyngeal wall on respiratory action. J. appl. Physiol. **21**, 454–462 (1966).

Teitelbaum, H. A., Ries, F. A.: A study of the comparative physiology of the glossopharyngeal nerve-respiratory reflex in the rabbit, cat and dog. Amer. J. Physiol. **112**, 684–689 (1935).

Tomori, Z.: Pleural, tracheal, and abdominal pressure variations in defensive and pathologic reflexes of the respiratory tract. Physiol. bohemoslov. **14**, 84–90 (1965).

— Widdicombe, J. G.: Muscular bronchomotor, and cardiovascular reflexes elicited by mechanical stimulation of the respiratory tract. J. Physiol. (Lond.) **200**, 25–50 (1969).

Toussaint, F. Y. P., Toussaint, F. Y.: Essais d'impregnation du système nerveux trachéo-broncho-pulmonaire chez l'homme et chez l'animal. Acta tuberc. belg. **50**, 179–197 (1959).

Whitteridge, D.: Multiple embolism of lung and rapid shallow breathing. Physiol. Rev. **30**, 475–487 (1950).

Widdicombe, J. G.: Respiratory reflexes from the trachea and bronchi of the cat. J. Physiol. (Lond.) **123**, 55–70 (1954a).

— Receptors in the trachea and bronchi of the cat. J. Physiol. (Lond.) **123**, 71–104 (1954b).

— The site of pulmonary stretch receptors in the cat. J. Physiol. (Lond.) **125**, 336–351 (1954c).

— The activity of pulmonary stretch receptors during bronchoconstriction, pulmonary oedema, atelectasis, and breathing against a resistance. J. Physiol. (Lond.) **159**, 436–450 (1961a).

— Respiratory reflexes in man and other mammalian species. Clin. Sci. **21**, 163–170 (1961b).

— Respiratory reflexes. In: Handbook of physiology, sect. 3, Respiration, vol. 1, pp. 585–630. Washington, D. C.: Amer. Physiol. Soc. 1964.

— Head's paradoxical reflex. Quart. J. exp. Physiol. **52**, 44–50 (1967).

— Kent, D. C., Nadel, J. A.: Mechanism of bronchoconstriction during inhalation of dust. J. appl. Physiol. **17**, 613–616 (1962).

— Nadel, J. A.: Reflex effects of lung inflation on tracheal volume. J. appl. Physiol. **18**, 681–686 (1963).

Chapter 4

Abdominal Visceral Receptors

By

BARRY F. LEEK, Edinburgh (Scotland)

With 16 Figures

Contents

Introduction	114
Electrophysiological Evidence	116
Slowly-Adapting Mechanoreceptors	116
Anatomical Location of Receptors	117
Receptors in Series with Smooth Muscle Cells	118
Nervous Discharge	119
Histological Location	122
The Consequences of an *in series* Location	123
Afferent Innervation	124
Function of *in series* Tension Receptors	126
Slowly-Adapting Tension Receptors Located in Mesenteries	129
Rapidly-Adapting Mechanoreceptors	130
Mesenteric Pacinian Corpuscles	130
Serosal Receptors	131
Muscularis Mucosae Receptors	133
Mucosal Receptors	134
Urethral Receptors	135
Chemoreceptors	135
Gastric pH Receptors	135
Acid-Sensitive Receptors	136
Alkali-Sensitive Receptors	136
Rôle of pH Receptors	137
Intestinal Chemoreceptors	138
Reflex Evidence	138
The Peristaltic or Myenteric Reflex	138
Ruminant Forestomach Mechanoreceptor Reflexes	140
On Salivation	140
On Primary Cycle Movements	141
On Secondary Cycle Movements	143
On the Cardia in Relation to Eructation	143
On Rumination	144
Abomasal Reflexes	144
Distension of the Abomasum	144
Acidification of the Abomasum	144
Gastro-Intestinal Mechanoreceptor Reflexes	145
Gastro-Intestinal Chemoreceptor Reflexes	147
Urinary Tract Reflexes	149
Some Sensory and Clinical Correlates	151

The Effects of Different Stimulus Intensities 151
    A Visceral Nociceptor Theory 152
Non-Painful Sensations 153
Painful Sensations 154
Conclusion 156
References 156

Introduction

In the normal state one is largely unaware of prevailing conditions in the abdominal viscera apart from the rather vague sensations which signal the fullness of the stomach, of the distal colon and of the bladder. This is all the more surprising, because the visceral nerves contain a preponderance of afferent (sensory) fibres. In the vagus nerves (Fig. 1) the proportion of afferent axons is 80–90% (DALY and EVANS, 1953; AGOSTINI et al., 1957), in the splanchnic nerves it is >50% (FOLEY, 1948) and in the pelvic nerves it is 30% (RANSON, 1921). Even these figures are probably too low, as DALY and EVANS (1953) and AGOSTINI et al. (1957) have demonstrated that many visceral afferent fibres are nonmyelinated and exist in larger numbers than hitherto suspected. On the purely numerical basis of their afferent:efferent axons ratio, the visceral nerves should be regarded, therefore, principally as sensory nerves and only secondarily as motor nerves. This is in contrast to the classical attitude derived from studies on the autonomic nervous system, which has unduly emphasized visceral motor pathways.

In contrast to the situation during normal visceral function, one is profoundly aware of certain types of visceral dysfunction, when they result in sensations of nausea and of abdominal pain. From time immemorial an objective analysis of the type, the location and the intensity of these sensations has been an essential part of clinical diagnostic technique and the earliest investigations of sensory mechanisms in the viscera were restricted to those mechanisms capable of reaching the level of consciousness; a classical account having been given by HURST (1911). Unfortunately, there are several limitations to the inferences which may be drawn about visceral receptors from studies of visceral sensation, e.g. (1) only a few receptor mechanisms are manifest consciously, (2) the sensation may be unrelated to the type of receptor stimulated (e.g. "burning" sensations are not due to thermoreceptor stimulation), (3) an applied "stimulus" may have either direct or indirect effects and may have more than one possible parameter of stimulation (e.g. distension of a viscus causes an increase both in volume and in tension), (4) "pain" thresholds vary from individual to individual and, even in the same person, they vary from time to time and (5) a sensation arising in one site may be referrable to an entirely different site. Nevertheless, clinical observations of those conditions for which the pathogenesis is well established have contributed some knowledge about the sensibility of various visceral structures and about the stimuli which excite them.

The magnitude of the afferent innervation of the abdominal viscera is so large, that it is clearly involved in functions other than signalling rather vague and infrequent derangements of visceral activity to the level of consciousness. Viscero-visceral reflexes occur mainly at subconscious levels and, since the turn of the century, considerable interest has developed in ascertaining the rôle of visceral

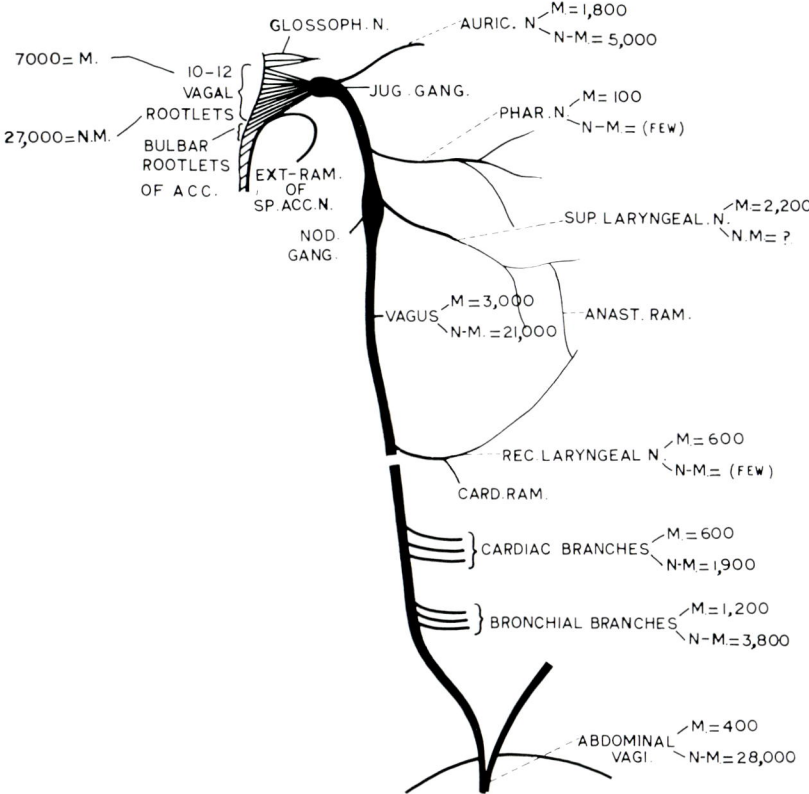

Fig. 1. Schematic diagram to show the number of medullated (M) and non-medullated (N-M) *sensory fibres* in the vagus nerve and its branches. The data were obtained from DUBOIS and FOLEY (1936), DALY and EVANS (1953) and AGOSTINI et al. (1957). In ruminants, the number of medullated fibres in the abdominal vagi (*i.e.* several thousands) is greater than the value shown here for non-ruminants (PAINTAL, 1963)

receptor mechanisms in the regulation of visceral activities, *i.e.* motility, secretions, and blood flow in the gastro-intestinal tract and in the uro-genital system. In some investigations of visceral reflexes, the reflexogenic zone and the effective stimulus have been determined fairly well and the underlying receptor mechanisms may be inferred with some confidence. These inferences apply, however, to a population of receptors, so that reflex studies are quite unsuitable for determining the location and the physiological properties of individual receptors. For these studies electrophysiological techniques are required.

ADRIAN (1933) was the first to record afferent impulses arising from the viscera: in this case from vagal afferent fibres innervating pulmonary inflation receptors. This technique was applied later to afferent fibres innervating the abdominal and pelvic viscera (TOWER, 1933; EVANS, 1936; TALAAT, 1937; GERNANDT and ZOTTERMAN, 1946). These early investigators obtained *multi-unit* records which made precise analysis either difficult or impossible. Precision came, however, with micro-dissection techniques which made *single unit* recording possible and, in this field, the work of Paintal and Iggo stands pre-eminent. The results obtained by electrophysiological techniques are complementary to those obtained by studies of viscero-visceral reflexes for, although they provide precise information about individual receptors, they have the demerits of any sampling system and they provide, by themselves, no indication of the significance of receptors in relation to reflexes or conscious sensation. Recent reviews of visceral receptor mechanisms based mainly on electro-physiological results have been given by Paintal (1963), Iggo (1966), Sharma (1967) and Iggo and Leek (1970).

In what follows, I have chosen to discuss visceral receptor mechanisms in categories determined by their means of investigation, but in a chronologically reverse order. In this way I shall endeavour, first to examine the physiological properties of various kinds of receptors, secondly, to examine the extent to which known receptors might fulfill the requirements of viscero-visceral reflexes and thirdly, to speculate on the relationship between the known receptors and those inferred as a result of visceral sensation and clinical observations.

Electrophysiological Evidence

The precise study of abdominal visceral receptors awaited the development of electrophysiological *single unit* techniques. The first records were obtained accidentally by Paintal (1954a) from vagal afferent fibres innervating gastric receptors which were excited chemically by phenyl diguanide injected into the aorta during an investigation of pulmonary receptors! Except where fine nerve filaments traverse a mesentery, it is necessary to subdivide a nerve trunk (*e.g.* a vagus nerve) repeatedly and to sample the afferent activity in each strand of nerve fibres as it is dissected away from the parent trunk. The dissection technique is described by Paintal (1954a *et seq*) and Iggo (1955 *et seq*). One cannot overemphasise the tedium of identifying and isolating small diameter (gastro-intestinal) nerve fibres with spikes of low amplitude and low frequency from a nerve trunk such as the vagus, in which the majority of nerve fibres (mainly pulmonary and cardiovascular units) have a much greater diameter and have spikes with a greater amplitude and frequency. The difficulties are, however, rewarded by results which are unobtainable by other techniques.

Most investigations have been concerned with mechanoreceptors rather than with chemoreceptors, because, for technical reasons, the parameters of a mechanical stimulus are easier to control than those of a chemical stimulus. The known mechanoreceptors fall into 2 categories: (1) those with a "slowly-adapting" response capable of detecting static and dynamic events and (2) those with a "rapidly-adapting" response capable of detecting dynamic events only. The reflex roles of these two categories of receptors may be expected to be quite different.

Slowly-Adapting Mechanoreceptors

These may be defined as receptors which, in response to a steady mechanical stimulus, give a discharge which is sustained for as long as the stimulus is applied. The sustained discharge is usually not regular, but may consist of an initial (dynamic) high frequency component which lasts a few seconds and is related to the rate of applying the stimulus. The discharge may then undergo a slight reduction over the course of 1–2 min (during receptive relaxation of the viscus) to attain a final level which may show rhythmic fluctuations about a mean value (Fig. 2).

Anatomical Location of Receptors. The detection of a mechanoreceptor zone results from observing the increased activity in single afferent units following, in the first instance, the distension of balloons located in the structures under

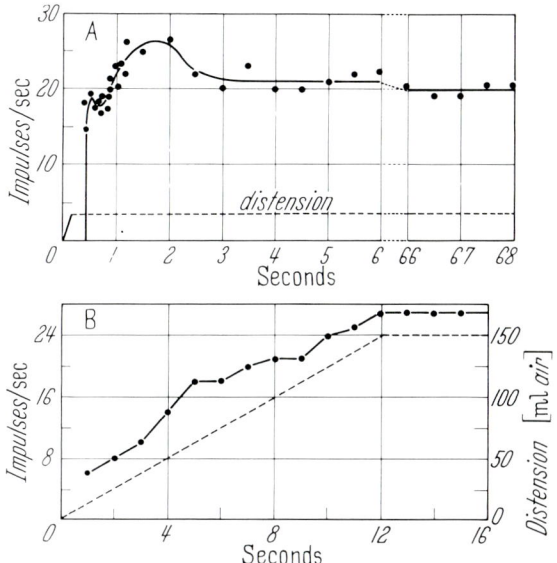

Fig. 2 A and B. Responses of two *slowly adapting* gastric tension receptors in cats to rapid (A) and slow (B) distension. In A, the rapid distension (*indicated by the dotted line*) produces in the afferent discharge an initial dynamic component, lasting 2–3 sec, and a static component which adapts slowly during the ensuing 1 min of maintained distension. In B, the slow distension produces an afferent discharge, the frequency of which is approximately proportional to volume of distension. At the end of the distension there is no dynamic component. The raised frequency occurring between 3 and 9 sec after the start of the distension may have been due to a superimposed intrinsic contraction of the region containing the tension receptor (PAINTAL, 1963)

investigation and, subsequently, manual exploration of the viscus, during which localized regions may be compressed, stroked, stretched, etc. Originally, PAINTAL (1954a) also used intravascular injection of phenyl diguanide (which excites various kinds of gastro-intestinal mechanoreceptors) to facilitate the detection of activity in vagal afferent units innervating these receptors. If the microdissection of the nerve strands is fine, this procedure is not necessary (IGGO, 1955 *et seq*; LEEK, 1969a). A necessary precaution is to ensure that the afferent activity being recorded actually arises from the viscus being distended and not from an adjacent structure which may be deformed by the distended viscus. PAINTAL (1954b, 1963) has described means of differentiating between these two possibilities when for example, gastric distension may excite nearby pulmonary receptors.

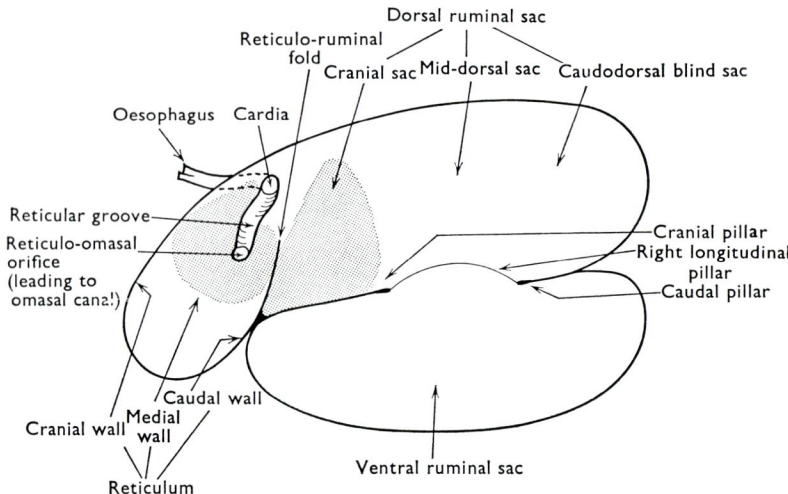

Fig. 3. A sagittal section through the reticulo-rumen of a sheep (*diagrammatic*) to show the regions where mechanoreceptors were located. The greatest densities of these receptors were found in the medial wall of the reticulum and of the cranial sac (*stippled areas*) (LEEK, 1969a)

Table 1. *The location of receptors innervated by 38 gastric afferent vagal units and the conduction velocities (c.v.) for 27 of these units. These locations are shown in Fig. 3* (LEEK, 1969a)

Site		No. of units	c.v. (m/sec)
Reticulum	cranial wall	2 ⎫	6, 8 ⎫
	medial wall	8 ⎬ 15	5, 7, 8, 10, 11 ⎬
	caudal wall	5 ⎭	14, 21, 24 ⎭
Cardia		1	19
Omasal canal		1	7
Reticular groove		4	5, 17
Reticulo-ruminal fold		4	12, 16
Cranial ruminal sac		9	9, 10, 11, 11, 13, 16, 10
Caudo-dorsal blind sac (dorsal wall)		1	15
Mid-dorsal ruminal sac		1	16
Right longitudinal pillar		1	23
Ventral ruminal sac		1	11
	Total = 38		Overall mean = 12.4

Receptors in Series with Smooth Muscle Cells

Using the manoeuvres of balloon distension and manual exploration, *slowly-adapting* mechanoreceptors have been located in the proximal oesophagus (ANDREW, 1957), the stomach of cats (PAINTAL, 1954a, b; IGGO, 1957a), the intestine of cats (IGGO, 1957a), the forestomach of goats (IGGO, 1955) and of sheep (LEEK, 1969a), the bladder of cats (IGGO, 1955), the uterus and broad ligament of rabbits (BOWER, 1959, 1966). Although they are apparently identical, these receptors have been variously termed "stretch receptors" by PAINTAL (1954a) and "tension receptors" by IGGO (1955). For reasons of accuracy given below, I prefer the latter description and I shall henceforth use it when implying slowly-adapting distension sensitive mechanoreceptors.

In cat stomach, PAINTAL (1954b) located 14 tension receptors: 7 being in the pyloric part, 4 were in the cardiac end of the lesser curvature and only 3 in the fundus. IGGO (1955) located 8 receptors: 2 being near the cardia and 6 in the pyloric antrum. In a later investigation by IGGO (1957a), another 16 tension receptors were located: 2 being near the cardia, 7 in the fundus and 7 in the pyloric antrum. Thus PAINTAL's (1954b) conclusion that the fundus may have a lower receptor density than other regions of the stomach may not be correct. In cat small intestine, IGGO (1957a, 1958) has located 11 tension receptor sites, of which 5 were in the duodenum and 6 were in the jejunum.

In the most recently evolved herbivorous mammals, "fermentation vat" facilities have been switched from the massive development of the hind-gut seen in earlier herbivores (*e.g. Equidae*) to an even more massive development taking the form of a non-glandular forestomach. This process is most pronounced in ruminant animals where the forestomach is clearly divided into three compartments (reticulum, rumen, and omasum), although parallel evolutionary events are seen within such widely differing groups as rodents, bats and marsupials (MOIR, 1965). In contrast to the stomach of a non-ruminant or to the ruminant glandular stomach compartment (the abomasum), the forestomach undergoes complex sequences of extrinsically controlled movements. For these movements, the forestomach has a profuse extrinsic innervation, both afferent and efferent, and the nerve fibres have a greater mean diameter than in the non-ruminant animals (IGGO, 1956). The cranial regions of the forestomach contain a high density of tension receptors. In sheep, LEEK (1969a) located 38 tension receptors (Fig. 3 and Table 1). The majority were situated in the medial wall of the reticulum (8) and of the cranial sac of the rumen (9). The remainder were found in the reticular groove (4), other parts of the reticulum (8), the omasal canal (1), the reticuloruminal fold (4) and other parts of the rumen (4). In view of the low density of receptors in the more caudal regions of the rumen, it seems unlikely that these regions could possess the reflexogenic significance postulated for secondary cycle ruminal movements by, for example, WEISS (1953) and STEVENS and SELLERS (1959). Other evidence to support this view is given by LEEK (1969b).

In cat bladder IGGO (1955) localized 2 of the 12 tension receptors studied. Both were situated near the neck of the bladder.

Nervous Discharge. Irrespective of their location, tension receptors appear to exhibit certain general properties on the basis of the afferent discharge recorded from single units. It is convenient to consider the discharge under two different conditions, (a) when the viscus containing the tension receptor is (apparently) quiescent and (b) when the viscus is undergoing a contraction. Under the former condition, there is generally a resting discharge, the mean spike frequency of which is (non-linearly) proportional to the resting tension in the viscus (PAINTAL, 1954b; IGGO, 1955). The receptors show threshold differences, some receptors having no resting discharge when the resting tension in the viscus is low. The resting discharge in most units is usually not regular but consists of intermittent bursts at the lower tensions and, as the tension is raised, there is a gradual transition,

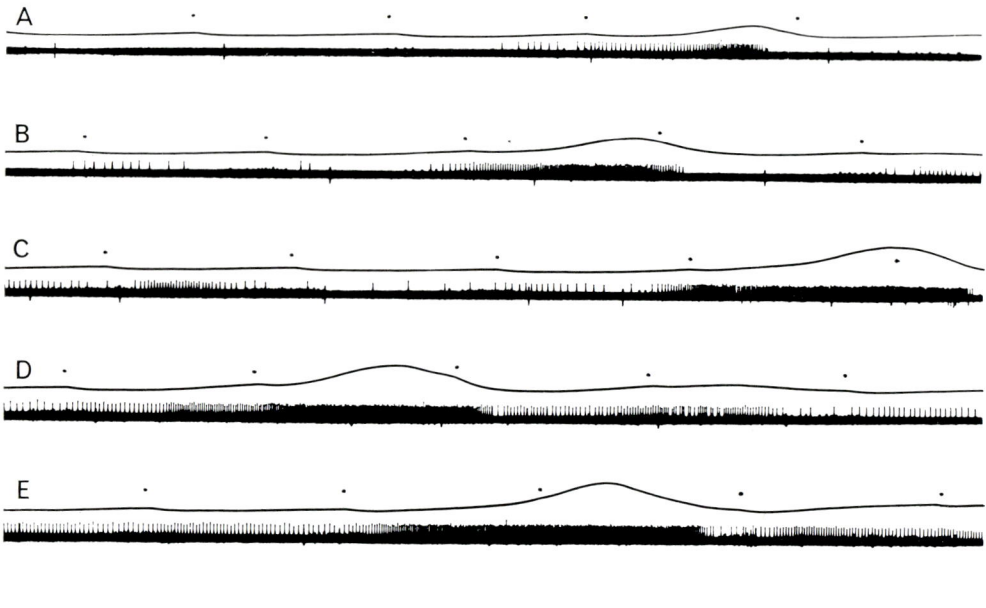

Fig. 4 A–E. The delayed effects upon the afferent discharges, recorded from a unit innervating reticular tension receptors, of increasing the volume of air in the reticular balloon. In A, B, C, D, and E, the reticular balloon contains 100, 200, 300, 400, and 600 ml air respectively. All the records were obtained from the same unit. Each record was obtained not less than 5 min after adding more air. In A there is negligible "resting discharge": a doublet of spikes (not shown) occur approximately every 7 sec. The reticular contraction has a small amplitude and the afferent discharge is correspondingly of relatively low frequency. As the level of distension is increased the "resting discharge" is also increased. At first, intermittent bursts of spikes appear which are unrelated to respiratory movements (B) and, upon further inflation, the interval between bursts is reduced and the peak discharge frequency reached during each burst is increased (C). Finally, the "resting discharge" becomes continuous, although it fluctuates at a rhythm related to respiration (D and E). At the higher levels of distension the durations and amplitudes of the reticular contractions are greater and associated with these, the afferent discharges are enhanced. In E, the peak frequency is greater than 100 spikes/sec. A small non-gastric "contaminant" spike, which extends below the base line, is also present but should be disregarded. The end of inspiration is marked with a dot (LEEK, 1969a)

so that a sustained discharge with rhythmical fluctuations is observed at the higher tensions (IGGO, 1955; LEEK, 1969a). This is illustrated in Fig. 4. IGGO (1955) suggested that these fluctuations might be due to localized movements in the viscus wall incapable of affecting the overall tension but sufficient to affect receptors located at the site of the movement. This seems a most likely explanation, particularly as the "intrinsic movements" of the cranial ruminal sac and of the reticulo-ruminal fold gave rise to palpable "ripples" which coincided with the fluctuations in the resting discharge being recorded at that time (LEEK, 1969a).

The change in the afferent discharge resulting from passively distending the viscus depends upon the rate of the distension (PAINTAL, 1954b; IGGO, 1955). If the distension is slow, there is a gradual increase in the afferent discharge, whereas, if the distension is sudden, there is a rapid rise in the afferent discharge to a high value, from which it steadily falls over the course of about 1 min to a new level, which is maintained (Fig. 2). It appears that the dynamic component of the afferent discharge seen during rapid distension may be attributable to at least two factors: (a) sudden distension evoking synchronously intrinsic movements over the whole wall and (b) the resistance to rapid stretch (or distension) provided by the visco-elastic elements in the wall. Furthermore, there is a considerable hysteresis effect, if one examines the relationship between the mean spike frequency of the resting discharge and tension in the viscus during incremental distension and that during decremental distension (LEEK, unpublished observation). Thus only in a very approximate way, do tension receptors provide an index of resting tension and/or volume in a viscus.

In all the situations where it was possible to observe the effect on the afferent discharge caused by a spontaneous or an induced contraction of the viscus containing the tension receptors, it was seen that the afferent discharge increased during the contraction in proportion to the tension which was developed (IGGO, 1955, 1957a; LEEK, 1969a). PAINTAL (1954b) overlooked this effect because he was unable to induce satisfactory contractions in the regions containing receptors. In a later review PAINTAL (1963) states that the gastric receptors which he described as "stretch receptors" (PAINTAL, 1954) were essentially the same as those described by IGGO (1955) as "tension receptors". The highest discharge frequencies are seen when the viscus contracts under isometric recording conditions (*i.e.* when the contraction causes a tension rise with no volume change). It is interesting that the spike frequency attained in this way, by developing tension activity, is much greater than that attained by distending the viscus passively to produce an equivalent tension (IGGO, 1955; LEEK, 1969a).

When the viscus is allowed to contract "isotonically" (*i.e.* when the contraction causes no change in tension but a reduction in volume), the extent of the lack of change in the afferent discharge is related to the extent to which the recording system approaches the perfectly isotonic condition (Fig. 5). Most systems involve moving a liquid against a constant pressure head and due to constrictions in the system and the viscosity of the liquid, there is considerable inertia at the beginning of the movement. Under these conditions (Fig. 6), the initial phase shows a small isometric component (*i.e.* increases in intraluminal pressure and tension receptor discharge) whereas the later phase, including the end of the phase of contraction, is almost isotonic with the resting state (*i.e.* the tension is equal to the resting state) and the afferent discharge is no greater and no less than during the resting state (IGGO, 1955). When the isotonic recording system is designed so that the viscus is filled with a gas (air) and the impediments to flow are minimal, a contraction is associated with negligible tension change and only a slight transient increase in the afferent discharge (Fig. 5; LEEK, 1969a). Whilst this approximately isotonic system is useful in the analysis of receptor properties, it is probable that, under normal physiological conditions, contractions of viscera are far from isotonic. Associated with the propulsion and the emptying of visceral contents, there are such impediments as fluid inertia and obstructions to flow. These would account for the tension changes, which are recorded by the various kinds of intraluminal manometric devices routinely used for detecting visceral movements.

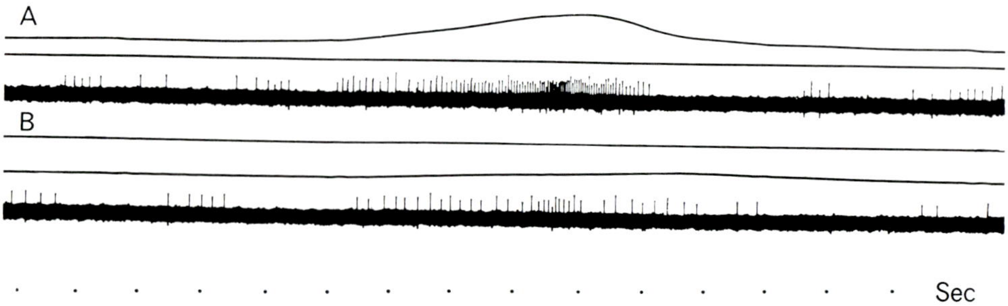

Fig. 5 A and B. The responses of an *in series* tension receptor in the reticulum during contractions occurring under *isometric* recording conditions (A) and *isotonic* recording conditions (B). In A, there is an increase in reticular pressure (*top trace in each record*) and hence an increase in tension in the reticular wall, no change in reticular volume (*middle trace in each record*) and a marked increase in the afferent discharge (*bottom trace in each record*). In B, there is negligible increase in reticular pressure and hence negligible increase in tension in the reticular wall, a reduction in volume (*shown as an upward deflection in the middle trace*) and only very slight increase in the afferent discharge (LEEK, 1969a)

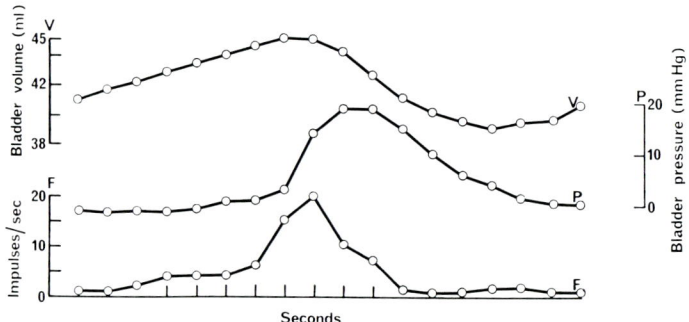

Fig. 6. Behaviour of an *in series* afferent unit during the reflex contraction of an innervated urinary bladder (cat, anaesthetized with chloralose). The urethra was cannulated and connected to a fluid-filled reservoir, at a head pressure of about 10 cm H_2O. The upper tracing (V) shows the bladder volume, the middle tracing (P) shows the intravesical pressure and the lower tracing (F) shows the discharge impulses in a single axon dissected from the pelvic nerve. A discharge of impulses starts slowly as the bladder is distended by the inflow of fluid, gradually increases with continued distension and reaches a peak as a reflex bladder contraction raises the intravesical pressure. Outflow of fluid then starts, the afferent discharge declines rapidly and the bladder pressure and volume fall more slowly. Peak discharge in the afferent fibres occurs at a nearly constant bladder volume while the bladder is contracting and the pressure is rising rapidly, *i.e.* the "isometric" phase (IGGO, 1966)

Histological Location. In which layer of the visceral wall are tension receptors located? IGGO (1957a) was able to remove the mucosa and sub-mucosa over the regions containing tension receptor activity without impairing this activity. Thus tension receptors so far tested have not been located in the mucosa or sub-mucosa. Apart from a few receptors described later which have been traced to the serosa and mesentery, the majority of tension receptors appear to be situated in the muscle layers. It is possible to modify and apply the principles which MATTHEWS

(1933) used in order to differentiate between receptors lying "in parallel" with and receptors lying "in series" with the extrafusal fibres of skeletal muscle. The "in parallel" receptors (*i.e.* muscle spindles) gave increased activity when the muscle was passively distended but gave decreased activity when the muscle shortened during an isotonic contraction. The "in series" receptors (*i.e.* Golgi tendon organs) gave increased activity both when the muscle was passively distended and when the muscle developed tension during an isometric contraction. Thus an "in parallel" receptor acts as a *length detector* and an "in series" receptor acts as a *tension detector*. The same ideas have been applied to visceral muscle (IGGO, 1955), by regarding a tangent at the circumference of the viscus as being equivalent to the linear situation existing in skeletal muscle. By this reckoning, since tension receptors are excited both passively by distension and actively by contraction, they must be *in series* with the contractile elements (the smooth muscle cells). The same conclusion was reached by LEEK (1969a) who, in addition, was able to study the behaviour of a receptor located in the cranial wall of the reticulum, to which access had been gained transthoracically. Light tactile stimuli applied to either the mucosal or the serosal surface were ineffective. Even quite forceful compression of the wall between a finger placed on the serosal surface or another finger on the mucosal surface caused little excitation of the receptor. However, slight tangential lengthening of the part of the wall containing the receptor zone caused marked excitation. Taking all the evidence together, it appears that tension receptors "in series" with smooth muscle cells are most readily excited by tangential lengthening and, because of this, I suggest tentatively that tension receptors may take the form of a flattened star with its points being attached to smooth muscle cells. A structure of this shape would be difficult to differentiate from other cells in the muscle layer and this may account for the failure of histologists, so far, to recognise these receptors. SHARMA (1967) has reviewed the accounts of various structures attributed to be visceral receptors and has concluded that none of the evidence is sufficiently convincing to be acceptable at this stage. However, STEVEN and MARSHALL (1970) briefly describe nerve endings embedded in collagenous whorls in ruminal muscle layers, which may be tension receptors.

The Consequences of an in series Location. The discovery that tension receptors are "in series" with smooth muscle cells provides an explanation for many of the phenomena described above. The rhythmical fluctuations in the resting discharge, which is observed when the viscus is in an apparently quiescent state, is almost certainly the result of localized intrinsic (myogenic) contractions. Neither the afferent discharge nor these intrinsic movements (in sheep reticulum) are affected by blocking agents effective at the preganglionic or postganglionic nerve endings, *e.g.* tetraethyl ammonium chloride and probanthine hydrochloride (LEEK, 1967). On the other hand, drugs which cause smooth muscle contractions (*e.g.* carbachol) and contractures (*e.g.* phenylephrine and other α-sympathomimetics) result in a greatly enhanced afferent discharge. Conversely, drugs (*e.g.* isoprenaline and other β-sympathomimetics) which inhibit intrinsic smooth muscle contractions cause the resting afferent discharge to decrease (LEEK, 1967; LEEK and VAN MIERT, unpublished observation). In interpreting the action of drugs, it must be emphasized that certain drugs do appear to exert their effects directly on the afferent nerve ending. PAINTAL (1964) considers that the excitatory effects of acetylcholine, phenyldiguanide, 5-hydroxytryptamine and nicotine on non-ruminant gastric tension receptor afferent discharges are due to the direct action of these drugs on the regenerative region of the non-medullated nerve endings (Fig. 7). In specialized receptors, however, the transducer action may be performed by non-neural components, in which case drug effects may be due to an action on these components rather than on the nerve ending itself.

It is noteworthy that tension receptor activity is also likely to be affected indirectly by any factor that influences smooth muscle activity. These may include (a) various humoral factors, such as hormones and blood electrolytes and

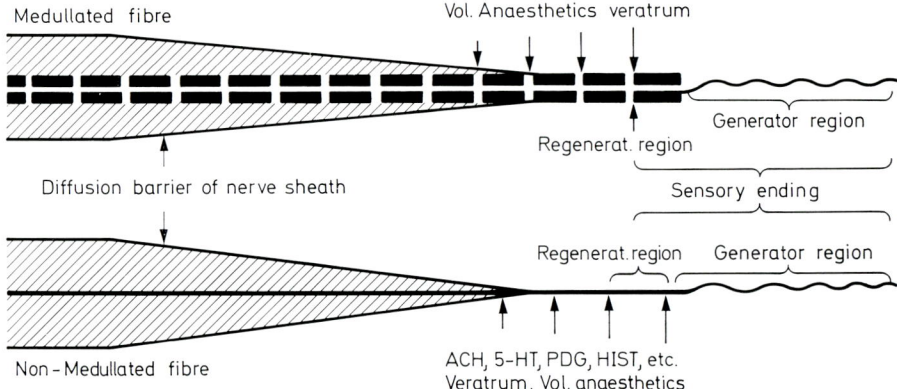

Fig. 7. Schematic diagrams of sensory endings of medullated and non-medullated nerve fibres showing the two parts of an ending and the probable site of action of drugs at the regenerative region, where there is no diffusion barrier. A greater variety of drugs affects the endings of non-medullated fibres because the fibres themselves are more susceptible to these drugs (PAINTAL, 1964)

(b) the physico-chemical nature of the contents of the viscera, in so far as these may excite peristaltic activity through intrinsic nerve networks ("myenteric reflex"). It has been suggested (LEEK, 1969a) that the reflex responses on reticulo-ruminal motility and on salivation of *lightly* stimulating the reticulo-ruminal mucosa (ASH and KAY, 1959) may have activated reticulo-ruminal smooth muscle cells through an intrinsic nervous mechanism, thereby leading to increased afferent discharges from "in series" tension receptors. IGGO (1957b) has demonstrated that even *light* tactile stimulation of the mucosa could excite tension receptors deep in the muscle layers but he did not consider whether this was a direct or an indirect effect.

Afferent Innervation. The gastro-intestinal tension receptors investigated by PAINTAL, IGGO and LEEK have been innervated by afferent fibres contained in the vagus nerves. PAINTAL (1954b) calculated the mean conduction velocity of these fibres in the cat to be 9 m/sec (6.5–13 m/sec). For similar fibres and also for fibres innervating intestinal receptors, IGGO (1958), using the collision technique for measuring conduction velocity, estimated (i) that the mean value was 1 m/sec (0.5–2.5 m/sec) and (ii) that conduction in distal regions of small myelinated nerve

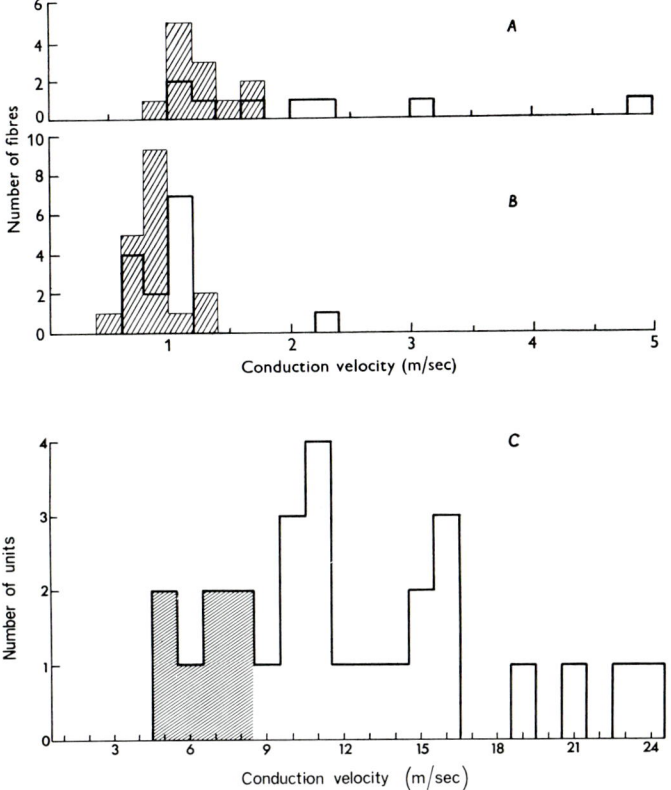

Fig. 8 A–C. The conduction velocities of gastro-intestinal afferent vagal units dissected from the cervical vagi of cats (A and B) and of sheep (C). A The discharge from gastric pH receptors. B The discharge from gastric and intestinal *in series* (distension-sensitive) tension receptors. C The discharge from reticulo-ruminal *in series* tension receptors. The hatched parts of A and B are the conduction velocities in the thoracic and abdominal regions of the units and the clear parts are the conduction velocities in the cervical region of the units. The conduction velocity and the diameter of an axon are lessened distally. Most of the axons had conduction velocities below 2.5 m/sec, that are characteristic of non-myelinated axons. The hatched part of C is the conduction velocities of those units (*in italics in Table 1*) which were observed or were presumed to have a path in the ventral (thoracic and abdominal) vagal trunk and the clear parts to those units (*not in italics in Table 1*) with a path in the dorsal vagal trunk. The overall mean is 12.4 + 1.0 m/sec (S.E. of mean) but the mean for the ventral vagal trunk is 6.6 + 0.5 m/sec and the mean for the dorsal vagal trunk is 14.5 + 1.0 m/sec. In contrast to cats, sheep have many more small myelinated axons in the thoracic/abdominal vagal trunks and the above conduction velocities typify this. (A and B from IGGO, 1966, and C from LEEK, 1969a)

fibres was slower than in the more proximal parts. From this he concluded that some of these small fibres probably lose their myelin sheath during their course through the thorax. IGGO's values are more acceptable than PAINTAL's, because the collision method of measuring the conduction velocity is more precise for small fibres and because the number of myelinated fibres in the cats' abdominal vagi is very small (Fig. 1).

Tension receptors in the reticulo-rumen are innervated by larger myelinated fibres (Fig. 8). Iggo (1955) obtained values of 12, 6, 5, and 2 m/sec for the four units in which he was able to measure the conduction velocities and to locate the receptors (in goats). In sheep, Leek (1969a) found an overall mean conduction velocity of 12.4 ± 1 m/sec (S.E. of mean), although afferent fibres coursing through the dorsal vagal trunk had a faster conduction velocity (14.5 ± 1 m/sec) than those through the ventral vagal trunk (6.6 ± 0.5 m/sec). As yet no explanations are available for these different values in the two trunks. These values for conduction velocity in the ruminant are consistent with the observation made by Iggo (1956a) that the abdominal vagi (of goats) contain several thousand small myelinated nerve fibres. Of these 80% had diameters of 2–4 µm and, hence, conduction velocities of 12–24 m/sec might be expected. That larger fibres are found in the ruminant vagi is presumably related to the fact that the complex sequence of movements necessary for mixing and propelling forestomach contents (see review by Sellers and Stevens, 1966) are wholly dependent upon an intact extrinsic afferent and efferent nerve supply. The afferent discharges from tension receptors in the reticulo-rumen provide both an excitatory drive to the gastric centres and a "feed-back" input during contractions (Leek, 1969b). In contrast, movements of the cat stomach and its ruminant equivalent, the abomasum, are largely independent of extrinsic nerves and these structures are innervated by small fibres, most of which are not myelinated (Agostini *et al.*, 1957).

Function of *in series* Tension Receptors. One is constantly being made conscious of the "fullness" of various visceral organs and, since *in series* tension receptors appear to be the principal sensory structures in the viscera, it may be assumed that these receptors, as one of their functions, signal something which is interpreted consciously as "fullness". *A priori* it may be thought that "fullness" is a reflection of the volume of a viscus, yet electrophysiological investigations have demonstrated that tension receptors give a very poor indication of volume and this is particularly true of the bladder. Theoretically, "volume" indication would be readily provided by "in parallel" receptors but these have not been shown to exist in the viscera. In a physical rather than a physiological situation, variations in the volume of a sphere lead to proportional variations in the tension measured at any point on the surface of the sphere. Thus tension measurements could be used as an indication of volume. In physiological situations, however, the relationship between tension and volume (Fig. 9) is not constant for several reasons, *e.g.* (i) in the reticulum of sheep, the tension exerted upon an intraluminal balloon is due to elastic elements "in parallel" with smooth muscle cells. Thus the tension in the muscle cells monitored by tension receptors is independent of and usually less than the tension exerted passively by elastic components of the connective tissue, (ii) in the bladder, slow increases in volume over low and moderate ranges do not lead to a significant increase in tension, because the smooth muscle of the bladder adapts (by receptive relaxation) to accommodate the increased volume, and (iii) by increasing the activity of smooth muscle cells, it is possible to increase the tension in a (closed) viscus without altering its volume.

One, therefore, concludes that, although undoubtedly the sensation of "fullness" is attributable to afferent discharges arising from in series tension receptors, "fullness" is only rather vaguely and variably related to the volume of a viscus. "Fullness" is a reflection of the actively developed tension in the wall of the viscus. Therefore, conditions which raise smooth muscle tone either through local mechanisms or through neural and/or humoral pathways may be expected to accentuate the sensation of fullness. A further aspect of this, which has not hitherto been considered, is the possible significance (in relation to sensation) of the very high afferent discharge frequencies produced, when a viscus contracts under relatively isometric conditions, at which times tension receptors would be

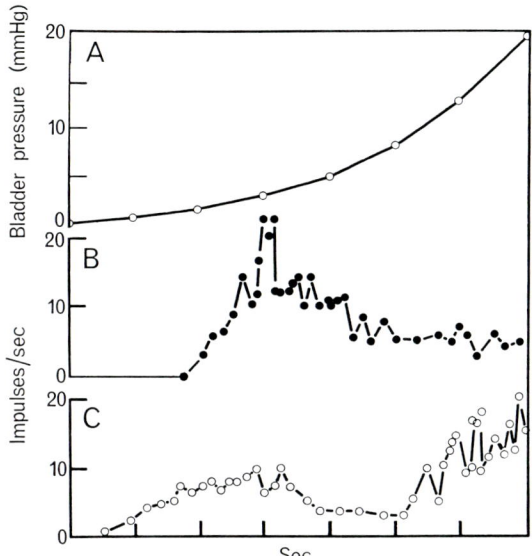

Fig. 9. The effect on the discharge of impulses in two afferent fibres, recorded from the same nerve filament, of distension of an innervated urinary bladder with fluid. Both units were silent before distension, and began to discharge impulses and reached peak activity at different times after the inflow of fluid started. Both of the afferent units were excited by contraction of the bladder and some of the individuality in the two responses may be due to localised contractions of the detrusor muscle. Cat, chloralose anaesthesia (IGGO, 1966)

acting as impaction monitors. The circumstances, which may be presumed to alter the conditions of a contraction from being *relatively isotonic* to being *relatively isometric*, would include (i) an increase in the viscosity of the contents, e.g. a change from liquid to more solid contents, (ii) an increase in the bulk of the contents and (iii) an obstruction to outflow of contents. Could it be, therefore, that the sensation of "fullness" arising from a viscus in which a contraction normally causes propulsion of its contents is derived not only from the tension of the quiescent viscus but also from the tension developed actively during a contraction: indicative of the degree of impaction ?

In the first description of gastric tension receptors, PAINTAL (1954b) ascribed to them a possible rôle on the peripheral mechanisms of satiation of hunger and thirst, based on their capacity to monitor gastric filling. This idea seems generally acceptable, although IGGO (1957a) has pointed out that the same receptors are also likely to give an increased afferent discharge when the stomach is empty and undergoing the powerful "hunger" contractions, described by CANNON and WASHBURN (1912). Incidentally, since the afferent discharge is conducted along C fibres, this would account for 0.5–1 sec of the delay observed between the contraction and the sensation of "hunger pangs".

Tension receptors, besides their involvement with visceral sensation, may fulfill an even greater role in relation to visceral reflexes. IGGO (1966) has shown the effect of suddenly obstructing the outflow of urine near the start of a bladder

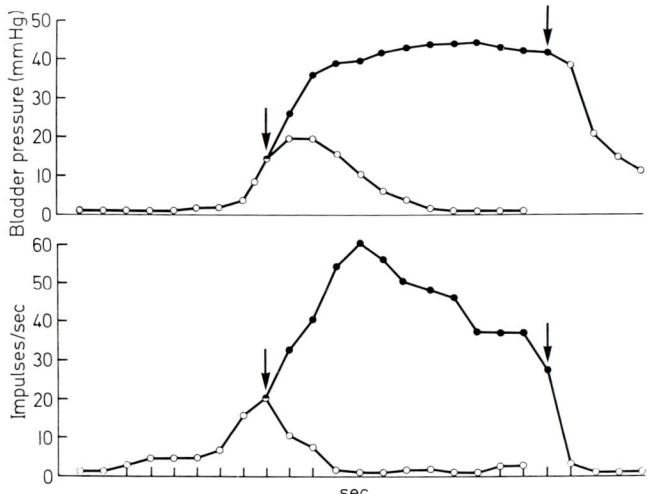

Fig. 10. The effect of obstructing the outflow of "urine" during a bladder contraction. The *open circles* ○————○ show the normal afferent discharge and bladder pressure changes when the urethra is not blocked. This unit is identical with the one shown in Fig. 6. The filled circles ●————● show the afferent discharge and bladder pressure changes during the period when the urethra is blocked (*at the first arrow*) and opened (*at the second arrow*). Urethral occlusion causes an enhanced and prolonged afferent discharge from *in series* tension receptors which, in turn, elicits reflexly a more powerful and protracted contraction of the bladder musculature. This illustrates one of the properties of an *in series* tension receptor, *i.e.* to act as an "impaction monitor" (IGGO, 1966)

contraction, thus converting the contraction from an isotonic to an isometric one (Fig. 10). A much greater afferent discharge developed and this led reflexly to a more powerful contraction of the bladder which was sustained for as long as the obstruction persisted: an effect previously observed by DENNY-BROWN and ROBERTSON (1933) and by MELLANBY and PRATT (1940).

The fullest analysis of a reflex involving visceral *in series* tension receptors has been made for the primary cycle movements of ruminant forestomach (LEEK, 1967, 1969b). The size of the reticulo-rumen is very large and its voluminous contents (*e.g.* about 70 kg in a cow) are mixed by a complex sequence of compartmental contractions co-ordinated by gastric centres in the medulla oblongata (IGGO, 1956a; BEGHELLI et al., 1963; HOWARD, 1970; HARDING and LEEK, 1970) with efferent fibres in the vagi (DUNCAN, 1953). The unitary discharge from afferent fibres innervating the smooth muscle of both the reticulum and the rumen

(LEEK, 1967; IGGO and LEEK, 1967a, b) has been recorded with the reticulum under different degrees of distension and with the reticulum contracting isotonically or isometrically. Thus it has been possible to examine both the input to the gastric centres derived from tension receptors under various conditions and the reflex modifications in the output from gastric centres due to varying the inputs. From these studies it was concluded that there exist two independent groups of reticular tension receptors, *i.e.* (i) those with a low threshold of excitability, which exert an excitatory effect on the gastric centres and (ii) those with a high threshold of excitability, which exert an inhibitory effect on the gastric centres (LEEK, 1967; IGGO and LEEK, 1967b).

The integrated afferent activity from these two groups of receptors exerts (i) a static effect, because during the quiescent period between contractions (approximately 1 min) it determines the brevity of this period and the magnitude of the subsequent efferent discharge and (ii) a dynamic effect, because the tension developed by this initial phase of the contraction reflexly affects the form of the efferent discharge responsible for the next phase and so on until the (primary cycle) contraction sequence ends. On the basis of these conclusions it is possible to account for the following observations: (i) as the reticular tension (during the quiescent period) is increased over a low and a moderate range (when only the low-threshold receptors would be excited), reticulo-ruminal contractions increase in frequency and in amplitude, whereas over the high range (when high-threshold receptors would also be excited) there is a progressive reduction in the frequency and in the amplitude of the contractions, (ii) after switching from isotonic to isometric recording conditions, there is no change in the frequency of the contractions but there is a change in the form of the efferent discharges and of the ensuing contractions. The initial tension developed by the isometric contraction reflexly excites the gastric centres and increases the next part of the efferent discharge but this, in turn, creates a much greater tension which reflexly inhibits the gastric centre and reduces the efferent discharge so that the tension declines to a level which is once again reflexly excitatory. The overall effect, therefore, is a reticular contraction which is prolonged, but has a lower peak amplitude. The isometric reticular contraction also reflexly inhibits the ruminal contraction.

Although the study of reticulo-ruminal reflexes may seem a very specialized field and one far divorced from human medicine, these reflexes have been electrophysiologically analysed more thoroughly than others in the gastro-intestinal tract and exhibit features which may subsequently be found to exist elsewhere, namely (i) tension receptors with different threshold properties may exert different or even opposing reflex effects and (ii) the afferent activity arising from tension receptors under static conditions and under dynamic conditions may have a different reflex significance.

Slowly-Adapting Tension Receptors Located in Mesenteries

A few examples of this type of receptor exist. They respond with a slowly-adapting discharge to traction on the mesentery, digital compression and distension of overlying organs. Unlike the in series receptors they do not discharge during contraction of the overlying organs. Indeed, during an isotonic contraction, a reduction in their afferent activity would be expected. In a way, they could behave as *in parallel* receptors and may, therefore, be thought to be a kind of *volume receptor*. At present this conclusion is doubtful as the numbers that have been found are so small. IGGO (1957b) describes one such receptor on the omentum attached to the greater curvature of the stomach and BOWER (1959) describes similar receptors in the broad ligament of the uterus.

Rapidly-Adapting Mechanoreceptors

Rapidly-adapting mechanoreceptors are those receptors which give an *"on"* and an *"off"* response when a steady mechanical stimulus is applied and later removed, but give no response while the stimulus is held steady. These receptors are, therefore, involved in monitoring dynamic events and are unresponsive to the steady state whatever its level. Rapidly-adapting receptors have been found in the mesentery (*e.g. Pacinian corpuscles*), beneath the serosa of the small intestine (*i.e.* the *"movement receptors"* described by BESSOU and PERL, 1966), in the intestinal muscularis mucosae (*i.e.* the *"mucosal mechanoreceptors"* described by PAINTAL, 1957) and in the urethra (*i.e.* *"flow receptors"* described by IGGO, 1956b and TODD, 1964, and *rapidly-adapting tension receptors* described by TODD, 1964). In addition, IGGO (1957b) has described receptors in the mucosa of cat stomach which serve principally as pH receptors and also give a rapidly-adapting response to certain types of mechanical stimulation. NIIJIMA (1962) has recorded rapidly-adapting activity from gastric mucosal receptors in response to mechanical stimulation, but he failed to observe whether they were also excited by solutions of high or low pH. These receptors may, therefore, have been the same as the chemoreceptors described by IGGO (1957b). It is interesting that, in the combined results of IGGO (1955) and LEEK (1969a) for the ruminant forestomach, only 3 of the 88 mechano-receptors gave a rapidly-adapting response and these were ill-defined.

Mesenteric Pacinian Corpuscles

The most obvious visceral receptors are the large Pacinian corpuscles located in the mesentery at its root and alongside the branches of the mesenteric arteries. Because of their large size and ease of access, Pacinian corpuscles have been the subject of extensive *in vitro* studies. They are *rapidly-adapting* mechanoreceptors which give both an *on* and an *off* response to a steady mechanical stimulus. They are capable of responding to sinusoidal vibrations up to 500 Hz. *In vitro* the transducer rôle of the cellular lamellae has been determined by comparing the properties of nerve endings before and after removal of the lamellae (LOEWENSTEIN and SKALAK, 1966) and a full discussion is given by LOEWENSTEIN in Vol. I of this Handbook. The development of generator potentials, the genesis of spikes in the initial segment, the saltatory conduction of spikes to the first node of Ranvier and the effect of humoral agents on these three sites are fully described elsewhere (GRAY and MALCOLM, 1950; GRAY and SATO, 1953; GRAY and DIAMOND, 1957; review by GRAY, 1959).

Interest in the *in vivo* activity of Pacinian corpuscles stems from the classical investigation of GAMMON and BRONK (1935). They recorded, in cats, *single* and *multi-unit* afferent activity in strands of the splanchnic and mesenteric nerves.

By sectioning one by one the fine nerve supply to each Pacinian corpuscle, activity which had a pulsatile periodicity was found to have its origin in these corpuscles. The frequency of the afferent discharge was found to be increased by intravenous perfusions of saline and of blood and *vice versa* the afferent activity was decreased by withdrawal of blood. Injections of adrenaline, likewise caused a rise in the systemic blood pressure but, in contrast, caused a reduction in the splanchnic blood flow and a concomitant reduction in the afferent discharge. GAMMON and BRONK (1935) therefore concluded that, unlike the baroreceptors in the carotid sinus and in the aortic arch, the mesenteric Pacinian corpuscular discharges were not determined by the systemic blood pressure but were indicative of the blood flow through the splanchnic bed. To test the reflex significance of these discharges, they perfused the mesenteric circulation at various pressures and failed to observe any reflex effect on the systemic blood pressure. Conversely, experimental perfusion of one half of the mesenteric circulation (*e.g.* raised flow rate) caused a compensatory reflex effect (*e.g.* vasoconstriction) in the other half. These authors considered that the reflex function of the mesenteric Pacinian corpuscles was therefore, to stabilize blood flow through the splanchnic bed.

More recently, LEITNER and PERL (1964) recorded the activity arising from receptors which were innervated by spinal nerves and which responded to cardiovascular changes and to adrenaline. In cats, they recorded from dorsal root fibres in the thorax (segments T_3–T_{12}) and found that the activity arose from Pacinian or paciniform corpuscles in the thorax and the abdomen. Those afferent fibres in the splanchnic nerves had conduction velocities in the range 18–66 m/sec. The discharge pattern of these corpuscles was not related to the mean or the diastolic blood pressure but depended on the pulse form.

Serosal Receptors

BESSOU and PERL (1966) have applied a "single fibre" technique to filaments dissected from the mesenteric nerves and they have observed the behaviour of 4 single afferent units which innervate receptors located beneath the serosa of cat intestine. Each so called *movement receptor* gave a rapidly-adapting response (less than 30 spikes/sec) to light mechanical stimulation of the mesentery or of the small intestine and they could follow sinusoidal vibrations of low frequencies only (never greater than 100 Hz). They were, therefore, quite different from Pacinian corpuscles. Using a fine probe it was possible to resolve the receptive field of each afferent unit into 1–5 sensitive points (smaller than 1 mm in diameter), each of which presumably corresponded to a terminal of a nerve branch. Each point lay upon or adjacent to a branch of the mesenteric artery at the place where it entered the intestinal musculature. The afferent fibres had conduction velocities of 2–21 m/sec (mainly 5–10 m/sec), from which it was inferred that they were small myelinated fibres, and on histological examination, fibres with diameters of 2–8 μ were found. Incidentally, this gives a conduction velocity/fibre diameter ratio of 1–2.5 which is considerably less than the generally accepted value of 6 (GASSER and GRUNDFEST, 1939).

These receptors signalled movement of the intestine, regardless of its origin, *viz.* (i) movement of the mesentery, (ii) distortion of the small intestine, and (iii) the dynamic phase of inflation and of deflation of an intraluminal balloon.

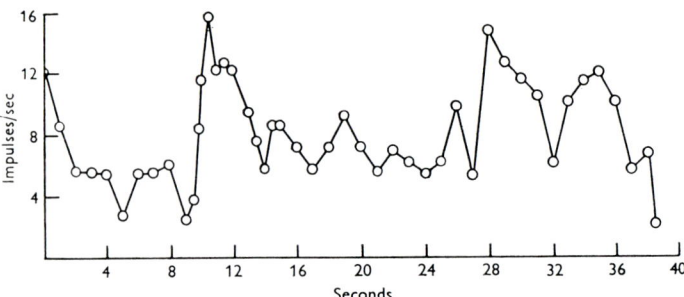

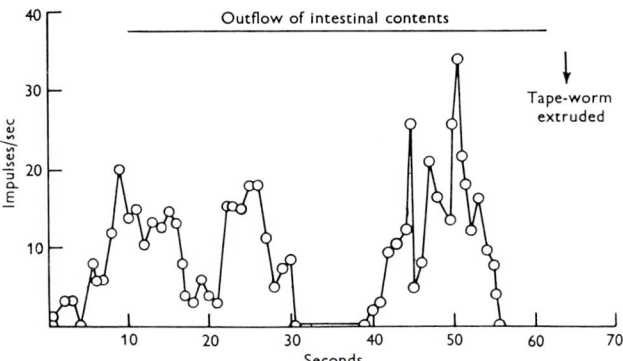

Fig. 11. Responses of a muscularis mucosae mechanoreceptor. *A* shows the unitary afferent activity after the introduction of 30% NaCl into the lumen of the small intestine. There is a progressive enhancement of the basal level of activity, on top of which are superimposed bursts of spontaneous activity (*e.g.* between 10 and 13 sec). *B* shows the activity at the end of a period of stimulation by 30% NaCl. The basal level is lower than in *A* but the spontaneous bursts are most prominent when peristaltic rushes occur (detected by observing the outflow of intestinal contents, *indicated by the horizontal line*, from a transection of the intestine aboral to the site of the receptor). In this record, the greatest activity was associated with a peristaltic movement which led to the extrusion of a tapeworm (PAINTAL, 1957)

The spike discharge pattern was related to the rate of change of intraluminal pressure and movements of the intestine which propelled its contents past the receptive zone, were particularly efficaceous in exciting a high frequency discharge. Although these receptors were discovered during investigations on spinal afferent fibres responsive to cardiovascular changes (LEITNER and PERL, 1964), it seems likely that they monitor events occurring in the intestine rather than in the blood vessels. As yet no reflex role for these receptors has been suggested.

GERNANDT and ZOTTERMAN (1946) recorded the electrical activity in fine branches of the mesenteric nerves and observed that pinching and stretching the intestine evoked activity in delta and C fibres. This activity was attributed to excitation of intestinal nociceptors. It seems likely, however, that such stimuli would have stimulated both *in series* tension receptors and serosal receptors of the types described above.

Muscularis Mucosae Receptors

PAINTAL (1957), using the "single vagal afferent fibre" technique, has detected rapidly-adapting receptors associated with muscular elements in the mucosal layer of cat small intestine. Unlike the intestinal *in series* tension receptors described by IGGO (1957a), PAINTAL's receptors did not respond to distension of the small intestine, were not located in the deep muscle layers and gave no sustained discharge in response to mechanical stimulation. Originally the fibres innervating these receptors were detected by injecting phenyl diguanide into the aorta to excite various gastro-intestinal mechanoreceptors and then by rejecting these fibres innervating receptors which were distension-sensitive, *i.e. in series* tension receptors which were excited by distension of intraluminal balloons. Later it was possible to recognise these units by their characteristic spontaneous discharge: 2–10 bursts/min, each burst usually lasting 1–6 sec with a peak spike frequency of 7–26 impulses/sec. The receptor zones were then located by manipulating the small intestine or by injecting a 30% (w/v) NaCl solution into the intestine. It was found that these receptors were present in all regions of the small intestine.

The spontaneous activity, which occurred rhythmically in bursts with few spikes in between, would superimpose itself on a discharge initiated artificially (Fig. 11). The spontaneous activity was also greatly augmented by the passage of intestinal contents during a peristaltic movement: solid contents (in the form of a tapeworm!) being more effective than the usual liquid chyle. Despite this enhancement during peristalsis, spontaneous activity was still present after giving atropine which rendered the intestine immobile and atonic. Conversely *strong* local (non-propulsive?) contractions frequently did not excite the receptors. Injecting 30% NaCl solution into the intestinal lumen was equally effective before and after administering atropine whereas the serosal application of the same solution was totally ineffective. These results allow one to conclude that the receptors are not located in the main muscle layers, but in a region near the mucosa. The receptors were not excited by other chemicals (*e.g.* sucrose, magnesium sulphate, etc.) at an equivalent osmolarity, except for KCl, which elicited a weak discharge. By direct observation, PAINTAL found that, of all the chemicals tried, only 30% NaCl and 30% KCl (to a lesser extent) caused shrinkage of the villi, even after atropinization, due to contraction of the muscularis mucosae. PAINTAL, therefore, concluded that the receptors were related to the muscularis mucosae and were probably aligned radially (*i.e.* along the core of the villus) rather than tangentially in a cross-section of the gut. 30% NaCl is, however, more than 30 times the isotonic concentration of NaCl and the *specificity* of its action on the muscularis mucosae must be viewed sceptically.

PAINTAL's conclusion is an interesting one and has several implications. In so far as the receptors give an augmented discharge during the propulsion of intestinal contents, they may be presumed to act as a kind of "flow" receptor. Apart from the "flow" input to the brain, there are the rhythmic bursts of activity associated with intrinsic movements of the muscularis mucosae and any factors which affect the frequency, duration and amplitude of these movements will thereby affect the "resting" input. It is known that sympathetic and parasympathetic nerves influence the movements and it may be assumed that certain chemical and mechanical factors in the intestinal contents may affect the movements through

Fig. 12 A–D. Responses of receptors in cat stomach to mechanical stimulation. A shows the response of an *alkali-sensitive* gastric mucosal chemoreceptor (fibre 40 in Table 2) to stroking the mucosa with a smooth probe, each stroke evoking a burst of spikes. After scraping off the superficial one-third of the mucosa (B), it was no longer possible to evoke a response to mechanical stimulation. C shows to the response of an *in series* gastric tension receptor to stroking the mucosa (*i.e.* first three bursts of impulses) and to lightly pressing the mucosa (*i.e.* continuous discharge). D shows the response of an *acid-sensitive* gastric mucosal chemoreceptor (fibre 52 in Table 2) to a very firm stretch of the mucosa. Note that the response to mechanical stimulation of the tension receptor deep in the muscle is greater than that of chemoreceptors in the mucosa. Time marks (1 sec) (Iggo, 1957b)

local nervous connections, *e.g.* the myenteric reflex. Thus the "resting" activity from the receptors may be highly variable, and the reflex significance both of the "resting" discharge and of the "flow" discharge has not been established.

Mucosal Receptors

Andrew (1957) has recorded the activity of receptors in the upper sphincter of the oesophagus which responded to light tactile stimulation of the mucosa. Iggo (1957b) has described receptors in the gastric mucosa which gave an irregular discharge of low frequency when (i) the mucosa was *firmly* stretched, (ii) the stomach was over-distended and (iii) a small area of the mucosa was stroked with a smooth probe or cotton wool. Unlike the *in series* tension receptors described earlier, these mucosal receptors were not excited either during a stomach contraction or by lightly pressing on the mucosa or after the mucosa had been removed. Besides giving weak mechanoreceptor responses, mucosal receptors were greatly excited by either high or low pH conditions (Fig. 12). Chemoreception appears to be the main rôle of these receptors, the mechanoreceptor activity being subsidiary, or even incidental, *e.g.* could the various mechanical manoeuvres have been effective through forcing intrinsic and extrinsic acids and alkalis into closer contact with the receptors? Two lessons worth learning from this investigation are (i) that chemoreceptors may give a response to mechanical stimuli and, therefore, need to be clearly distinguished from true mechanoreceptors (a point overlooked by NIIJIMA, 1962) and (ii) that light mechanical stimulation of the mucosa may excite a greater response from receptors deep in the muscle layers than from those in the mucosa itself.

Urethral Receptors

IGGO (1956b) recorded the activity of one "flow receptor" which behaved quite unlike tension receptors *in series* with the muscular tissue of the bladder (IGGO, 1955). This unit was located in the urethra and was only excited by the flow of fluid along the urethra: passive, steady distension being ineffective. The spike frequency of the discharge was related to the rate of inflow of fluid along the urethra and stopped when the inflow was turned off. It reached 300 spikes/sec in contrast to the tension receptor discharge for which the maximum was 40 spikes/sec. This type of receptor was also described subsequently by TODD (1964). "Flow receptors" were located at various points along the urethra but were particularly evident near to the external urethral sphincter. Their discharge was irregular but sustained for as long as the flow lasted. The flow appeared to be turbulent and TODD concluded that the resulting high frequency tension variations were responsible for evoking the receptor activity.

In addition to the "flow receptors" TODD (1964) recorded from receptors which gave a rapidly-adapting response when urethral distension was applied or removed but not when the distension was held steady or when there was flow along the urethra. Histological examination of urethral tissue showed the existence of small lamellated (paciniform) structures which might subserve the rôle of flow receptors and/or rapidly-adapting tension receptors. No reflex function for the latter receptors has been given so far, but the concept of flow receptors was fundamental to BARRINGTON's theory of micturition reflexes and its subsequent modification by GARRY et al. (1959).

Chemoreceptors

Gastric pH Receptors

Despite the considerable volume of literature implicating various modalities of chemical sensitivity in the gastro-intestinal tract, the only "single fibre" studies of chemoreceptor activity seem to be those of IGGO (1957b) on gastric pH receptors in cats (Table 2). Activity in 19 single units was recorded from afferent vagal fibres having conduction velocities of 1–5 m/sec. The receptors were found in all regions of the stomach and removal of the mucosa (sometimes even only the most superficial part of the mucosa) caused loss of receptor activity, thereby implying a superficial mucosa locus with receptive fields as large as 5 cm^2. These receptors were not excited by any of the many chemicals tried except for *either* strong acids *or* strong alkalis depending on whether the receptor was *acid-sensitive* or *alkali-sensitive*.

Table 2. *The response of gastric mucosal nerve endings to mechanical and chemical stimulation* (IGGO, 1957b)

Fibre No.	Stroke	Stretch	Contraction	Compression	Acid	Alkali	Ringer-Locke	Threshold
15 m	+	+ E	—	—	—		—	
27 m	+	—		—				
28	+	—		—			—	
29 m	+	—			—	+	—	
36	+	—					—	
37 m	+	—		—				
40 m	+	—		—	—	+	—	0.01 N NaOH
46	+	—	—	—	—	+	—	pH 8.0
48	+	—			+	—	—	
49	+	—			—	+		pH 8.0
52 m	+	+ E			+	—		pH 2.2
53 m	+	+ E	—		—	+	—	pH 9.0
55 m	+	—	—		—	+	—	pH 9.3
56 m	+		—			+	—	
58 m					—	+	—	pH 9.0
59 m	+				+	—	—	
60 m	+				—	+	—	0.01 N NaOH
61 m	+				+	—		pH 3.0
62					+		—	pH 1.5

+ = response; — = no response; E = over-distension or firm stretch of the mucosa; m = mucosal origin confirmed by scraping of the mucosa.

In a current investigation of gastric pH receptors, HARDING and LEEK (unpublished observations) also have observed independent acid-sensitive receptors and alkali-sensitive receptors, which were located in the fundic and pyloric regions of sheeps' stomach and which were innervated by vagal fibres with conduction velocities of 1–6 m/sec. The acid-sensitive receptors respond to hydrochloric acid after a short latency and to volatile fatty acids after a longer latency. These results have led to a tentative conclusion that diffusion of acids over a distance of up to 200 μm through an aqueous zone is necessary before the receptors are excited. This implies that the receptors are not located on the exposed luminal surfaces but that they may be located perhaps in the gastric pits near to the isthmus region of the peptic glands.

Acid Sensitive Receptors. Acids applied to the exposed gastric mucosa excited activity in 5 units. The threshold was not greater than pH 3 at which point the afferent discharge was erratic and brief, whilst at lower pH values the discharges were sustained and had higher spike frequencies. Repeated application of strong solutions (0.1–0.2 N HCl) sometimes damaged the mucosa and destroyed the receptors. Intraluminal perfusion of an intact stomach yielded activity in 1 unit when the perfusion fluid was 0.1 N HCl (Fig. 13). Greater activity was elicited by 0.2 N HCl and there was no response to 0.03 N HCl or Ringer-Locke solution. Receptors sensitive to acids were not excited by alkalis.

Alkali Sensitive Receptors. Applications of 0.1 N NaOH to the exposed gastric mucosa excited 9 units to discharge at frequencies of 30–50 spikes/sec for as long

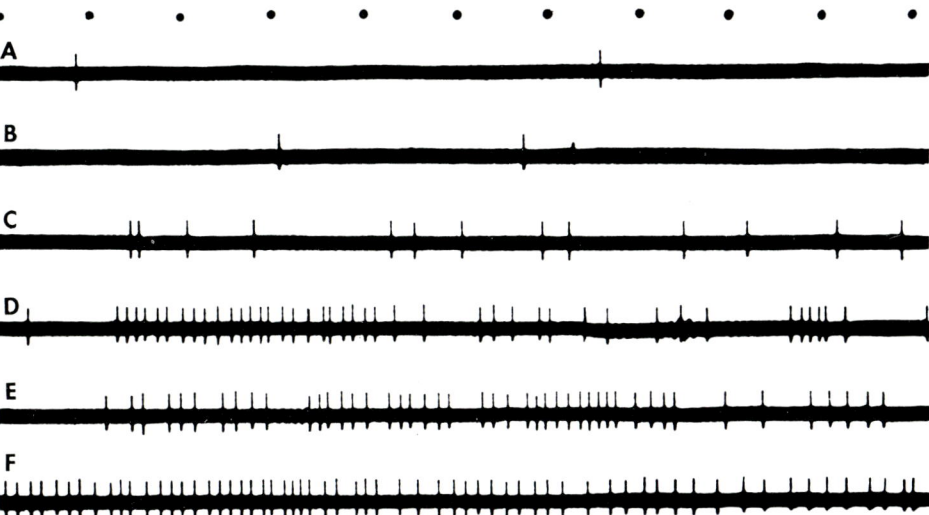

Fig. 13 A–F. The response of an *acid-sensitive* gastric mucosal chemoreceptor (fibre 52 in Table 2) to the application of various fluids on to the exposed mucosa: A distilled water, B buffer solution at pH 5, C at pH 3, D at pH 2 and E continued in F 0.1 N-HCl. Time marks (1 sec) (IGGO, 1957b)

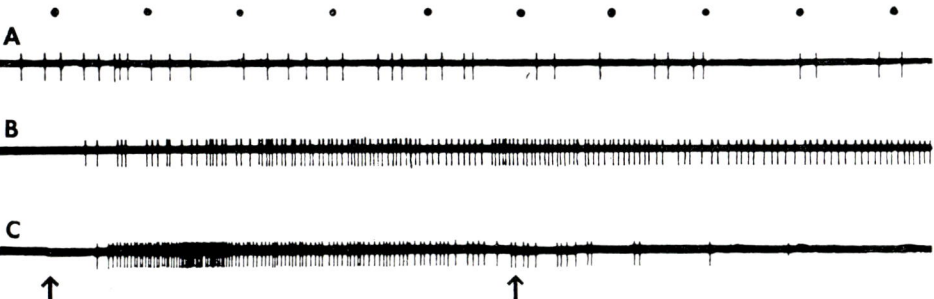

Fig. 14 A–C. The response of an *alkali-sensitive* gastric mucosal chemoreceptor (fibre 49 in Table 2) to the application of various fluids to the exposed mucosa: A borate buffer at pH 9.3, B 0.1 N NaOH and C 0.1 N NaOH applied at the first arrow and neutralized by the application of 0.1 N HCl at the second arrow. Time marks (1 sec) (IGGO, 1957b)

as the solution remained on the mucosa (Fig. 14). Exact threshold pH values could not be established but the more excitable units gave responses at pH 8. These receptors did not respond to the application of strong acids but both the acid-sensitive and alkali-sensitive receptors responded to certain kinds of coarse mechanical stimulation as described earlier.

Rôle of pH Receptors. A variety of reflexes, discussed later, implicate acid-sensitive receptors in the stomach. The possible significance of alkali-sensitive receptors in the stomach is a mystery. Under normal conditions gastric pH rarely exceeds pH 7. Could it be that acid-sensitive and alkali-sensitive receptors are a general feature of all regions of the gastro-intestinal tract and that the existence of alkali-sensitive receptors in a acid environment such as the stomach is merely vestigial? Certainly the results of experiments on gastro-intestinal reflexes implicate pH receptors in the small intestine as well as in the stomach.

Intestinal Chemoreceptors

SHARMA and NASSET (1962) have examined the chemosensitivity of the small intestine by obtaining *multi-unit* records from fine branches of the mesenteric nerves in acute experiments on anaesthetized cats and in chronic experiments on conscious dogs with Thiry-Vella loops. Enhancement of the discharge in these nerves was induced by perfusion of the gut lumen with glucose and various amino-acids.

The idea of receptors metering the end products of digestion is an interesting one and open to considerable speculation. The receptors may have a role in relation to feeding, particularly in relation to specific appetites. NASSET (1964) has postulated that a homeostatic mechanism exists at the gastro-intestinal level for the short term regulation of the amino-acid content of the gut lumen. This conclusion results from his observation that the amino-acid mixture in the jejunum remains relatively constant irrespective of the dietary amino-acid composition. Perhaps "chemical taste" is a general property of various regions of the alimentary canal, as it is clearly more widespread than its classical confinement to the tongue and adjacent structures.

Reflex Evidence

When an effect can be induced by applying a stimulus to some part of the abdominal viscera, it may be possible to infer the type and location of the receptors being stimulated, providing always that the stimulus parameter can be defined adequately and that the design of the experiment includes satisfactory controls. The latter proviso is particularly important in visceral situations, where smooth muscle cells may, in part, react directly to certain types of stimuli (*e.g.* the myogenic response to visceral distension). Where the reaction is indirect, the pathway involved may be either humoral or nervous and a nervous pathway may be intrinsic (*e.g.* the peristaltic or myenteric reflex) or extrinsic (*e.g.* vago-vagal reflexes). In this chapter, only a few of the effects induced by humoral mechanisms will be considered, because, although "receptors" may be inferred, it is not always clear whether or not the receptor is either a part of the cells or has a local neural link with the cells which are responsible for secreting the humoral transmitter substances, *e.g.* fats, etc., acting in the duodenum, cause the release of enterogastrone which inhibits gastric motility. Do the fats act directly on the cells which release the enterogastrone or is there a separate chemoreceptor mechanism interspersed between the stimulus and the secretory cell?

The Peristaltic or Myenteric Reflex

Peristaltic movements of the gastro-intestinal tract are influenced by vagal activity, but are not wholly dependent upon it. They arise through intrinsic reflexes involving neural pathways running between receptors and the muscle layers within the walls of the viscera. BAYLISS and STARLING (1899) in their original description of peristaltic movements noted that they were induced by locally applied stimuli such as distension. Subsequently, BÜLBRING et al. (1958) showed that the integrity of the mucous membrane was essential for initiating peristalsis. Asphyxiation, local anaesthesia or removal of the mucosa abolished the peristaltic reflex and BÜLBRING et al. (1958) concluded that there were distension-sensitive receptors situated in the mucosa. Later it was concluded that

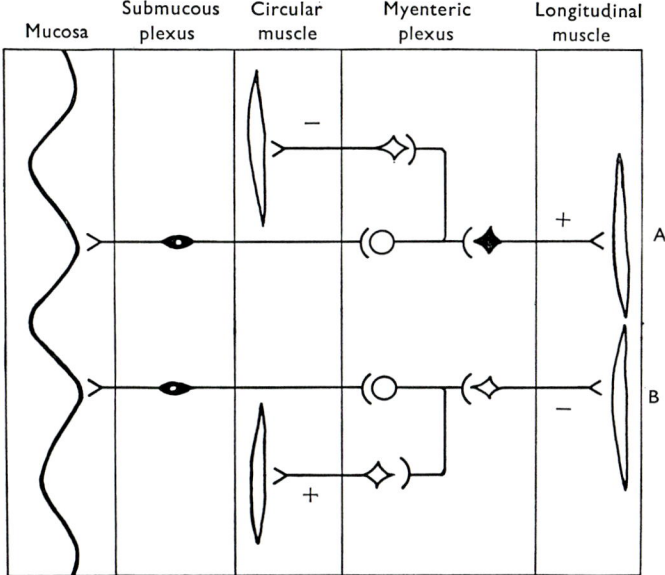

Fig. 15. Diagram of a possible arrangement of sensory, motor (+) and inhibitory (−) nerves in the intrinsic plexus which may cause reciprocal activation of the two muscle coats. Both sensory cells are distension-sensitive but that which excites the longitudinal muscle (A) has a lower threshold than that which excites the circular muscle (B) (KOTTEGODA, 1969)

chemoreceptors might also initiate peristalsis as a result of the effects observed when the mucosa was exposed to 5-HT (BÜLBRING and LIN, 1958), substance P (BELESLIN and VARAGIC, 1958) and phenyl diguanide (BÜLBRING and LIN, 1958; BÜLBRING and CREMA, 1958).

The distension-sensitive receptors appear to be of two kinds, viz. (i) low threshold receptors and (ii) high threshold receptors (KOSTERLITZ and LEES, 1964). KOTTEGODA (1970) has summarized the evidence on which this conclusion is founded. It is postulated that low levels of distension excite the low-threshold receptors and cause the preparatory phase of peristalsis; namely, contraction of the longitudinal muscle together with reciprocal inhibition of the circular muscle (Fig. 15). This results in a zone where the small intestine now has a greater cross-sectional area and tension, which thereupon stimulates the high threshold receptors. KOTTEGODA (1970) refers to it as "an increase in radial distension" but I prefer to regard it as an increase in the circumference of the intestine. This is because suitably orientated receptors would thereby undergo tangential lengthening, which was found to be *the effective parameter* of a distension stimulus in the ruminant forestomach (LEEK, 1969a). The high threshold receptors in the small intestine are however, believed to be in the mucosa unlike their location "in series" with muscular elements in the ruminant forestomach. Activity in the high threshold receptors is responsible for initiating the "emptying phase", namely, an aborally moving wave of contraction of the circular muscle coupled with reciprocal relaxation of the longitudinal muscle (KOTTEGODA, 1969, 1970).

The concept of a viscus having two sets of mechanoreceptors, both of which respond to some kind of stimulus but possess different thresholds of stimulation and elicit opposing effects, has also been postulated for the ruminant forestomach (LEEK, 1967; IGGO and LEEK, 1967b) and may be a feature of other visceral organs, *e.g.* the bladder. Although BÜLBRING and LIN (1958) postulated the existence of chemoreceptors to account for the effects of 5-HT *etc.*, it does seem more likely that these very active substances may not excite specific chemoreceptors, but that they may sensitize mechanoreceptors in the manner discussed by PAINTAL (1964). Indeed, normal peristalsis may depend on a certain degree of sensitization by 5-HT, because it is continuously being liberated into the lumen of the intestine in amounts proportional to the filling pressure (BÜLBRING and LIN, 1958). At present there is no evidence to suggest that the mechanoreceptors which are involved in the peristaltic reflex are also innervated by afferent fibres coursing in the extrinsic nerves to the gut. Neither the tension receptors "in series" with contractile elements in the deep muscle layer of the intestine (IGGO, 1957a) nor the receptors in the muscularis mucosae (PAINTAL, 1957) would fulfill the rôle of the mucosal receptors postulated for the peristaltic reflex.

Ruminant Forestomach Mechanoreceptor Reflexes

Various forms of mechanical stimulation of specific areas of the reticulo-rumen are capable of producing pronounced reflex effects on salivation, primary and secondary cycle movements of the forestomach compartments and on the evocation of the complex acts comprising rumination. Using sheep with large ruminal fistulae, ASH and KAY (1959) were able to apply mechanical stimuli manually to the mucosal surface of the reticulum and rumen. The most potent reflexogenic areas were in the region of the reticular groove, the medial surface of the reticulum and the reticulo-ruminal fold.

On Salivation. Gently rubbing or stretching the mucosa caused an increase in the output of saliva and in the frequency of reticulo-ruminal (primary cycle) contractions. Occasionally it induced rumination (ASH and KAY, 1959). It was concluded at that time that these effects may have arisen from the stimulation of mechanoreceptors located in the mucosa. Now it seems more likely that the effects are attributable to the afferent activity arising from tension receptors "in series" with the deep muscle layers, described by IGGO (1955) and LEEK (1969a). In halothane-anaesthetized sheep with spontaneous reticulo-ruminal movements present and a cannula in the parotid duct one may record an increased outflow of saliva following each reticular contraction when it is an isometric one, but no increased outflow when it is an isotonic one (LEEK, unpublished observation).

This observation is in accordance with the idea that "in series" tension receptor activity may reflexly promote salivation, as these receptors give a markedly enhanced discharge during an isometric reticular contraction but not during an isotonic one.

Although *light* mucosal stimulation was capable of eliciting the effects described by Ash and Kay (1959), this does not rule out the involvement of receptors located deeper than the mucosal layer, because (i) Iggo (1955) has shown that in the cat's stomach light mucosal stimulation was more excitatory on the tension receptors "in series" with muscle in the deep layers than on mucosal (chemo-)receptors capable of responding to a mechanical stimulus, and (ii) Leek (1969a) has suggested that *light* mucosal stimulation might excite mucosal mechanoreceptors subserving an intrinsic reflex, similar to the peristaltic reflex, thereby causing an increase in the activity of muscle cells in the deep layers which, in turn, would excite *in series* tension receptors.

The potent reflexogenic area delineated by Ash and Kay (1959) resembles the area of greatest tension receptor density observed by Leek (1969a) and shown in Table 1. Thus there is circumstantial evidence for the view that tension receptors fulfill the role of reticulo-ruminal mechanoreceptors capable of reflexly enhancing salivation.

On Primary Cycle Movements. The ruminant forestomach compartments (reticulum, rumen, and omasum) undergo a complex sequence of (primary cycle) movements for mixing their contents and transferring a fraction to the abomasum, which is the acid and pepsin secreting compartment and is equivalent to the stomach of the monogastric animals (see review by Sellers and Stevens, 1966). The primary cycle movements, unlike abomasal and stomach movements, are extrinsically activated and depend upon the integrity of gastric centres in the medulla oblongata and of vagal nerves, which provide both afferent and efferent pathways (see review by Leek, 1969b). It is concluded that the gastric centres require a tonic, net excitatory afferent drive, for which the input arising from mechanoreceptors in the reticulo-rumen is the most important.

Moderate levels of distension of the reticulum (Iggo, 1951, 1956a; Titchen, 1958; Dussardier, 1955; Stevens and Sellers, 1959; Iggo and Leek, 1967a, b) and of the reticulo-ruminal fold (Titchen, 1958) enhanced primary cycle movements. This effect was not affected by local anaesthesia of the reticular mucosa (Stevens and Sellers, 1959). The enhanced movements were associated with increased unitary discharges recorded from efferent cervical vagal fibres (Iggo and Leek, 1967a), which arise from gastric centre motoneurones located in and near the rostral part of the dorsal vagal nucleus (Harding and Leek, 1970). From these observations one concludes that there exists in the reticulum distension-sensitive receptors located at a greater depth than the mucosa. These receptors are the *in series* tension receptors examined by Iggo (1955) and Leek (1969a). Because these receptors respond even to low levels of distension they have been termed "low-threshold" tension receptors (Iggo and Leek, 1967b).

High levels of distension of the reticulum reduce the amplitude and ultimately the frequency of primary cycle movements (Fig. 16). Associated with these is a reduction in the unitary discharge in efferent gastric vagal fibres (Iggo and Leek, 1967b). Similarly, when high tensions develop during an isometric reticular contraction, there is a reflex reduction in the efferent discharges in fibres activating those structures involved in the next part of the primary cycle sequence. LEEK (1967) and IGGO and LEEK (1967b) have, therefore, concluded that, in addition

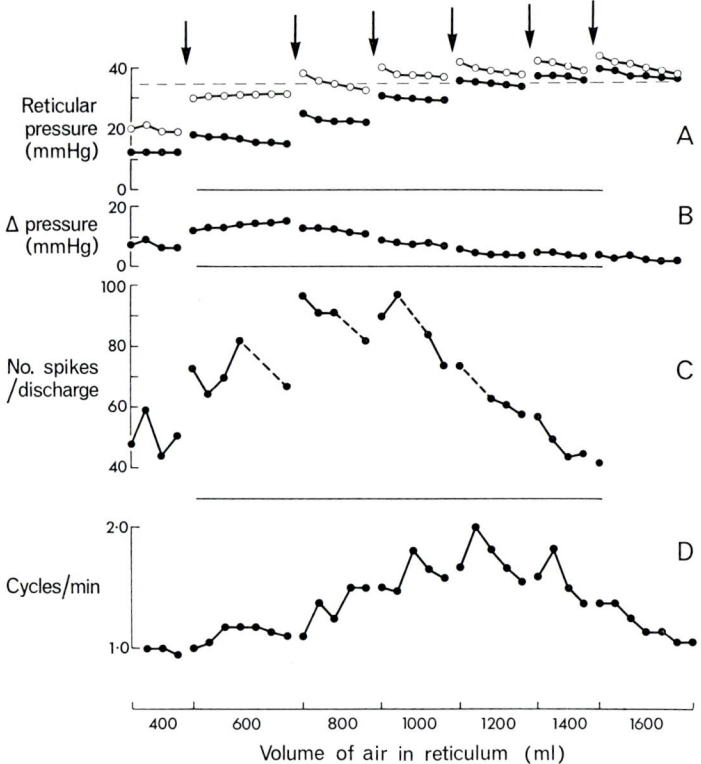

Fig. 16 A–D. The effect of progressively distending a sheep reticulum: reflex excitation at low levels of distension and reflex inhibition at high levels of distension. A balloon in the reticulum was inflated with increments of 200 ml air added immediately before each contraction marked with an arrow. In A ○——○ indicates the maximal reticular pressure developed during a contraction and ●——● indicates the reticular pressure during the quiescent part of the gastric cycle. B is the pressure rise developed during each contraction (i.e. the difference between ○——○ and ●——● in A). C is total number of spikes in the discharge of *one efferent* vagal reticular fibre associated with each contraction. D is the frequency of reticular contractions. The horizontal dotted line in A ($\equiv$ 35 mm Hg) represents the tension level when high-threshold reflex inhibitory effects (due to high-threshold tension receptors) are superimposed on low-threshold reflex excitatory effects (due to low-threshold tension receptors). At distensions of less than 800 ml no inhibition is seen. At volumes of 800–1200 ml the pressure developed during the contraction (○——○ in A) exceeds the high-threshold value (i.e. 35 mm Hg) and there is reflex inhibition of the amplitude of the contraction (shown in B as a diminution in pressure differential), and a reduction in the peak frequency of the efferent unitary discharge (not shown here). The number of spikes/discharge (C) does not diminish until volumes of >1000 ml, because (although the peak frequency is reduced) the low-threshold reflex excitation causes a prolongation of the discharge (and the contraction) which more than offsets the reduced peak frequency at 800–100 ml. Above 1200 ml, the frequency of the reticular contractions (D) is reduced, because it is mainly determined by the resting tension in the reticulum (●——● in A) and, as this now exceeds the high-threshold value, reflex inhibition results (IGGO and LEEK, 1967b)

to the "low-threshold" receptors which exert a reflex excitatory effect, there exist "high-threshold" receptors which exert a reflex inhibitory effect. Although the investigations of IGGO (1955) and LEEK (1969a) have not demonstrated two distinct populations of tension receptors, it is clear that some receptors have a higher threshold than others. The "high-threshold" receptors also utilise afferent pathways in the vagus as splanchnic nerve section has no effect on these reflex responses.

The idea of a two-threshold system of receptors, which qualitatively, but not quantitatively, are excited by the same kind of stimulus but produce opposite reflex effects, has already been discussed for the peristaltic reflex. Such an arrangement of tension receptors in the reticulum would account not only for the reflex effects on efferent gastric vagal discharges and reticulo-ruminal movements, described by IGGO and LEEK (1967b) but also for the inhibition of salivation caused by extreme reticular distension (KAY and PHILLIPSON, 1959) and for the inhibition of salivation and of reticulo-ruminal movements induced by very forceful squeezing and stretching of the reticular wall (ASH and KAY, 1959).

On Secondary Cycle Movements. Gaseous insufflation of the rumen induces secondary cycle movements in both conscious and decerebrate sheep (WEISS, 1953; review by DOUGHERTY, 1961; REID and TITCHEN, 1965). Excessive insufflation inhibits the movements. WEISS (1953) concluded that gaseous distension of the *caudo*-dorsal blind sac was responsible for the reflex elicitation of secondary cycle movements. These movements were very much more frequent during insufflation of sheep standing *tail* uppermost on a 30° ramp (which impairs eructation and increases ruminal distension) than during insufflation of sheep standing *head* uppermost on a 30° ramp which facilitates eructation and reduces ruminal tension. I have repeated WEISS's experiment with a modification, in which insufflation was so arranged that the gas pressure in the rumen remained the same, irrespective of the position of the sheep and the frequency of eructation. Under these conditions the tail-high position on the ramp gave rise to only slightly more secondary cycle movements than the head-high position. I attribute this results to the existence of a reflexogenic zone in the *cranial* part of the rumen which is subjected to a greater degree of *fluid* (rumen contents) distension in the tail-high position than in the head-high position, the gas pressure being the same in each situation. This conclusion accords well with the electrophysiological evidence that the rumen has numerous tension receptors in the medial wall of its cranial sac and very few tension receptors in its more caudal regions (LEEK, 1969a, b). ASH and KAY (1959) also found the caudal regions to be relatively inexcitable.

On the Cardia in Relation to Eructation. Using cine-radiographic techniques, DOUGHERTY and MEREDITH (1955) have analysed the movements of structures in the cranial regions of the reticulo-rumen during eructation induced by insufflation of gas. The movements, including the opening of the cardia, are evoked reflexly by gaseous distension of these cranial structures and occur even after ablation of most of the rumen (DOUGHERTY et al., 1958). Presumably the reflex phenomena are elicited by the excitation of *in series* tension receptors in the reticulum or in the reticular groove. More interesting, however, is the demonstration by DOUGHERTY *et al.* (1958) that these reflex events are inhibited by the presence of liquids (such as rumen contents, water or liquid paraffin) around the region of the cardia. It is concluded that there must be receptors near the cardia which can distinguish between gas and liquid but their location and the means by which they make this differentiation is not known. A similar mechanism appears to exist in the rectum, as most animals seem capable of distinguishing between flatus and even very fluid faeces.

On Rumination. This term is used to cover all the events taking place in association with the outwardly-visible procedure of "cud-chewing". Rumination is a complex phenomenon controlled from centres in the hypothalamus (ANDERSSON, 1951). It appears to require a state of drowsiness combined with a peripheral sensory input from the forestomach. Ruminants fed finely-milled foodstuffs to not ruminate (SCHALK and AMADON, 1928; FREER et al., 1962), whereas high-roughage diets enhance rumination and forceful tactile stimulation of the areas around the reticular groove and the reticulo-ruminal fold also evokes it (SCHALK and AMADON, 1938; DOWNIE, 1954; ASH and KAY, 1959). It is not clear what receptors evoke rumination, although tension receptors may be involved.

Abomasal Reflexes

The abomasum is embryologically and physiologically homologous with the stomach of non-ruminant animals.

Distension of the Abomasum reflexly inhibits primary cycle movements of the reticulo-rumen (PHILLIPSON, 1939; TITCHEN, 1958). After sectioning the splanchnic nerves, distension of the abomasum reflexly enhances primary cycle movements (TITCHEN, 1958). Thus, there appears to be two sets of distension-sensitive receptors in the abomasum: one set with afferent pathways in the vagal nerves which excite the gastric centres and another set with afferent pathways in the splanchnic nerves which inhibit the gastric centres. There are no electrophysiological records of abomasal receptors but it is likely that the abomasum is similar to the stomach of the cat, where IGGO (1957a) observed "in series" tension receptors. Presumably, such receptors could elicit the above reflex effects.

Acidification of the Abomasum with hydrochloric acid, under certain circumstances, may reflexly enhance primary cycle movements of the reticulo-rumen (TITCHEN, 1958; IGGO and LEEK, 1967b). The threshold for this effect is about pH 1. One infers that the reflex effects are due to the stimulation of acid receptors, similar to those studied in cat stomach by IGGO (1957b) but, in these, the threshold was around pH 2.5 (Fig. 13). Although pH 1 is considerably lower than the normal pH of abomasal contents (pH 2–4), it is not necessarily outwith the physiological range, because in the lumen of the peptic glands themselves pH 0.8 is attainable. One may hypothesize that, if acid receptors were located at the junction of the peptic glands with the mucosal surface, pH 1 could occur and acid receptors at this site could signal the extent of an imbalance between acid secretion and acid neutralization by abomasal contents.

ASH (1961a, b) has demonstrated that volatile fatty acids (the principal end-products of ruminal fermentation processes) have a potent excitatory action on gastric acid secretion. This, in itself, does not constitute evidence for neural receptors sensitive to volatile fatty acids, because the latter may exert their action directly on gastrin-secreting cells.

The rôle of abomasal receptors in reflexly affecting primary cycle movements is probably related to the regulation of abomasal fillings with contents from the forestomach. Too rapid filling would tend to distend the abomasum and to raise the pH through neutralizing gastric hydrochloric acid, the reflex consequences of which would be a reduction in primary cycle movements and a corresponding reduction in the rate of transference of forestomach contents to the abomasum.

Gastro-Intestinal Mechanoreceptor Reflexes

A variety of reflex effects following distension of the stomach have been described and a selection of these are given below.

CRAGG and EVANS (1960) have described (i) a cardio-vascular pressor response in the cat and rabbit after distension of the oesophagus, cardia, stomach, and pylorus and (ii) gastric (receptive) relaxation following distension of the pylorus. These effects were abolished by vagotomy and mimicked by stimulating the central end of one sectioned vagus nerve. These reflex effects could very well involve *in series* tension receptors of the kind observed in cats by IGGO (1957a). The same receptors may be responsible for the increased nervous activity recorded by microelectrodes implanted in hypothalamic "satiety centres" and the decreased nervous activity recorded from the "feeding centres" in response to gastric distension (ANAND and PILLAI, 1967).

Evidence of gastric tension receptors involved in the gastric phase of pancreatic secretion may be inferred from the investigation of BLAIR *et al.* (1966). In chloralosed cats already giving a pancreatic secretion in response to endogenous or exogenous secretin, it was possible to increase the amylase content of the secretion by distending the pyloric antrum or the fundus. These effects were unaltered after splanchnotomy, but vagotomy abolished the response to distension of the fundus which therefore involved a vago-vagal pathway and distension-sensitive receptors. The enhanced secretion of pancreatic amylase resulting from distension of the antrum persisted after splanchnotomy and vagotomy but could be abolished by atropine or cocainization of the antral mucosa, from which it was inferred that there were distension-sensitive receptors (in the mucosa?) activating a local cholinergic nervous pathway linked to a humoral pathway.

DANIEL and WIEBE (1966) observed inhibition of duodenal motility as a result of gastric distension. This response was not affected by either vagotomy or by pyloric section but it was blocked by reserpine and by guanethidine. It is concluded that distension-sensitive gastric receptors activate an extrinsic reflex using sympathetic (splanchnic) pathways for both the afferent and the efferent limbs.

HOPKINS (1966), in a reappraisal of some old results, concludes that, following test meals, the volume of the stomach contents declines not exponentially (as was formerly thought) but linearly proportional to the square root of the volume. This conclusion offers a most interesting physical explanation for a physiological situation. To use a rather crude analogy, if fluid flows out of an inelastic container (*e.g.* water flowing out of a bucket with a hole in the bottom) the volume of the fluid remaining in the bucket will decrease exponentially with time, whereas if fluid flows out of an elastic container (*e.g.* water flowing out of a balloon with a hole in it) the volume of the fluid remaining in the container will be related to the tension in the elastic container. Surface tension (T) is proportional to the surface area of a container and HOPKINS (1966) has argued that, if the stomach were regarded as a cylinder [of volume (v), length (l), and radius (r)] rather than as a sphere, then

$$T \propto 2\pi r \cdot l$$

and since

$$v = \pi r^2 \cdot l \quad \text{or} \quad \sqrt{v} = r\sqrt{\pi \cdot l}, \quad \text{and} \quad r = \sqrt{v}/\sqrt{\pi \cdot l}$$

$$T \propto 2\pi (\sqrt{v}/\sqrt{\pi \cdot l})l \quad \text{or} \quad T \propto \sqrt{v}.$$

Whilst the experimental results do fit the theory, there may be reasons for this besides the hypothesis that the stomach behaves as if it were an elastic cylinder. One important point which this study has brought out is that gastric emptying rate is likely to be related to gastric tension. For the physics of the situation it matters little whether the tension affects gastric emptying entirely passively, due to the elastic elements in the stomach wall or whether the tension elicits active processes for gastric emptying through reflexes which originate from gastric tension receptors and affect (i) the degree of receptive relaxation of the gastric musculature (CRAGG and EVANS, 1960; JANSSON, 1969), (ii) pyloric antral motility (ARMITAGE and DEAN, 1966) and possibly (iii) duodenal motility (DANIEL and WIEBE, 1966).

Altogether there seems to be circumstantial reflex evidence that "gastric emptying" involves extrinsically innervated gastric tension receptors of the kind described by IGGO (1957a). The further involvement of these receptors (but not necessarily, of their extrinsic afferent pathways) in evoking gastric secretion is clear from the experiments of DRAGSTEDT *et al.* (1953). Gastric secretion was induced either by passive distension of the pyloric antrum or, more profusely, by the tension developed actively during antral contractions, particularly if the outflow of material was impeded, so that conditions approached the isometric state. *In series* tension receptors operating through local pathways were implicated in this phenomenon and in other phenomena reviewed by DRAGSTEDT and WOODWARD (1968).

The effects of distending the gall bladder and the bile ducts on gastrointestinal movements in anaesthetized and in spinal dogs have been described by NAKAYAMA and MORI (1967). The stimulus was distension to a pressure of 50–160 mm Hg (which seems rather excessive) and the afferent pathways lay in the splanchnic nerves. Distension of the gall baldder caused inhibition of the stomach and of the small intestine. The large intestine was usually inhibited, although it was sometimes excited. Distension of the bile ducts inhibited the small intestine and either inhibited or excited stomach movements. SHERRINGTON (1906) describes a reflex effect on blood pressure resulting from bile duct distension. NEWMAN and PAUL (1966) have distended the gall bladder, recorded multi-unit discharges in splanchnic nerve fibres and observed excitatory and inhibitory effects on the nervous activity recorded from the anterior lobe of the cerebellum. At present, no electrophysiological "single fibre" studies of distension-sensitive receptors in the biliary tract exist.

The sensibility of the small intestine has been investigated by DOWNMAN et al. (1948). The small intestine was exposed and the reflex effects of various stimuli applied to the intestine were observed on the pupil of chloralosed cats and on the movements and blood pressure of spinal decerebrate cats. The effective stimuli were mechanical (e.g. pinching, squeezing, cutting, and scratching), thermal/noxious (i.e. heat above 46°C) and chemical (e.g. 1.3% KCl, 1 N, or 0.1 N HCl, 10% NaCl). Local cooling and freezing and also gently rubbing the mucosa were ineffective. The mucosal layer was split away from the overlying muscle layer with minimum disturbance to the innervation of either layer and, under these conditions most forms of mechanical, thermal, and irritant chemical stimulation of the mucosa were ineffective, whereas the muscle layers and the serosa remained responsive, particularly to mechanical stimuli. It was found, in addition, that the adjacent mesentery was *very* sensitive to all forms of stimulation. The mucosa thus appears to be reflexogenically less excitable than the deeper regions: a conclusion which accords well with that derived from electrophysiological studies described earlier.

Inhibition of pyloric antral motility can be induced passively by distension of the duodenum or actively when the duodenum undergoes forceful contractions (DANIEL and WIEBE, 1966). This is an extrinsic reflex, as it is usually still present after section of the pylorus. It is abolished by bilateral vagotomy and chemical sympathectomy, from which it is concluded that the vagi provide the afferent pathways and the splanchnic nerves provide the efferent pathway. With this reflex, the efficacy either of passive distension or of forceful contractions as the stimulus provides strong evidence for the involvement of *in series* tension receptors of the type studied in the duodenum by IGGO (1957a).

Gastro-Intestinal Chemoreceptor Reflexes

Theoretically, chemoreceptor activity may be elicited either by a structural property or by a physical property of a chemical substance capable of acting as a stimulus. In the former instance only a few substances with identical molecular components are capable of fitting a template on the receptor surface and of inducing receptor activity. Such specific chemoreceptors are prevalent in the olfactory mucosa and, to a lesser extent, in the tongue, subserving the sensations of smell and taste respectively. More prevalent in the gastro-intestinal tract are nonspecific chemoreceptors which are excited by a physical property possessed by any number of chemical substances, e.g. acid receptors and osmoreceptors. In certain situations, however, some of these receptors are in effect not entirely nonspecific, because interposed between the chemical substances and the chemoreceptors are "diffusion barriers" which cause the chemoreceptors to have a selective, if not a specific, response to certain groups of related chemicals.

A reflex which implicates acid receptors in the pyloric antrum is described by ANDERSSON and OLBE (1964). Dogs, each with a PAVLOV pouch and an isolated, innervated antrum, were prepared. Sham-feeding caused gastric secretion (due to efferent vagal activity). This effect could be eliminated by acidifying the antrum to give pH 2 or less. This level of acidity would be sufficient to excite acid receptors in the gastric mucosa of cats (Fig. 13) (IGGO, 1957b).

Acid receptors in the duodenum are implicated in the reflex inhibition of gastric secretion and motility when acid conditions (of pH 3.5 or less) prevail in the duodenum (THOMAS, 1957). This is an extrinsic reflex (the "enterogastric reflex") using vagal pathways. It is blocked either by local anaesthetic applied to the mucosa (SIRCUS, 1958) or by vagotomy (CODE and WATKINSON, 1955; THOMAS, 1957). HUNT and KNOX (1968, 1969) have made a particularly elegant investigation of this reflex in man. Their technique was based on measuring the gastric emptying rate of test meals which contained various concentrations of hydrochloric, acetic, lactic, phosphoric, citric, butyric, and hexanoic acids. Acids of high molecular weight were less effective than those of low molecular weight for slowing gastric emptying but the oil/water partition coefficients, the dissociation constants (pK) and the valencies of the acids did not influence their efficacy. Only the acids and not their salts were effective, so that the receptor was responding to hydrogen ion concentration and was therefore, an *acid receptor*. Comparing the effectiveness of various kinds of acid at different concentrations, HUNT and KNOX (1969) concluded that the receptors were behaving, as if they were titrating the acid to pH 6.5, possibly using bicarbonate as the neutralizing base. Thus carbonic acid would be formed and act as the stimulus for the acid receptor, irrespective of the kind of acid either present in the duodenum or responsible for the formation of the carbonic acid. There appears to be a diffusion barrier between the duodenum and the titration site, which would account for the greater efficacy of low molecular weight acids compared with high molecular weight acids. This diffusion barrier seems to be aqueous in type, because fat soluble acids showed no advantage. So far no acid receptors in the duodenum have been sought by electrophysiological techniques.

The idea that osmotic factors acting on the duodenum inhibit gastric secretion stems from the work of DAY and KOMAROV (1939) and this idea gives rise to the concept of *osmoreceptors*. SIRCUS (1958) showed that sugars, peptones, and fats (only after being digested) inhibited gastric secretion in proportion to the osmotic pressure they exerted in the duodenum, even after removal of the antrum. Unlike the enterogastric reflex derived from acid receptors in the duodenum, the osmoreceptor mechanisms was not blocked by local anaesthetics applied to the duodenal mucosa and, on this basis, SIRCUS (1958) concluded that a humoral pathway was involved. In man, HUNT and PATHAK (1960) provide evidence of osmoreceptors in the intestine which inhibit gastric emptying. Test meals of various substances and osmotic activities were ingested and, after a fixed interval, the stomach was emptied by stomach tube in order to ascertain the rate of gastric emptying. For most simple molecules and ions, the rate of emptying was related to their osmotic activity, except in the case of acids (for which the independent mechanism described above exists) and of sodium ions, urea and glycerol (for which facilitated transport across the osmotic membrane was suggested). From a comparison of the rates of emptying of water (fast), glucose (slow), and starch (fast initially then slow, as for glucose), HUNT (1960) concluded that the osmo-receptors were located at or beyond the site of hydrolysis of starch, *i.e.* distal to the pylorus.

A more precise location was deduced from a later investigation using the same techniques (ELIAS *et al.*, 1965) whereby it was concluded (i) that the osmoreceptors lie deep or close to the site of production of the enzymes sucrase and maltase,

(ii) that a barrier for fructose exists which can however, be saturated and may therefore, take the form of a "metabolic sink" and (iii) that, since there was no effective distinction upon gastric emptying rate between sucrose and a fructose-glucose mixture, the barrier for fructose lies between the hydrolysing enzymes and the osmoreceptors. Although there is good evidence for the existence of osmoreceptors, it must be emphasized that these receptors may or may not be linked to extrinsic afferent nerves. In the latter instance, the receptor tissue may also be secretory and produce a hormone such as enterogastrone. In the vascular system, osmoreceptor activity evoked by different concentrations of dextran and of salt solutions can be recorded in certain afferent nerves, *e.g.* the hepatic nerve of the rabbit during liver perfusion (ANDREWS and STRATMAN, 1968) but no electrophysiological evidence of osmoreceptor activity in the gastro-intestinal tract exists.

There are numerous examples of "cause and effect" relations in which the cause (or stimulus) is fairly specific and presumably depends upon a certain configuration of that part of the molecule which attaches itself to receptive sites on the surface of the receptor. In the pyloric antrum there appear to be receptors which respond specifically to the products of protein digestion. Some of these receptors are probably capable of direct action in the form of secreting hormones (*e.g.* gastrin) without the intervention of neural links. Other receptors use a local nerve network to activate secretory cells, *e.g.* in the gastric phase of pancreatic secretion (BLAIR *et al.*, 1966). Similar receptors in the duodenum employ extrinsic reflex pathways in the vagus nerves, *e.g.* in the enterogastric reflex (THOMAS, 1957). Many of the other substances, such as long chain fatty acids and monosaccharides, also seem to act upon specific receptors but their effects involve mainly humoral pathways and will not be considered further here. SHARMA and NASSET (1962) have demonstrated that glucose can stimulate chemoreceptors innervated by medium and large fibres and amino-acids can stimulate others innervated by small fibres.

Volatile fatty acids, particularly acetic acid, appear to exert reflex effects which are not dependent on their pH effects, *per se*. The existence of receptors which are excited specifically by volatile fatty acids is postulated to account (i) for the reflex inhibition of reticulo-ruminal movements by volatile fatty acids acting on receptors in the forestomach and abomasal mucosa (ASH, 1959), (ii) for the excitation of gastric acid secretion by volatile fatty acids in the abomasum (ASH, 1961 a, b) and (iii) for the depression of food intake following injections of acetic acid directly into the rumen (BAILE and PFANDER, 1966). This may form the basis of a chemosensitive mechanism for regulating food intake in the ruminant (BAILE, 1968). Apart from the *multi-unit* records of SHARMA and NASSET (1962) no specific chemoreceptors have been demonstrated electrophysiologically.

Urinary Tract Reflexes

Bladder filling and periodical emptying, though subject to influences from higher centres in the brain, can occur in a co-ordinated way even in low decerebrate and in spinal animals. The co-ordination is largely due to the sensory input derived from bladder receptors during the filling (or collecting) phase and to sensory "feed-back" from bladder and urethral receptors during the emptying (or micturition) phase. In this system, the receptors inferred from the investigation of reflexes are similar to those made evident by the electrophysiological techniques described earlier.

During bladder filling there are progressive increases in its volume, in its tension and in the frequency and amplitude of intrinsic contractions of its musculature. The reflex phenomena associated with this phase include (i) a reduction of the efferent activity in sacral (parasympathetic) neurones (GROAT and RYALL, 1969), (ii) an increase in the efferent activity in sympathetic nerves (EDVARDSEN, 1968a) and (iii) an increase in the tone of the external urethral sphincter (GARRY et al., 1959). The first two of these effects result in receptive relaxation of the bladder and this is particularly evident during the last third of the filling phase (EDVARDSEN, 1968a) when the intrinsic contractions would be most pronounced. This is good circumstantial evidence for the view that reflex receptive relaxation results from the excitation of *in series tension receptors* of the type described by IGGO (1955) with afferent fibres in sacral (parasympathetic) nerve (EDVARDSEN, 1968a). The increased external sphincter tone which occurs during bladder filling is also likely to result from the activation of the *in series* tension receptors but, in addition, flow of liquid along the urethra when the bladder is empty will also reflexly increase sphincter tone (GARRY et al., 1959). This results presumably from the activation of *flow receptors* in the urethra, described by IGGO (1956b) and TODD (1964).

At a critical point of bladder filling, reflex emptying is initiated. The critical point is determined by bladder tension and is independent of bladder volume. Rapid bladder filling with inadequate reflex receptive relaxation or lack of receptive relaxation following peripheral sympathectomy (EDVARDSEN, 1968b) causes the critical tension to be reached whilst bladder volume is still relatively small. The question now arising is: how do tension receptors bring about (i) receptive relaxation when bladder tensions are below a critical value and (ii) bladder contraction when bladder tensions are above the critical value? This situation is analogous to that discussed earlier, in which moderate levels of reticular tension reflexly enhance ruminant forestomach movements, whereas high levels of reticular tension reflexly inhibit the movements (IGGO and LEEK, 1967b). A parallel also exists in the somatic system whereby moderate tension applied to an extensor muscle leads to the "stretch reflex" (*i.e.* contraction through the excitation of "low threshold" receptors, *viz.* muscle spindles) and higher tensions lead to the "clasp-knife reflex" (*i.e.* reflex relaxation, through the excitation of "high threshold" receptors, *viz.* GOLGI tendon organs). In both the bladder and the ruminant reticulum, two theoretical possibilities exist to account for the opposing reflex effects of moderate and of high tensions, *viz.* (i) low-threshold receptors giving one reflex effect and high-threshold receptors giving a dominant and opposing reflex effect or (ii) a system of interneurones arranged so that low-threshold pathways are used by low-frequency afferent inputs and high-threshold pathways, which cause opposing and dominant reflex effects, are opened up by high-frequency afferent inputs. Because of analogous effects in the somatic system and in the peristaltic reflex, the idea of two independent sets of receptors with different threshold and opposing reflex effects is currently favoured.

The critical bladder tension which initiates the emptying phase, reflexly evokes a bladder contraction and reflex opening of the external urethral sphincter (BARRINGTON's 1st and 5th reflexes; GARRY *et al.*, 1959). Flow of urine (or any liquid) along the urethra, under conditions of extreme passive distension of the bladder or during bladder contractions, causes (i) no increase in the already decreased external sphincter tone *i.e.* the "flow receptor" reflex is now rendered ineffective (GARRY *et al.*, 1959) and (ii) a reinforcement of the efferent parasympathetic fibre discharge causing an enhanced bladder contraction (BARRINGTON's 2nd reflex; DE GROAT and RYALL, 1969). These reflex effects are attributable to *in series* tension receptors (described by IGGO, 1955), because they are reinforced by the increased tension developed actively during the bladder contraction even though the bladder volume is decreasing (as in Fig. 10). Any tendency to obstruct bladder emptying would be expected to increase even further the force and the duration of the bladder contraction. This has been demonstrated, in fact, in cats (MELLANBY and PRATT, 1939, 1940; IGGO, 1955).

Besides their reflex actions associated with bladder filling and emptying, tension receptors in the bladder are capable of evoking potent vasoconstriction of hind-limb blood vessels. These reflexes are particularly evident in subjects with spinal cord injuries (WHITTERIDGE, 1956).

Some Sensory and Clinical Correlates

Sensory studies and clinical observations have provided some rather general information about visceral receptors, although its interpretation requires care. Here there is an attempt to examine the correlation between some of the receptor mechanisms made evident by electrophysiological and reflex experiments and some inferred from sensory and clinical investigations.

The Effects of Different Stimulus Intensities

In general, *low levels* of stimulation evoke local reflex effects subserving local regulatory functions, *higher levels* of stimulation evoke more widespread (and even opposing) reflex effects sometimes accompanied by non-painful sensations, whilst *even higher levels* of stimulation evoke profound reflex and behavioural effects coupled with the sensation of pain. At least three theoretically possible mechanisms could account for these effects: (i) a single population of receptors which respond to the same stimuli and increase their afferent discharges in proportion to the stimulus strength. Low frequency inputs would evoke local reflex response and higher frequency inputs would, in addition, open up high-threshold inter-neuronal circuits and give rise to more widespread (and even the opposite) reflex effects. Very high frequency inputs would open up interneuronal circuits of even higher threshold, some of which would project to the cerebral cortex and give rise to the sensation of pain, (ii) two or more populations of receptors which respond by the *same physiological mechanism* to the same kind of stimulus but possess different thresholds; the high-threshold receptors being responsible for the diffuse reflex effects and pain sensations, (iii) two or more populations of receptors which possess different thresholds and respond to the same kind of stimulus by *different physiological mechanisms*. In contrast to (ii), high intensity stimulation would result in tissue damage either directly or indirectly (through tissue ischaemia/ anoxia *e.g.* during extreme distension of a viscus). The damaged tissues would release humoral agents which would excite specific "pain" receptors (nociceptors) and might reduce the threshold of (*i.e.* sensitize) other receptors.

There is no conclusive evidence in favour of any one of the above possibilities. Electrophysiological and reflex investigations have demonstrated "two thresholds" effects but whether this is attributable to two separate populations of receptors or whether it is due to separate low and high threshold interneuronal pathways is not certain. The first of these two alternatives is generally assumed, particularly when two separate afferent pathways can be demonstrated. Variation in the threshold of interneuronal links is particularly relevant with regard to pain sensations, as their thresholds are especially labile, being much affected by such factors as previous experience ("conditioning"), general alertness and psychological states.

A Visceral Nociceptor Theory

Electrophysiological studies have demonstrated that different receptors responsive to the same kind of stimulus have different thresholds but, so far, these studies have not differentiated the receptors into distinctly separate groups, each with its own reflex pathways and effects. It is generally accepted that noxious stimuli must produce some degree of tissue damage before nociceptors are excited and give rise to nociceptive reflexes and "pain" sensation. Various humoral agents have been postulated to be released from damaged tissues and to bring about these effects, *e.g.* Substance P, bradykinin, histamine, 5-HT. POTTER *et al.* (1962) have compared the minimum dose of these agents which, administered intra-arterially, caused "vocalization" in chloralose-anaesthetized dogs. "Vocalization" was interpreted as being indicative of the excitation of nociceptors, although these agents do sensitize receptors of all kinds (PAINTAL, 1964). "Vocalization" was evoked most readily by substance P and bradykinin. POTTER *et al.* (1962) have raised the interesting possibility that these or similar agents might be released by an axon reflex mechanism in sufficient quantities to excite adjacent nociceptors, when other types of receptors are subjected to high intensity stimulation. If correct, this theory would readily account for the fact that, for example, in a "spasmodic colic" affecting structures dually innervated by vagal and splanchnic nerves, many tension receptors with vagal afferent fibres would be forcibly stimulated despite the afferent pathway for pain sensations being restricted to the splanchnic nerves, in accordance with the experimental studies of MOORE (1938) and the neurosurgical experiences of WHITE and SWEET (1969). The ultimate sensory and reflex effects of nociceptor activity show considerable variation. This is partly due to the action of important descending spinal influences, which modify the accessibility of afferent nociceptor inputs to ascending sensory tracts. These mechanisms are discussed in Volume II of this Handbook.

Non-Painful Sensations

Moderate distension either of the stomach, of the small or large intestine or of the bile duct gives rise to a feeling of (i) "fullness"; referrable to the stomach even though the distension may affect only the intestine or the bile duct (Hurst, 1911; Hoelzel, 1947) and (ii) (temporary) satiation. In the stomach the effects are due to tension and are not due to volume. The sensations decline as the tension diminishes due to receptive relaxation even though the volume of the distension remains the same. For the same reason, rapid filling of the stomach gives rise to the above effects when the final volume of the gastric contents is much smaller than that necessary to elicit the sensations when the stomach filling rate is slow (Hurst, 1911). All these effects could be accounted for by tension receptors (*not* "volume" or "stretch" receptors) fulfilling a distension-sensitive role. The effects also provide circumstantial but not conclusive evidence that the tension receptors are *in series* with the smooth muscle cells: because the effects diminish as the tension falls even though the volume is unchanged, the receptors are located in series with the structural elements responsible for determining the tension. As the stomach is capable of reflex receptive relaxation, it is concluded that tension is not determined passively by elastic elements but is determined actively by contractile elements and, therefore, that distension-sensitive receptors are *in series* with the smooth muscle cells. This conclusion accords with that of Iggo (1957a), based on electrophysiological evidence. Another theoretically possible location for the receptors exists but this has not been found by electrophysiological investigation *i.e.* that the receptors are located near the mucosa, sandwiched (and therefore subjected to compression) between the muscle layer (which provides the tension) and the intragastric contents (which resist the tension). Unfortunately, some authors refer to distension causing "*radial* tension" whereas electrophysiological evidence suggests that the effective tension is *tangential*.

At present it is a little difficult to reconcile the involvement of *in series* gastric tension receptors in the sensation of stomach fullness and appetite satiation with their involvement in the sensation of epigastric "hunger pangs" derived from "strong contractions" of an empty stomach, described by Cannon and Washburn (1912). These powerful contractions occur under conditions of fasting and of insulin-induced hypoglycaemia. Undoubtedly *in series* gastric tension receptors would be excited, yet the sensation which ensues is one of "emptiness" in contrast to the sensation of "fullness" described above. For the moment, one can only assume that the conscious, opposing interpretations placed on apparently similar afferent inputs during hunger contractions and during gastric distension may be dependent upon (i) integrative mechanisms in the brainstem and (ii) the direction of the relative imbalance between the "hunger" and the "satiety" centres.

The bladder shows many of the effects observed for the stomach. This is not surprising, as electrophysiological investigations have demonstrated that the same kind of mechanoreceptors (*i.e. in series* tension receptors) predominates in both organs (Iggo, 1955, 1957a). With *slow filling* rates a large volume of contents is necessary to give the sensation of bladder fullness and to initiate reflex bladder emptying whereas much smaller volumes elicit these effects when (i) the filling rate is rapid, (ii) reflex receptive relaxation is abolished and (iii) intrinsic movements of the bladder wall are enhanced. As for the stomach, these effects are explicable on the basis of *in series* tension receptor activation, as the other theoretical possibility, *mucosal pressor-receptors* have not been shown to exist, by electrophysiological means.

Painful Sensations

Irrespective of the lack of knowledge about pain mechanisms, a few general features of visceral pain in relation to sensory receptors are worth discussing. Many of the painful visceral conditions fall into one of the following categories:

(i) Irritation (including inflammation) of the mucosa or of the serosa.
(ii) Torsion or traction on mesentery.
(iii) Passive distension of a viscus.
(iv) Powerful contractions of a viscus.
(v) Impactions.

Low levels of irritation (*e.g.* by toxins, irritant purgatives, or "catarrhal" inflammation) often produce hypermotility without arousing conscious sensations apart from a increased frequency of the need to defaecate, *e.g.* in a mild gastro-enteritis, or to micturate, *e.g.* in a mild cystitis. In the latter instance, the irritants probably directly or indirectly excite intrinsic contractions of the bladder musculature which in turn, excites *in series* tension receptors at *apparently* lower bladder tensions and leads to micturition of subnormal volumes of urine. Higher levels of irritation cause appreciable tissue damage and give rise to painful sensations accompanied by nausea or vomiting (in those species capable of doing so). These are features of (i) ulcerative conditions affecting the stomach and the duodenum, the reticulum (due to foreign bodies) and the abomasum, (ii) the more severe forms of gastro-enteritis (including appendicitis and colitis), cystitis, and endometritis (including the changes associated with dysmenorrhoea). The sensations experienced by a fistulated subject when there was irritation of the gastric mucosa have been described by WOLF and WOLFF (1947).

Torsion of or traction on mesenteries elicits nociceptive reflex responses even in anaesthetized or in decerebrate animals. Electrophysiological investigations have not identified the receptors responsible for these responses, although reflex studies, such as those of DOWNMAN *et al.* (1948), have demonstrated the high level of sensibility existing in the mesenteries. These responses are a common feature of surgical manoeuvres of the viscera in anaesthetized subjects and are accompanied by painful sensations in conscious subjects. Mesenteric receptors are presumably responsible for, at least, the initial pain arising from the various forms of herniations, intussusceptions, prolapses, and torsions, *e.g.* the particularly severe torsion encountered in horses when the colon rotates on its mesenteric axis. In man, traumatic or inflammatory involvement of the visceral peritoneum gives rise to painful sensations vaguely located to the midline, whereas that of the parietal peritoneum produces pain which is more discretely localised to the affected region of the abdomen and which may be accompanied by "referred pain" in some extra-abdominal region having a common dermatomal origin with the affected part of the abdomen.

Passive distension of a viscus may be due to the excessive accumulation of gases or of liquids. The gases arise from microbial fermentation processes but their accumulation results from subnormal elimination rates rather than from supranormal production rates *per se*. This is clearly demonstrable in ruminants, which normally can eliminate rumen gases at 5–10 times their maximum production rate. Despite this, acute gaseous distension of the reticulo-rumen (*i.e.* "bloat") is a common clinical condition, which results from the failure to move the gas layer (which is in the dorsal part of the rumen) cranially to the region of the cardia (which is in the dorso-medial part of the reticulum). Distension of the cardiac region by gases (*but not by liquids or by foams*) reflexly evokes eructation (*vide* p. 143). Bloat results from diets which contain or, following fermentation, produce factors that either depress extrinsic reflex movements of the reticulo-rumen (so that the gas layer is not moved cranially) or form a stable foam (so that the gas does not separate out from the rumen liquids). Similar considerations probably hold for "flatulent colics" encountered elsewhere, particularly in humans and in horses. Distension-sensitive receptors (tension receptors) have been demonstrated electrophysiologically in the forestomach and in the gastro-intestinal tract and it is these receptors presumably which are involved in the discomfort or the pain caused by gaseous visceral distension. As yet there are no electrophysiological studies on the inferred receptors which are capable of distinguishing between gas and fluid distension, *e.g.* between rumen gas and rumen fluids or between flatus and faeces in the distal colon and rectum.

Distension of a viscus with liquids may be assumed to involve distension-sensitive receptors. Such distension gives rise to sensations and reflex responses which are not mimicked by fluid distension of the peritoneal cavity, *e.g.* in ascites. This is circumstantial evidence for the conclusion derived from electrophysiological studies (p. 122 *et seq*) that the distension-sensitive receptors are tension receptors (which respond to tangential forces) rather than pressor receptors (which would respond to transmural compression or radial forces). For reasons discussed earlier (p. 120 and 149) the onset and severity of the sensations and the clinical signs depend on the rate of the fluid distension *vis-à-vis* the rate of either active or passive receptive relaxation of the viscus. This again suggests that the tension receptors occupy a situation *in series* with the contractile elements of the viscus, *e.g.* in the bladder. Another interesting consideration concerns the "dumping syndrome" which may be a sequel to partial gastrectomy and shows signs which include epigastric pain and nausea following the intake of food and (particularly hyperosmotic) fluids. Although various causes (*e.g.* reduced blood volume, serotonin release) have been suggested to account for the clinical signs (see review by MENGUY, 1964), it does seem possible that distension-sensitive receptor mechanisms may be involved, at least in part. In the duodenum and proximal jejunum, there would be present abnormally large volumes of contents, which would excite distension-sensitive receptors (tension receptors): a conclusion in accordance with the parallel observation that increased peristalsis found in the small intestine in this syndrome. Enhanced peristalsis would also excite rapidly-adapting mucosal receptors (p. 134) and serosal ("movement") receptors (p. 131). Experimentally, the symptoms of the "dumping syndrome" have been reproduced by distension of the jejunum (MACHELLA, 1949; GLAZEBROOK and WELBOURN, 1952).

Powerful contractions of an abdominal viscus produce clinical signs known as "spasmodic colic", *i.e.* recurrent, short periods of sharp pain. These contractions are a frequent accompaniment to inflammatory conditions affecting the small and large intestines and the bladder in man. Spasmodic colic referrable to the large intestine is a particularly common occurrence in horses but it is uncommon in cattle. Presumably, *in series* tension receptors are excited by these powerful contractions and are involved in the pain process. It is not clear why the contractions should be so powerful or why they should be confined to spasms. Humoral agents are probably active in these conditions and this may account for the symptomatic relief that may be obtained in man by causing a hyperaemia, *e.g.* by lying in a hot bath. A possible connection through an axon reflex, whereby tension receptor activity in vagal fibres may elicit nociceptor activity in splanchnic fibres has been discussed above (p. 152).

A similar association between tension receptors and pain sensations may exist in clinical conditions caused by impactions. The location of tension receptors *in series* with muscle cells make them particularly effective as "impaction monitors" (p. 127). Their activity seems likely to have some role in the painful sensations associated with impactions, such as those due to (i) calculi in the bile ducts, urethra and ureters, (ii) impairment of bladder emptying through calculi and prostatic occlusions, (iii) impairment of colonic emptying because of impacted faeces (particularly, in man, horses, and dogs) or of prostatic enlargement (in dogs), and (iv) contraction of the gravid uterus during stage I of labour and during dystocia caused by inadequate dilatation of the cervix.

Conclusion

The electrophysiological studies of visceral mechanoreceptors provide an insight into the sensory mechanisms of those reflexes and non-painful sensations which involve mechanical stimuli. A considerable number of "reflex" investigations implicate chemoreceptors of various kinds but their electrophysiological identification is sadly lacking. Visceral pain, though so common, is not explicable as long as we are unable to identify nociceptors and to determine the possible link between nociceptors and other types of receptors.

Acknowledgement

I am indebted to Professor A. Iggo for his help during the preparation of the manuscript.

References

Adrian, E. D.: Afferent impulses in the vagus and their effect on respiration. J. Physiol. (Lond.) **79**, 332–358 (1933).

Agostini, E., Chinnock, J. E., Burgh Daly, M. de, Murray, J. G.: Functional and histological studies of the vagus nerve and its branches to the heart, lungs and abdominal viscera in the cat. J. Physiol. (Lond.) **135**, 182–205 (1957).

Anand, B. K., Pillai, R. V.: Activity of single neurones in the hypothalamic feeding centres: effect of gastric distension. J. Physiol. (Lond.) **192**, 63–77 (1967).

Andersson, B.: The effect and localisation of electrical stimulation of certain parts of the brain stem in sheep and goats. Acta physiol. scand. **23**, 8–23 (1951).

Andersson, S., Olbe, L.: Inhibition of gastric acid response to sham feeding in Pavlov pouch dogs by acidification of antrum. Acta physiol. scand. **61**, 55–64 (1964).

Andrew, B. L.: Activity in afferent nerve fibres from the cervical oesophagus. J. Physiol. (Lond.) **135**, 54–55 P (1957).

References

Andrews, W. H. H., Stratman, C. J.: Afferent nerve impulses in the hepatic nerve of perfused rabbit livers. J. Physiol. (Lond.) **195**, 32–33 P (1968).

Armitage, A. K., Dean, A. C. B.: The effects of pressure and pharmacologically active substances on gastric peristalsis in a transmurally stimulated rat stomach-duodenum preparation. J. Physiol. (Lond.) **182**, 42–56 (1966).

Ash, R. W.: Inhibition and excitation of reticulo-rumen contractions following the introduction of acids into the rumen and abomasum. J. Physiol. (Lond.) **147**, 58–73 (1959).

— Acid secretion by the abomasum and its relation to the flow of food material in the sheep. J. Physiol. (Lond.) **156**, 93–111 (1961a).

— Stimuli influencing the secretion of acid by the abomasum of sheep. J. Physiol. (Lond.) **157**, 185–207 (1961b).

— Kay, R. N. B.: Stimulation and inhibition of reticulum contractions rumination and parotid secretion from the forestomach of conscious sheep. J. Physiol. (Lond.) **149**, 43–57 (1959).

Baile, C. A.: Regulation of feed intake in ruminants. Fed. Proc. **27**, 1361–1366 (1968).

— Pfander, W. M.: A possible chemosensitive regulatory mechanism of ovine feed intake. Amer. J. Physiol. **210**, 1243–1250 (1966).

Bayliss, W. M., Starling, E. H.: The movements and innervation of the small intestine. J. Physiol. (Lond.) **24**, 99–143 (1899).

Beghelli, V., Borgatti, G., Parmeggiani, P. L.: On the role of the dorsal nucleus of the vagus in the reflex activity of the reticulum. Arch. ital. Biol. **101**, 365–384 (1963).

Beleslin, D., Varagic, V.: The effect of cooling and of 5-hydroxytryptamine on the peristaltic reflex of the isolated guinea-pig ileum. Brit. J. Pharmacol. **13**, 266–270 (1958).

Bessou, P., Perl, E. R.: A movement receptor of the small intestine. J. Physiol. (Lond.) **182**, 404–426 (1966).

Blair, E. L., Brown, J. G., Harper, A. A., Scratcherd, T.: A gastric phase of pancreatic secretion. J. Physiol. (Lond.) **184**, 812–824 (1966).

Bower, E. A.: Action potentials from uterine sensory nerves. J. Physiol. (Lond.) **148**, 2–3 P (1959).

— The characteristics of spontaneous and evoked action potentials recorded from the rabbit's uterine nerves. J. Physiol. (Lond.) **183**, 730–747 (1966).

Bülbring, E., Crema, A.: Observations concerning the action of 5-hydroxytryptamine on the peristaltic reflex. Brit. J. Pharmacol. **13**, 444–457 (1958).

— Lin, R. C. Y.: The effect of intraluminal application of 5-hydroxytryptamine and 5-hydroxytryptophan on peristalsis; the local production of 5-HT and its release in relation to intraluminal pressure and propulsive activity. J. Physiol. (Lond.) **140**, 381–407 (1958).

— — Schofield, G. C.: An investigation of the peristaltic reflex in relation to anatomical observations. Quart. J. exp. Physiol. **43**, 26–37 (1958).

Cannon, W. B., Washburn, A.: An explanation of hunger. Amer. J. Physiol. **29**, 441–454 (1912).

Code, C. F., Watkinson, G.: Importance of vagal innervation in the regulatory effect of acid in the duodenum on gastric acid secretion. J. Physiol. (Lond.) **130**, 233–252 (1955).

Cragg, B. G., Evans, D. H. L.: Some reflexes mediated by afferent fibres of the abdominal vagus in the rabbit and cat. Exp. Neurol. **2**, 1–12 (1960).

Daly, M. de B., Evans, D. H. L.: Functional and histological changes in the vagus nerve of the cat after degenerative section at various levels. J. Physiol. (Lond.) **120**, 579–595 (1953).

Daniel, E. E., Wiebe, G. E.: Transmission of reflexes arising on both sides of the gastroduodenal junction. Amer. J. Physiol. **211**, 634–642 (1966).

Day, J. J., Komarov, S. H.: Glucose and gastric secretion. Amer. J. dig. Dis. **6**, 169–175 (1939).

Denny-Brown, D., Robertson, E. G.: The state of the bladder and its sphincters in complete transverse lesions of the spinal cord and cauda equina. Brain **56**, 397–463 (1933).

Dougherty, R. W.: The physiology of eructation in ruminants. In: Digestive physiology and nutrition of the ruminant. London: Butterworth 1961.

— Habel, R. E., Bond, H. E.: Esophageal innervation and the eructation reflex in sheep. Amer. J. vet. Res. **19**, 115–128 (1958).

Dougherty, R. W., Meredith, C. D.: Cinefluographic studies of the ruminant stomach and of eructation. Amer. J. vet. Res. **16**, 96–100 (1955).

Downie, H. G.: Photokymographic studies of regurgitation and related phenomena in the ruminant. Amer. J. vet. Res. **15**, 217–223 (1954).

Downman, C. B. B., McSwiney, B. A., Vass, C. C. N.: Sensitivity of the small intestine. J. Physiol. (Lond.) **107**, 97–106 (1948).

Dragstedt, L. R., Oberhelman, H. A., Zubiran, J. M., Woodward, E. R.: Antral motility as a stimulus for gastric secretion. Gastroenterology **24**, 71–78 (1953).

— Woodward, E. R.: Mistakes in peptic ulcer sugery in the early years: a physiologic analysis. Ann. Surg. **167**, 886–897 (1968).

Dubois, F. S., Foley, J. O.: Experimental studies on the vagus and spinal accessory nerves in the cat. Anat. Rec. **64**, 285–307 (1936).

Duncan, D. L.: The effects of vagotomy and splanchnotomy on gastric motility in the sheep. J. Physiol. (Lond.) **119**, 157–169 (1953).

Dussardier, M.: Controle nerveux du rythme gastrique des ruminants. J. Physiol. (Paris) **47**, 170–173 (1955).

Edvardsen, P.: Nervous control of urinary bladder in cats. I. The collecting phase. Acta physiol. scand. **72**, 157–171 (1968a).

— Nervous control of urinary bladder in cats. II. The expulsion phase. Acta physiol. scand. **72**, 172–182 (1968b).

Elias, E., Gibson, G. J., Greenwood, Linda F., Hunt, J. N., Tripp, J. H.: Location of an alimentary osmoreceptor. J. Physiol. (Lond.) **180**, 28–29 P (1965).

Evans, J. P.: Observations on the nerves of supply to the bladder and urethra of the cat, with a study of their action potentials. J. Physiol. (Lond.) **86**, 396–414 (1936).

Foley, J. O.: The functional types of nerve fibres and their numbers in the great splanchnic nerve. Anat. Rec. **100**, 766–767 (1948).

Freer, M., Campling, R. C., Balch, C. C.: Factors affecting the voluntary intake of food by cows. 4. The behaviour and reticular motility of cows receiving diets of hay, oat straw and oat straw with urea. Brit. J. Nutr. **16**, 279–295 (1962).

Gammon, G. C., Bronk, D. W.: The discharge of impulses from Pacinian corpuscles in the mesentery and its relation to vascular changes. Amer. J. Physiol. **114**, 77–84 (1935).

Garry, R. C., Roberts, T. D. M., Todd, J. K.: Reflexes involving the external urethral sphincter in the cat. J. Physiol. (Lond.) **149**, 653–665 (1959).

Gasser, H. S., Grundfest, H.: Axon diameters in relation to the spike dimensions and the conduction velocity in mammalian A fibres. Amer. J. Physiol. **127**, 393–414 (1939).

Gernandt, B., Zotterman, Y.: Intestinal pain: an electrophysiological investigation on mesenteric nerves. Acta physiol. scand. **12**, 56–72 (1946).

Glazebrook, A. J., Welbourn, R. B.: Some observations on the function of the small intestine after gastrectomy. Brit. J. Surg. **40**, 111–117 (1952).

Gray, J. A. B.: Initiation of impulses at receptors. In: Handbook of physiology, sect. 1, Neurophysiology. Washington: Amer. Physiol. Society 1959.

— Diamond, J.: Pharmacological properties of sensory receptors and their relation to those of the autonomic nervous system. Brit. med. Bull. **13**, 185–188 (1957).

— Malcolm, J. L.: The initiation of nerve impulses by mesenteric Pacinian corpuscles. Proc. roy. Soc. B **137**, 96–114 (1950).

— Sato, M.: Properties of the receptor potential in Pacinian corpuscles. J. Physiol. (Lond.) **122**, 610–636 (1953).

Groat, W. C. de, Ryall, R. W.: Reflexes to sacral parasympathetic neurones concerned with micturition in the cat. J. Physiol. (Lond.) **200**, 87–108 (1969).

Harding, R., Leek, B. F.: Differentiation between motoneurone and interneurone activity recorded from the medullary gastric centres of sheep. J. Physiol. (Lond.) **209**, 42–43 P (1970).

Hoelzel, F.: Use of non-nutritive materials to satisfy hunger. Amer. J. dig. Dis. **14**, 401–404 (1947).

Hopkins, A.: The pattern of gastric emptying: a new view of old results. J. Physiol. (Lond.) **182**, 144–149 (1966).

Howard, B. R.: The dorsal nucleus of the vagus as a centre controlling gastric motility in sheep. J. Physiol. (Lond.) **206**, 167–180 (1970).

References

Hunt, J. N.: The site of receptors slowing gastric emptying in response to starch in test meals. J. Physiol. (Lond.) **154**, 270–276 (1960).
— Knox, M. T.: A relation between the chain length of fatty acids and the slowing of gastric emptying. J. Physiol. (Lond.) **194**, 327–336 (1968).
— — The slowing of gastric emptying by nine acids. J. Physiol. (Lond.) **201**, 161–179 (1969).
— Pathak, J. D.: The osmotic effects of some simple molecules and ions on gastric emptying. J. Physiol. (Lond.) **154**, 254–269 (1960).
Hurst, A. F.: The sensibility of the alimentary canal. London: Oxford University Press 1911.
Iggo, A.: Spontaneous and reflexly elicited contractions of reticulum and rumen in decerebrate sheep. J. Physiol. (Lond.) **115**, 28–29 P (1951).
— Tension receptors in the stomach and the urinary bladder. J. Physiol. (Lond.) **128**, 593–607 (1955).
— Central nervous control of gastric movements in sheep and goats. J. Physiol. (Lond.) **131**, 248–256 (1956a).
— Afferent fibres from the viscera. XX Int. Physiol. Congr. 458–459 (1956b).
— Gastro-intestinal tension receptors with unmyelinated afferent fibres in the vagus of the cat. Quart. J. exp. Physiol. **42**, 130–143 (1957a).
— Gastric mucosal chemoreceptors with vagal efferent fibres in the cat. Quart. J. exp. Physiol. **42**, 398–409 (1957b).
— The electrophysiological identification of single nerve fibres with particular reference to the slowest conducting vagal afferent fibres in the cat. J. Physiol. (Lond.) **142**, 110–126 (1958).
— Physiology of visceral afferent systems. Acta neuroveg. (Wien) **28**, 121–134 (1966).
— Leek, B. F.: An electrophysiological study of single vagal efferent units associated with gastric movements in sheep. J. Physiol. (Lond.) **191**, 177–204 (1967a).
— — An electrophysiological study of some reticulo-ruminal and abomasal reflexes in sheep. J. Physiol. (Lond.) **193**, 95–119 (1967b).
— — Sensory receptors in the ruminant stomach and their reflex effects. In: Physiology of digestion and metabolism in the ruminant. Newcastle upon Tyne: Oriel Press 1970.
Jansson, G.: Vago-vagal reflex relaxation of the stomach in the cat. Acta physiol. scand. **75**, 245–252 (1969).
Kay, R. N. B., Phillipson, A. T.: Responses of the salivary glands to distension of the oesophagus and rumen. J. Physiol. (Lond.) **148**, 507–523 (1959).
Kosterlitz, H. W., Lees, G. M.: Pharmacological analysis of intrinsic intestinal reflexes. Pharmacol. Rev. **16**, 301–339 (1964).
Kottegoda, S. R.: An analysis of possible nervous mechanisms involved in the peristaltic reflex. J. Physiol. (Lond.) **200**, 687–712 (1969).
— Peristalsis of the small intestine. In: Smooth muscle. London: Edward Arnold 1970.
Leek, B. F.: An electrophysiological analysis of the reflex regulation of reticulo-ruminal movements. Ph. D. Thesis, Edinburgh 1967.
— Reticulo-ruminal mechanoreceptors in sheep. J. Physiol. (Lond.) **202**, 585–609 (1969a).
— Reticulo-ruminal function and dysfunction. Vet. Rec. **84**, 238–243 (1969b).
Leitner, L. M., Perl, E. R.: Receptors supplied by spinal nerves which respond to cardio vascular changes and adrenaline. J. Physiol. (Lond.) **175**, 254–274 (1964).
Loewenstein, W. R., Skalak, R.: Mechanical transmission in a Pacinian corpuscle. An analysis and a theory. J. Physiol. (Lond.) **182**, 346–378 (1966).
Machella, T. E.: The mechanisms of the post-gastrectomy "dumping" syndrome. Ann. Surg. **130**, 145–159 (1949).
Matthews, B. H. C.: Nerve endings in mammalian muscle. J. Physiol. (Lond.) **78**, 1–53 (1933).
Mellanby, J., Pratt, C. L. G.: The reactions of the urinary bladder of the cat under conditions of constant pressure. Proc. roy. Soc. B **127**, 307–322 (1939).
— — The reactions of the urinary bladder of the cat under conditions of constant volume. Proc. roy. Soc. B **128**, 186–201 (1940).
Menguy, R.: Motor function of the alimentary tract. Ann. Rev. Physiol. **26**, 227–248 (1964).
Moir, R. J.: The comparative physiology of ruminant-like animals. In: Physiology of digestion in the ruminant. London: Butterworths 1965.
Moore, R. M.: Some experimental observations relating to visceral pain. Surgery **3**, 534–555 (1938).

Nakayama, S., Mori, T.: Effects of distension of the gall bladder and bile ducts on the movement of the stomach and intestine in the dog. Jap. J. Physiol. **17**, 458–465 (1967).

Nasset, E. S.: The nutritional significance of endogenous nitrogen secretion in non-ruminants. In: The role of gastrointestinal tract in protein metabolism. Oxford: Blackwell 1964.

Newman, P. P., Paul, D. H.: The representation of some visceral afferents in the anterior lobe of the cerebellum. J. Physiol. (Lond.) **182**, 195–208 (1966).

Niijima, A.: Afferent impulses in the vagal and splanchnic nerves of toads' stomach and their rôle in sensory mechanism. Jap. J. Physiol. **12**, 25–44 (1962).

Paintal, A. S.: A method of locating the receptors of visceral afferent fibres. J. Physiol. (Lond.) **124**, 166–172 (1954a).

— A study of gastric stretch receptors. Their role in the peripheral mechanism of hunger and thirst. J. Physiol. (Lond.) **126**, 255–270 (1954b).

— Responses from mucosal mechanoreceptors in the small intestine of the cat. J. Physiol. (Lond.) **139**, 353–368 (1957).

— Vagal afferent fibres. Ergebn. Physiol. **52**, 75–156 (1963).

— Effect of drugs on vertebrate mechanoreceptors. Pharmacol. Rev. **16**, 341–380 (1964).

Phillipson, A. T.: The movements of the pouches of the stomach of sheep. Quart. J. exp. Physiol. **29**, 395–415 (1939).

Potter, G. D., Guzman, F., Lim, R. K. S.: Visceral pain evoked by intra-arterial injection of substance P. Nature (Lond.) **193**, 983–984 (1962).

Ranson, S. W.: Afferent paths for visceral reflexes. Physiol. Rev. **1**, 477–522 (1921).

Reid, C. S. W., Titchen, D. A.: Reflex stimulation of movements of the rumen in decerebrate sheep. J. Physiol. (Lond.) **181**, 432–448 (1965).

Schalk, A. F., Amadon, R. S.: The physiology of the ruminant stomach, study of the dynamic factors. N. Dak. agric coll. Exp. Stn. Bull. **216**, 1–64 (1928).

Sellers, A. F., Stevens, C. E.: Motor functions of the ruminant forestomach. Physiol. Rev. **46**, 634–661 (1966).

Sharma, K. N.: Receptor mechanisms in the alimentary tract: their excitation and functions. In: Handbook of physiology, sect. 6, Alimentary canal. Washington: Amer. Physiol. Soc. 1967.

— Nasset, E. S.: Electrical activity in mesenteric nerves after perfusion of gut lumen. Amer. J. Physiol. **202**, 725–730 (1962).

Sherrington, C. S.: The integrative action of the nervous system. New Haven: Yale University Press 1906.

Sircus, W.: Studies on the mechanisms in the duodenum inhibiting gastric secretion. Quart. J. exp. Physiol. **43**, 114–133 (1958).

Steven, D. H., Marshall, A. B.: Organization of rumen epithelium. In: Physiology of digestion and metabolism in the ruminant. Newcastle: Oriel Press 1970.

Stevens, C. E., Sellers, A. F.: Studies of the reflex control of the ruminant stomach with special reference to the eructation reflex. Amer. J. vet. Res. **20**, 461–482 (1959).

Talaat, M.: Afferent impulses in the nerves supplying the urinary bladder. J. Physiol. (Lond.) **89**, 1–13 (1937).

Thomas, J. E.: Mechanics and regulation of gastric emptying. Physiol. Rev. **37**, 453–474 (1957).

Titchen, D. A.: Reflex stimulation and inhibition of reticulum contractions in the ruminant stomach. J. Physiol. (Lond.) **141**, 1–21 (1958).

Todd, J. K.: Afferent impulses in the pudendal nerves of the cat. Quart. J. exp. Physiol. **49**, 258–267 (1964).

Tower, S. S.: Action potentials in sympathetic nerves elicited by stimulation of frog's viscera. J. Physiol. (Lond.) **78**, 225–245 (1933).

Weiss, K. E.: Physiological studies on eructation in ruminants. Onderstepoort J. vet. Res. **26**, 251–283 (1953).

White, J. E., Sweet, W. H.: Pain and the neurosurgeon. Springfield, Illinois: C. C. Thomas 1969.

Whitteridge, D.: The effects of distension of viscera. In: Lectures on the Scientific Basis of Medicine, vol. 4, p. 305–310. University of London: The Athlone Press 1956.

Wolf, S., Wolff, H. G.: Human gastric function: an experimental study of a man and his stomach. New York-London: Oxford University Press 1947.

Chapter 5

Central Thermoreceptors and Thermoregulation

By

Richard F. Hellon

With 11 Figures

Contents

1 Discovery and Localization in the Hypothalamus	162
2 Responses to Heating and Cooling the Hypothalamus	163
3 Responses to Heating and Cooling the Spinal Cord	165
4 Natural Variations in Temperature in the CNS	167
5 Quantitative Responses and Models of Temperature Regulation	167
6 Electrical Activity in Hypothalamus and Spinal Cord	170
(a) Preoptic and Anterior Hypothalamic Recordings	171
(b) Midbrain Recordings	175
(c) Peripheral Inputs to Hypothalamus and Midbrain	176
(d) Spinal Cord Recordings	179
(e) Transmitter Substances and Temperature Sensitivity	180
(f) Fever	181
7 Summary and Conclusions	182
References	182

There are two groups of thermal receptors whose individual characteristics are beginning to be understood: those on various parts of the body surface and those which are being found in the central nervous system. Although our prime concern is with the second group, it will be necessary in discussing the role of these central thermoreceptors in the regulation of body temperature to refer also to the surface receptors. The latter are considered in detail elsewhere in this Handbook.

This chapter will deal first with the discovery and properties of the temperature receptors which have been found in the brain stem and spinal cord. Then we will consider how these receptors are currently thought to be concerned in the regulation of body temperature. Finally the most common disturbance of this regulation will be discussed: the fever of infectious diseases.

It is not intended to survey the whole subject of thermoregulation, but rather to concentrate on investigations where direct involvement of central thermoreceptors has been demonstrated. More comprehensive reviews will be found in the publications of Euler (1961), Hardy (1961), Bligh (1966), Hammel (1968), Benzinger (1969), and Hardy et al. (1970). The discussion will be confined to results obtained from mammals; most human experiments have been omitted, as have those involving lesion in the CNS of animals. The fascinating question of the evolution of temperature regulation has also been left out, although there are indications that species such as lizards, turtles and some fish possess thermosensitivity in the region of the third ventricle.

1. Discovery and Localization in the Hypothalamus

The concept that the brain exerts a controlling influence on body temperature dates from the second half of the 19th century. Various investigators had found that damage to certain parts of the brain, particularly the corpus striatum, resulted in an increase in heat production and a persistent rise in body temperature. However this evidence was difficult to interpret and indeed a contemporary textbook doubted whether these "heat centres" had any physiological function (PEMBREY, 1898). It was in the early years of the present century that the first clear demonstrations of the thermosensitivity of the brain were made. KAHN (1904) and MOOREHOUSE (1911) raised the temperature of the carotid blood in rabbits and dogs and obtained vasomotor and respiratory responses which they attributed to warming of the brain. The first direct thermal stimulation of the cranial contents seems to have been made by BARBOUR (1912). He applied strong warming or cooling to the "corpus striatum" of rabbits and produced changes in body temperature: warming lowered the body temperature and cooling raised it. These results were confirmed in conscious cats by PRINCE and HAHN (1918a), who also showed that vasoconstriction and vasodilatation could be induced in the skin of rabbit legs by respectively cooling or warming of the brain (PRINCE and HAHN, 1918b). The sensitive region was more precisely localized by HASAMA (1929) who discovered that the "regio subthalamica" was concerned.

When stereotaxic techniques were applied to this problem by RANSON and his colleagues a much more exact delineation was possible, especially with the added refinement of localized high-frequency heating between pairs of bilateral electrodes. They were able to show in cats that the reactive region from which polypnoea and panting could be elicited by heating was within a few millimetres of the midline and lay between the crossing of the anterior commissure and the optic chiasma (MAGOUN et al., 1938). The region is indicated in Fig. 1. Other parts of the brain did not produce these responses and the authors concluded: "To us the obvious interpretation of these results is that in these experiments we have been setting in play by artificial heating the same mechanisms in the same regions of the brain which are activated in the normal animal when the temperature of the blood rises above normal." The validity of this conclusion however is still controversial and will be discussed later (see Section 4). These results have been repeatably confirmed in a variety of unanaesthetized animals and will also be considered in a later section.

The first investigator to systematically cool the hypothalamus was STRÖM (1950a, b), although the earlier work already mentioned and that of SHERRINGTON (1924) indicated that shivering and vasoconstriction could be produced by gross cooling of the brain. In anaesthetized cats or conscious dogs STRÖM was able to elicit vasoconstriction in skin blood vessels when these were already in a dilated state, but he found no evidence of shivering. More definite evidence was produced by KUNDT et al. (1957) and HENSEL and KRÜGER (1958) who used a chronically

implanted thermode in the hypothalamus of conscious cats. They found that vasoconstriction in the ear vessels, shivering and a rise in rectal temperature could all
be generated by cooling of about 1°C. Placement of the thermode 5 mm away
from the hypothalamus abolished these effects.

Thus by the end of the 1950's it was firmly established that both warming
and cooling of just the anterior hypothalamus by 1 or 2°C could mobilize effector
mechanisms for heat loss or heat conservation, respectively. Later investigations,
which will now be considered, have been concerned with extending these results

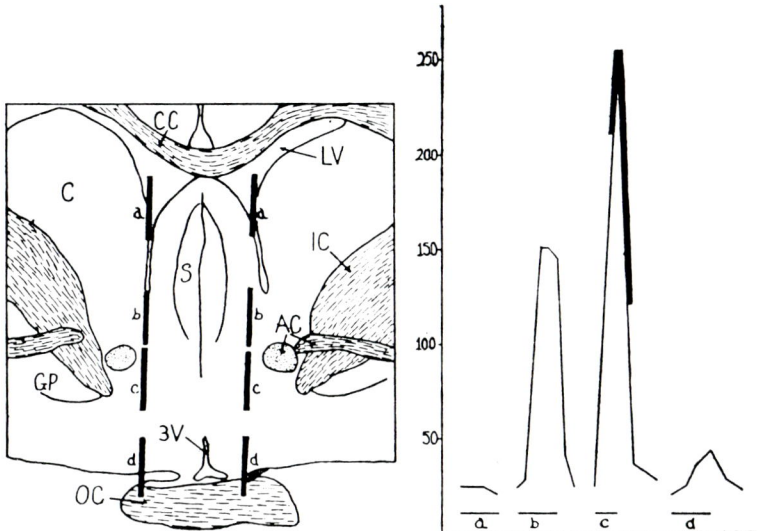

Fig. 1. Left: transverse section of cat brain in the preoptic area showing the four positions
(a, b, c, d) of a pair of electrodes used for r–f heating. Right: respiratory rates (per min)
obtained during heating in the four positions; panting is shown by the heavy line; the duration
of period "a" is 5 min, other times are in proportion. (From MAGOUN et al., 1938)

to other species; with quantifying the relationship between hypothalamic temperature and the animal's responses; and with the interaction of hypothalamic
temperature and external temperature.

2. Responses to Heating and Cooling the Hypothalamus

Three methods have been used to raise or lower the local temperature in the
hypothalamus: (a) chronically-implanted thermode tubes, either in the midline
or surrounding the area, through which water is passed; (b) highfrequency heating with implanted electrodes; (c) heating or cooling the carotid blood supply.
All three have their limitations. The first two methods inevitably give rise to
complex thermal gradients in the brain (FUSCO et al., 1961; CUNNINGHAM et al.,
1967) which make it difficult to define exactly how much local temperature is
being displaced. The third method is only effective in species with a prominent
internal carotid supply such as rabbits and primates.

In this section discussion will be restricted to experiments made on unanaesthetized animals.

In experiments on cats, KRÜGER et al. (1959) caused some vasoconstriction with a lowering of hypothalamic temperature by 1.5°C, but an reduction of 4°C was needed to diminish cutaneous blood flow by 50% and induce shivering. FREEMAN and DAVIS (1959) used heating and cooling and found that responses could only be obtained from the anterior hypothalamus—the posterior region was not effective. Anterior hypothalamic heating promoted vasodilatation of the footpads and a stretched-out posture but no panting; cooling gave rise to vasoconstriction and a huddled posture, but no shivering. However shivering caused by external cooling could be suppressed by heating the anterior hypothalamus.

Comparable experiments on cattle have demonstrated that a rise in skin temperature, respiratory frequency and in the rate of moisture loss from the skin can all be brought about when hypothalamic temperature is raised from 38.5°C to 40–41°C (INGRAM and WHITTOW, 1962; INGRAM et al., 1963).

Extensive work by ANDERSSON and his collaborators has been done on goats. Essentially they have given similar results to those cited for other species (ANDERSSON and LARSSON, 1961; ANDERSEN et al., 1962). In addition they have reported that hormonal responses from the thyroid and suprarenal glands can be evoked. Hypothalamic cooling caused an increase in urinary excretion of catecholamines (ANDERSSON et al., 1963a) and in plasma levels of protein-bound iodine (ANDERSSON et al., 1963b, 1965).

Similar changes in catecholamine excretion have been found in pigs by BALDWIN et al. (1969). These authors (BALDWIN and INGRAM, 1968) have also shown in pigs that local heating of the hypothalamus can increase respiratory frequency, but only in a warm environment. Cooling could reduce panting in the heat or increase oxygen consumption in a cold environment.

DOWNEY et al. (1964) cooled the internal carotid blood of rabbits while measuring oxygen consumption. They found a linear relation between lowering of brain temperature and increase in oxygen consumption which indicated that changes of less than 0.5°C in brain temperature would be enough to increase metabolism. In heating experiments on rabbits EULER (1964) raised hypothalamic temperature with an array of electrodes for high-frequency heating. An increase of 0.2, 0.3, or 0.5°C caused a vasodilatation of ear vessels and increase in respiratory rate so that body temperature fell. The extent of this fall was dependent on the elevation in hypothalamic temperature.

Pigs, rats and primates have been used in behavioural experiments where the animal had been trained to control its environment. For example it can switch on bursts of infra-red heating in a cold environment or choose the length of time its surroundings remain at high or low temperature. When hypothalamic temperature is raised or lowered then these species will "work" respectively for less or more external heat (SATINOFF, 1964; CARLISLE, 1966; MURGATROYD, 1966; BALDWIN and INGRAM, 1967; CORBIT, 1969; ADAIR, 1970; GALE et al., 1970).

Some of the most significant work in this field has been done on dogs by HARDY and HAMMEL. Its importance stems from their use of a calorimeter to relate quantitatively the animal's responses and the displacement of hypothalamic temperature. Besides showing that a dog's thermoregulatory mechanisms can be manipulated in a similar manner to that just described for other species (HAMMEL et al., 1960), they provided some clear evidence that there was an interaction between responses to hypothalamic warming and the prevailing ambient temperature (FUSCO et al., 1961). In cool (14°C), neutral (26°C), and warm (29°C) environments, the effect of raising hypothalamic temperature was to increase heat loss and drive the body temperature down, but the mechanisms which were brought into play, depended on the environment. At 14°C the dog responded by stopping shivering and reducing heat production. At 26°C heat production was also depressed, while at 29° vigorous panting elicited. When the "thermal clamp" was removed from the hypothalamus, its temperature fell to the lowered level of the rest of the body and shivering resulted. At this time thermal gradients were not present in the hypothalamus and so it was possible to define its temperature with some precision. In all environments the hypothalamus was 1°C below its usual level, but in the neutral environment the heat production increased to 3.0–3.5 kcal/kg/hr, in the cool environment to 4.5–5.0 kcal/kg/hr. These varying responses to hypothalamic heating and the subsequent hypothermia, indicate that there is probably an interaction between temperature receptors in the skin and the neurons which are being stimulated in the brain stem. Further consideration of this interaction and the model derived from it will be given in a later section.

All the evidence discussed so far allows certain conclusions to be drawn: (i) that there are neurons in the brain which can respond to a temperature increase or decrease of 1°C or less; (ii) that these neurons are only present in the anterior region of the hypothalamus; (iii) that these neurons can activate both physiological and behavioural mechanisms for temperature regulation; (iv) that there is an interaction between these neurons and information coming from peripheral temperature receptors in the skin.

3. Responses to Heating and Cooling the Spinal Cord

Some authors (e.g. THAUER, 1935) have long maintained that a spinal animal is capable of a degree of thermo-regulation. In the last few years it has become evident that many of the thermoregulatory responses which can be generated by manipulating hypothalamic temperature can also be produced by warming or cooling the spinal cord. The species which have been used include dogs, rabbits, oxen, guinea-pigs and pigeons. The early work in this field was done under anaesthesia, but as in the discussion of hypothalamic responses, only the later experiments performed without anaesthetics will be considered here.

Warming the cord has been found to activate heat loss mechanisms and suppress those for heat production. Skin vasodilatation, panting, salivation, and the inhibition of shivering have been observed in dogs (JESSEN et al., 1967; JESSEN, 1967), rabbits (IRIKI, 1968), oxen (HALES and JESSEN, 1969), guinea-pigs (BRÜCK and WÜNNENBERG, 1966), and pigeons (RAUTENBERG, 1969). Conversely reducing cord temperature can bring about shivering in dogs (JESSEN et al., 1968) and pigeons (RAUTENBERG, 1969). Some results showing the responses which occur on heating the spinal cord are presented in Fig. 2. The data were obtained from a conscious dog with an implanted thermode over the dorsal surface of its cord. The heating period lasted twenty minutes and caused a dramatic increase in

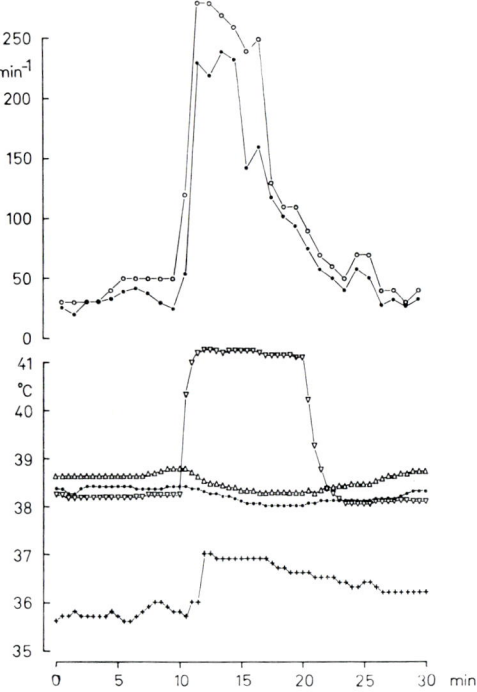

Fig. 2. Respiratory and vasomotor effects in a conscious dog during warming of the spinal cord. The curves from above downwards show: respiratory rate, average (●), and maximum (○); hypothalamic temperature (△); rectal temperature (●); peridural temperature over the cord (▽); paw surface temperature (+). (From Jessen, 1967)

respiratory rate and vasodilatation in the paw. However the high respiration did not persist for the whole duration of heating, possibly because hypothalamic, and rectal temperatures were falling.

There are further similarities between responses from the hypothalamus and the spinal cord. As with the hypothalamus, ambient temperature interacts with cord responses (Rautenberg, 1969; Brück and Wünnenberg, 1967) and unrestrained animals show postural changes which are conductive to greater or lesser heat loss. Finally Jessen (1970) has reported that warming both the hypothalamus and the spinal cord has an additive effective on a dog's respiratory heat loss compared with warming the two sites separately; there is a similar additive effect on oxygen consumption when cord and hypothalamus are cooled together (Jessen et al., 1968).

As in the hypothalamus, thermal gradients must be created when an elongated tissue like the spinal cord is heated or cooled with a U-shaped thermode placed extra-durally on the dorsal surface. These gradients will arise across the thickness of the cord as well as along its length. The neurons which are responding to temperature seem to lie in the ventro-lateral region of the cord (Wünnenberg and Brück, 1968; Simon and Iriki, 1970), but direct measurement of temperature in this region during thermode heating or cooling does not seem to have been made.

4. Natural Variations in Temperature within the Central Nervous System

We have seen that the artificial thermal stimulation of the anterior hypothalamus or the spinal cord by probably 0.5°C or less can activate behavioural and physiological thermoregulatory mechanisms. The evidence clearly points to the presence of specialized thermodetector neurons in these parts of the CNS and, as will be discussed in the following section, a quantitative relationship can be demonstrated between the stimulation of these neurons and the animal's responses measured calorimetrically. At the moment it is pertinent to ask what natural changes have been found in CNS temperature which might be related to responses governing body temperature.

Variations in hypothalamic temperature of up to 1°C have been observed in cats (SEROTA, 1939; FORSTER and FERGUSON, 1952; KUNDT et al., 1957; ADAMS, 1963), dogs (RAWSON et al., 1965), monkeys (RAWSON and HAMMEL, 1963; HAMILTON, 1964), rats (ABRAMS and HAMMEL, 1964, 1965), and pigs (BALDWIN and INGRAM, 1968). These variations were not clearly related to any thermoregulatory activity and yet, as we have seen, an artificial displacement of brain temperature by this amount is sufficient to evoke such activity. HAYWARD and BAKER (1969) emphasize that although arterial temperature is a major factor in determining brain temperature, in some species who have a carotid rete (e.g. cat and sheep) cerebral arterial temperature can be reduced by venous blood bathing the rete.

There do not seem to have been any measurements of natural variations in spinal cord temperature and it is clearly important that this should be done.

The important point is made by FORSTER and FERGUSON (1952) and HAMMEL et al. (1963) that responses to ambient cooling or heating such as shivering or panting could occur without a demonstrable change in hypothalamic temperature. To quote BLIGH (1966): "This contrast between the effectiveness of quite small local hypothalamic temperature changes in inducing thermoregulatory responses under experimental conditions and the ineffectiveness of considerably greater changes under conditions comparable to these of normal life, is in urgent need of investigation." One explanation for this seeming paradox may lie in the activation of other thermoreceptors, particularly those in the skin, and this point will now be considered.

5. Quantitative Responses and Models of Temperature Regulation

The regulation of body temperature from a physical point of view consists of a balancing of the factors concerned with heat loss against the factors concerned with heat production. However, such a system could come into equilibrium at a variety of internal temperatures depending on the circumstances at the time. Since in homeotherms this does not seem to be the case, it follows that there is a built-in mechanism or controller which fixes the deep body temperature close to a particular level. This level varies from one species to another and is presumably genetically determined. The site of this regulated temperature is still a matter for conjecture, but much of the evidence points to the preoptic and anterior hypothalamic region which is also the site for the temperature controller itself (HAMMEL, 1968).

Several recent advances in our understanding of temperature regulation have come by applying the concepts and theories of control engineering. Thus in an object producing and losing heat continuously there would be a central sensing element and a "reference" or "set-point" temperature towards which the system would tend to drive. The control of the various avenues of heat loss and production would be some function of the difference between the set-point temperature and the actual temperature of the sensing element. This difference is known as the "load error". Such an arrangement would result in temperature fluctuations around the set-point which would depend on the sensitivity of the control system and the thermal inertia of the object. As discussed in Section 4 it seems unlikely that this normally happens in experimental animals. These fluctuations could be reduced by the action of surface receptors which would immediately activate responses before central temperature had begun to deviate. However these surface receptors could not by themselves constitute the input to the regulating system, if only because they would be affected by the very responses which they initiate, such as vasodilatation or sweating. At the simplest level then, a regulating mechanism might be based on advance warning by surface receptors of an environmental change coupled with central control exercised by the hypothalamus and probably the spinal cord.

A series of experiments by HAMMEL and his colleagues, mainly on conscious dogs, have been concerned with the quantitative analysis of the responses evoked by warming and cooling the hypothalamus and by the modification of these responses by the ambient temperature (HAMMEL et al., 1963a, 1963b; HELLSTRØM and HAMMEL, 1967). They first showed that the evaporative heat loss by panting or the heat production of shivering were linearly related to the raising or lowering of hypothalamic temperature. The type of result obtained is shown in Fig. 3 from which it will be seen that there are threshold or set-point temperatures for each response which must be exceeded before the dog begins to respond. A general equation to describe these curves takes the form:

$$R - R_0 = \alpha_R (T_{HY} - T_{SET_R}). \quad \text{where} \quad R - R_0 \geqslant 0. \tag{1}$$

Here R is the thermoregulatory response, R_0 is the basal level of that response, α_R is the proportionality constant for the response, T_{HY} is the actual hypothalamic temperature and T_{SET_R} is the threshold or set-point temperature for the response. The difference term in brackets represents the load error or actuating signal. By performing the experiments at different environmental temperatures the values of T_{SET_R} but not of α_R were found to be altered. As shown in Fig. 3, cooling the exterior shifted the set-points for heat production and evaporation to the right, whereas a warm environment caused both set-points to move to the left. The implication was that information from skin thermoreceptors was changing the magnitude of T_{SET_R}, so providing a new set-point. Such a mechanism could explain how a dog can pant in a hot environment without any rise in T_{HY}. This shifting set-point hypothesis has been discussed in detail by HAMMEL (1965, 1968). He has proposed that the action of many other factors besides skin temperature can be explained as if they were acting by altering set points. These include the level of core or extra-hypothalamic temperature, the state of arousal, exercise, sleep and fever. However, except in experiments in which hypothalamic temperature has been systematically displaced and the subjects' responses followed, this hypothesis is difficult to test experimentally. The reason for this is that T_{SET} can only be an idealized temperature which in reality would be represented by the summed activity of an untold number of hypothalamic and other neurons. Nevertheless it remains one of the most useful and versatile of present models and, as will be described below, neurophysiological evidence in its favour is beginning to appear.

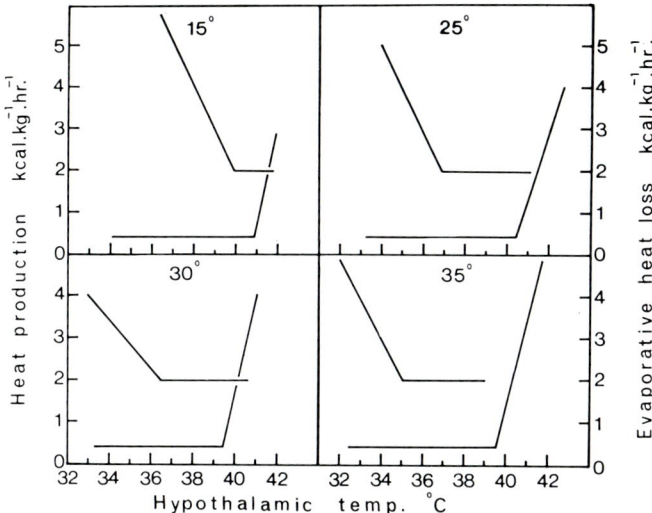

Fig. 3. Heat production and evaporative heat loss from a conscious dog during manipulation of hypothalamic temperature at air temperatures of 15, 25, 30, and 35°C. Curves with negative slopes show heat production; curves with positive slopes show evaporative heat loss. (Adapted from HELLSTRØM and HAMMEL, 1967)

Recent preliminary evidence from JESSEN (1970) shows that warming the spinal cord in conscious dogs gives curves very similar to those in Fig. 3. When the hypothalamus of the same dog was warmed the slopes (α_R) of the two curves were similar, indicating that the cord and hypothalamic temperatures are equally powerful stimulators of the panting response. The simultaneous warming of both regions shifted the sloping part of the curve to the left in the same way as environmental warming had done in HAMMEL's experiments. These results and those mentioned earlier (Section 3) emphasize the importance of the spinal cord as a site for central thermoreceptors which will have to be included in any future model of temperature regulation.

A different type of model has been proposed by STOLWIJK and HARDY (1966) which is based on a mathematical analysis of human metabolic, vasomotor, and sweating responses. The general equation for each response takes the form:

$$R - R_0 = \alpha_R \, (T_{HY} - T_{SET_{HY}}) \, (T_S - T_{SET_S}) \qquad (2)$$

Symbols are as in Eq. (1), except T_S representing average skin temperature. There are set-points for both hypothalamus and average skin temperatures; these are fixed and the same for all thermoregulatory responses. Clearly this model would function quite differently from HAMMEL's. For instance, the curves shown in Fig. 3 would have variable slopes at different skin temperatures because of the product nature of Eq. (2). However STOLWIJK and HARDY obtained good agreement when comparing computed solutions of the model equations and actual data taken from human experiments.

Another proposal which also uses a product type of equation has been put forward by BRÜCK and WÜNNENBERG (1970) and is based on observations made on guinea pigs. In the cold these animals generate heat by conventional muscular shivering if they are adult and warm-adapted, or by non-shivering thermogenesis in interscapular brown fat if they are adult and cold-adapted or new-born. The heat produced by the brown fat is carried directly by venous channels to warm the underlying cervical cord (SMITH and ROBERTS, 1964) which in guinea pigs has been shown to possess thermoreceptive properties (BRÜCK and WÜNNENBERG, 1966). BRÜCK and WÜNNENBERG propose that shivering is controlled by the product of the deviations of cervical cord and skin temperatures from their respective set-points, while non-shivering thermogenesis is driven by a corresponding product derived from hypothalamic and skin temperatures. It will be interesting to test this model with calorimetric measurements of heat production and also for other thermoregulatory responses.

6. Electrical Activity in Hypothalamus and Spinal Cord

It has been necessary to describe in some detail the responses evoked by stimulation of the central thermoreceptors and the theories and models which have thus been advanced, in order to make a more meaningful discussion of the properties of individual neurons in thermoregulation.

The first attempt to correlate local temperature in the brain stem with electrical recording was made by EULER (1950). He recorded D.C. potentials in the antero-medial hypothalamus of cats while brain temperature was changed by heating or cooling carotid blood. Within this region, but not outside it, potentials changes were highly correlated and synchronous with the imposed temperature changes. In some recordings the two were positively correlated and in others the correlation was negative. A high temperature sensitivity of up to 10 mV/°C was found. These potentials seem unlikely to be the summed responses of individual action potentials and their origin remains uncertain.

Over the past decade, about twenty papers have appeared which report various correlations between the rate of firing of single neurons in the brain stem or spinal cord and temperatures within and without the body. These experiments have been done on various mammals (cats, rabbits, dogs, and rats) and a lizard (CABANAC et al., 1967).

Before considering these papers in detail it would be appropriate to consider the various limitations of this method. Most investigations have been done under anaesthesia, but as we shall see this may not be too serious a drawback, provided the barbiturates are avoided. Extracellular microelectrodes will tend to record from neurons with larger somata and thus a biassed sample may be obtained. Another type of bias may arise from the fact that many neurons in the hypothalamus have a slow spontaneous firing rate and so may be missed or mechanically stimulated during the descent of a microelectrode (CROSS and GREEN, 1959). At the moment there is no sure way of deciding whether the particular neuron being studied is actually responding to temperature or whether it is being synaptically excited from a true detector cell, although EISENMAN and JACKSON (1967) (see p. 174) have suggested one way of making this distinction. The anatomical pathways to and from the hypothalamus are as yet poorly defined, so there can be no certainty about the position of an observed neuron in the neural circuitry. As HAMMEL (1968) has emphasized: "There exists the enormous handicap of exploring in a region of the CNS where neurons subserving many kinds of regulatory processes comingle. Even more troublesome is the comingling of neurons subserving several kinds of thermoregulatory activity, rendering uncertain the position of the neuron under study in the chain of neurons activating the response."

Nevertheless, despite these serious drawbacks, the technique is increasing our understanding of temperature regulation, albeit with faltering steps and tentative answers. Since there is yet no way of measuring the integrated output from a neural controller, unit recording offers the only present method for analysing the detailed processes in the brain stem and spinal cord.

(a) Preoptic and Anterior Hypothalamic Recordings

The first report of single unit activity in the hypothalamus appeared in 1961 (NAKAYAMA et al.) to be followed by a full paper in 1963 (NAKAYAMA et al.). Using cats under urethane anaesthesia, they made microelectrode explorations in the preoptic and anterior hypothalamic areas up to 2 mm from the midline, and heated or cooled this region with high frequency current or water-perfused thermodes. They found that the firing rate of 80% of neurons was not affected by local temperature in the range of 32–42°C. Some of the results are shown in Fig. 4 which shows how remarkably stable these cells were over this wide temperature span. Similar findings have been made by others (HARDY et al., 1964; EISENMAN and JACKSON, 1967; WIT and WANG, 1968a). NAKAYAMA et al. found that the remaining 20% of neurons showed varying degrees of temperature-sensitivity, their activity increasing linearly with rising local temperature. The results from some of these warm-sensitive cells are given in Fig. 5 and their sensitivity measured on a Q_{10} basis was between 5 and 15. None of the cells encountered showed "cold-sensitivity", that is had their firing rate increased by lowering local temperature.

The presence of these warm-sensitive neurons seems very likely to account for the responses to thermal stimulation which have already been described in Section 2.

It is possible that the insensitive cells (Fig. 4) could provide set-point information and that the output to the effector mechanisms depends on the difference between the firing rates of these insensitive cells and those which do respond to local temperature. With present techniques however it is not yet possible to test this suggestion.

The finding of cold-sensitive neurons in the anterior hypothalamus was first described in dogs under chloralose/urethane anaesthesia by HARDY et al. (1964).

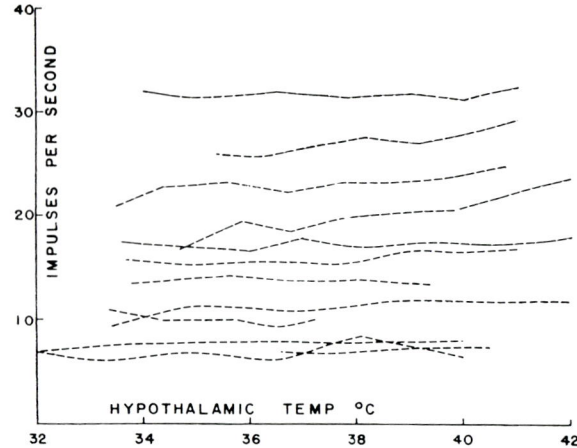

Fig. 4. Firing rate of twelve temperature-insensitive neurons in the preoptic-anterior hypothalamic region of cats under urethane. (From NAKAYAMA et al., 1963).

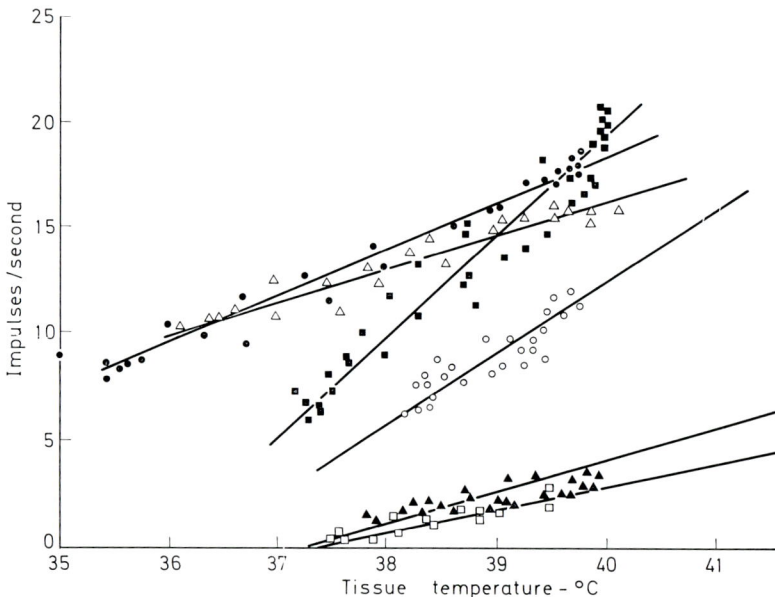

Fig. 5. Firing rate of six warm-sensitive neurons in cats under urethane. (From HARDY, 1969a)

Fig. 6 is taken from their paper and shows how the activity of these cells increased as temperature was reduced by a few degrees. In these experiments warm-sensitive and insensitive units were also found. The proportions of insensitive, warm-sensitive and cold-sensitive neurons were 60, 30, and 10% respectively, and the three types seemed to be randomly distributed in the area explored.

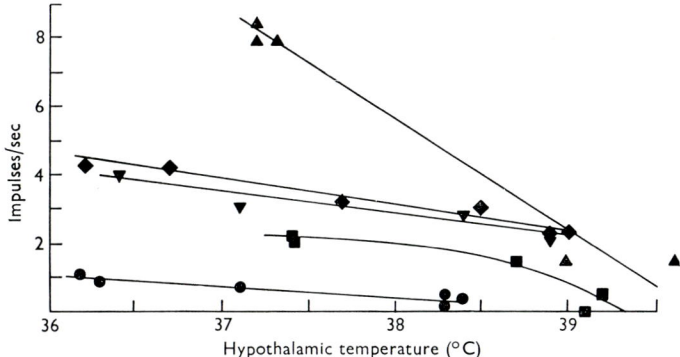

Fig. 6. Firing rate of five cold-sensitive neurons in dogs under chloralose/urethane. (From HARDY et al., 1964)

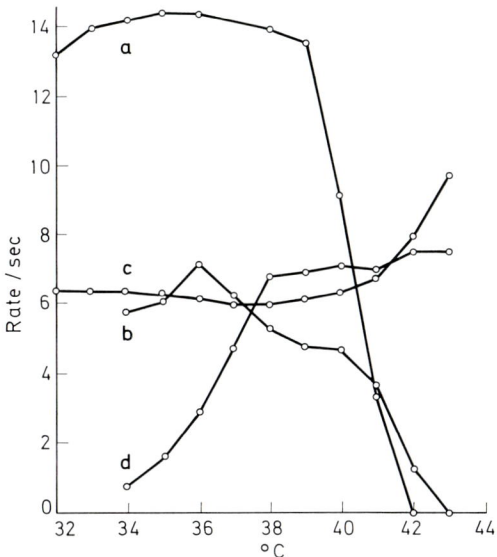

Fig. 7. Firing rates of warm- and cold-sensitive neurons showing non-linear responses; cats under urethane. (From EISENMAN and JACKSON, 1967, courtesy of Academic Press)

To what extent this cold-sensitivity is due to particular membrane properties, to a cascade neural network (HAMMEL, 1968), or to inhibition by warm-sensitive cells cannot be stated yet. Indeed the same doubts exist about the warm-sensitive neurons. All that can be done is to draw an analogy with the end organs which have these specialized sensitivities and which are well documented for the skin and tongue (e.g. HENSEL, 1968).

All the temperature-sensitive neurons which have been described so far show a linear relation between firing rate and local temperature over a range of 8 or

10°C. Further explorations by EISENMAN and JACKSON (1967) in cats under urethane and HELLON (1967) in conscious rabbits showed the existence of neurons with non-linear characteristics. Over part of the temperature range these units were thermally insensitive and only above or below a certain threshold level did they show a response which was either warm-sensitive or cold-sensitive. Fig. 7 illustrates the responses of some of these neurons which are more commonly found than those with the linear curves. It might be predicted that there would be four types of the non-linear curves: those which were unaffected by cooling but excited or inhibited by warming and those which were unaffected by warming but excited or inhibited by cooling. In fact all four types have been reported in the papers mentioned and in the later ones of WIT and WANG (1968a) and NAKAYAMA and HARDY (1969). There are two points which should be made about these cells which have a threshold temperature. First, the position of the threshold is in many cases very close to the normal brain temperature for the species. For example the eight cells recorded in conscious rabbits (HELLON, 1967) all had threshold temperatures in the range 38.5–39.5°C and the usual rabbit brain temperature is close to 39°C. Thus a small change in the temperature of the hypothalamus will excite or inhibit these neurons which had previously been firing at a steady rate. Secondly the non-linear curves in Fig. 7 bear a strong resemblance to the plots in Fig. 3 of the effector responses to hypothalamic heating and cooling. It therefore is quite possible that these neurons play an important part in generating these thermal responses to local temperature.

EISENMAN and JACKSON (1967) have proposed that the cells showing linear responses with positive slopes are the actual thermodectors while those giving non-linear responses with positive or negative slopes are inter-neurons synaptically connected to the thermodetectors. Their chief evidence for this is that the behaviour of the linear cells was largely unaffected by the short-acting barbiturate, methohexital, when this was given to an unanaesthetized decerebrate preparation. In contrast the non-linear neurons were markedly depressed. This suggests that the thermosensitivity of the linear units may depend on some special characteristic of their membrane and that the anaesthetic acts on a synaptic link between these neurons and the non-linear ones.

This raises the question of how far those unit studies which have been made under anaesthesia may be extrapolated to the normal animal. EISENMAN and JACKSON compared the types of units found in cats under urethane with those in isolated forebrain preparations without anaesthetic and found the same types of response in both situations. MURAKAMI et al. (1967) systematically studied the action of several anaesthetics and neuromuscular blocking drugs on the temperature-sensitive neurons in encéphale isolé preparations of dogs. In general, pentobarbitone, ether, chloralose/urethane, gallamine, and decamethonium reduced the firing rate and usually the temperature sensitivity of the cells. However there were some which were unaffected by chloralose/urethane. Urethane alone was not tested. The only experiments on normal animals seem to be those of the writer (HELLON, 1967) using rabbits and these give results essentially similar to those from animals under urethane. It seems that urethane is to be preferred if an anaesthetic must be used, a conclusion which is supported by the recordings of CROSS and DYER (1970) from the posterior hypothalamus.

(b) Midbrain Recordings

Efforts to find temperature-sensitive neurons in other parts of the brain stem have usually failed (NAKAYAMA et al., 1963; ANAND et al., 1966) although there is evidence that strong heating of the medulla can evoke increased respiration and cardiovascular effects (HOLMES et al., 1960; CHAI et al., 1965). However, in several papers where anterior hypothalamic units have been studied, there are indications that the units are not responding to temperature changes in their own locality but in a more caudal region where the imposed fluctuations in anterior hypothalamic temperature will be delayed and attenuated. As a result the curves of

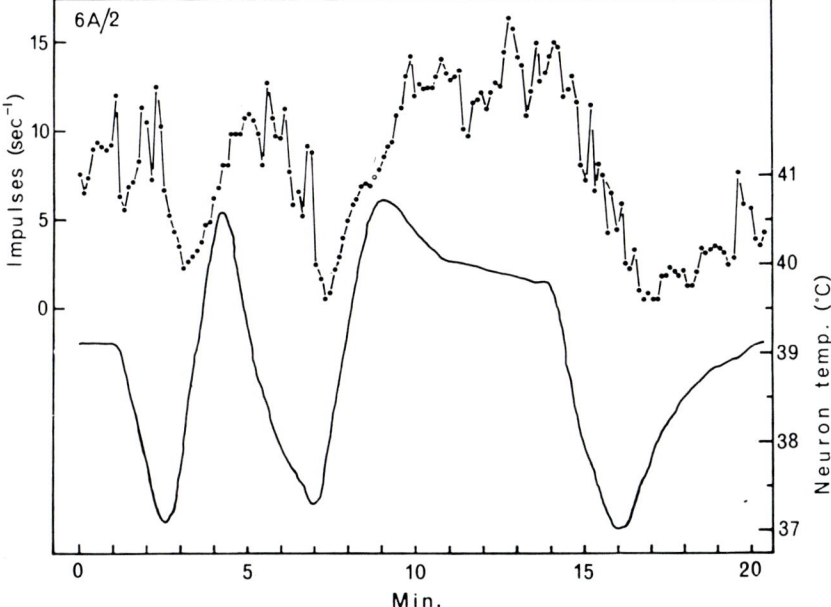

Fig. 8. Firing rate of a warm-sensitive neuron in a rabbit (urethane) showing how its responses lag behind the temperature changes measured at its own location. (HELLON, unpublished data)

firing rate and temperature at the neuron position are out of phase. An example of this sort of recording is given in Fig. 8 which shows how the responses of a unit lag behind the temperature swings at its own location in the anterior hypothalamus by about 30 sec. The implication is that this cell is a secondary neuron which is not itself temperature-sensitive and that the true temperature detector probably lies more caudally in the midbrain. Two studies have been made to specifically investigate the temperature properties of cells in the midbrain (CABANAC and HARDY, 1969; NAKAYAMA and HARDY, 1969). A technique was used which allowed independent control of temperature in the preoptic and midbrain regions of rabbits anaesthetized with urethane while these regions were explored with microelectrodes. They found that 23% of cells in the midbrain reticular formation at the level of the red nucleus and pons were excited by cooling in their own locality but not by warming. None of them was sensitive to preoptic changes of temperature.

When recording in the preoptic area units were seen to respond to midbrain temperature as well as to temperature at their own position. Fig. 9 is an example of a midbrain neuron which was excited by cooling in its own neighbourhood but unaffected by preoptic temperature.

The physiological significance of these midbrain temperature units is not yet certain. HARDY (1969b) has found that cooling the midbrain alone can evoke an increase in oxygen consumption but not so effectively as preoptic cooling.

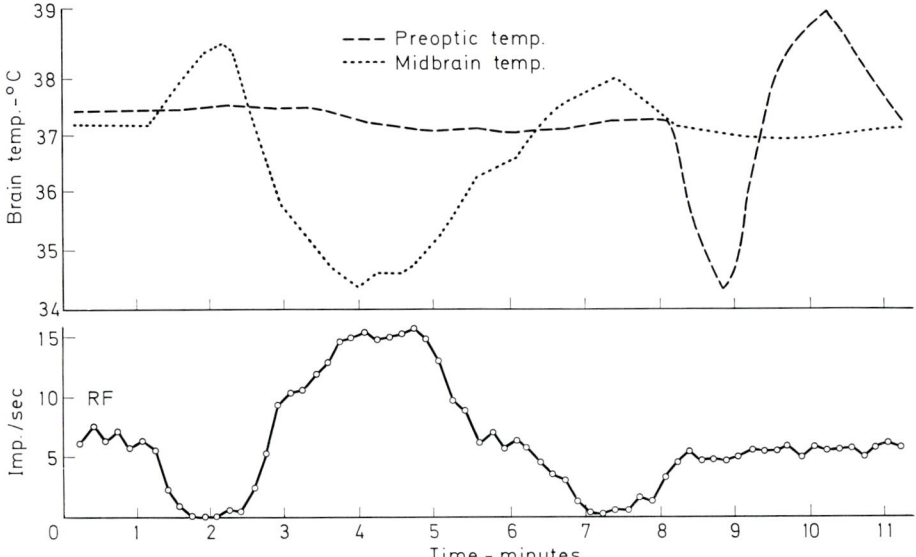

Fig. 9. Firing rate of a cold-sensitive neuron in the midbrain of a rabbit showing its responses to midbrain temperature but not to preoptic temperature. (From NAKAYAMA and HARDY, 1969)

(c) Peripheral Inputs to Hypothalamus and Midbrain

It has already been pointed out that there is good evidence for an interaction between thermoregulatory responses generated from the hypothalamus and environmental temperature. This implies that receptors in the skin are in some way interacting with neurons in the hypothalamus. The first attempts at recording unit activity in the anterior hypothalamus during changes in skin temperature were not successful (HARDY et al., 1964; MURAKAMI et al., 1967). Hot or cold towels were applied to the nose and shaved skin of dogs. Even in encéphale isolé preparations it was not possible to show any change in hypothalamic firing rate to peripheral temperature. It is not clear why these studies should have given negative results because two more recent papers have demonstrated that thermal information from the periphery can be relayed to the anterior hypothalamus. One possible reason is that widespread stimulation of fairly large skin areas is necessary. WIT and WANG (1968a) heated the whole body surface of cats (under urethane) with infrared lamps. They found five units situated 3–5 mm from the midline which increased their activity after the radiation had been on for several minutes.

Using rabbits, which were only sedated with urethane, HELLON (1970a) made unit recordings in a wind tunnel where the air stream temperature could be raised or lowered. Cells were found which were slowed or accelerated, mainly by cooling the air temperature from 25° to 10°C with a latent period of a few seconds to 90 sec. Fig. 10 shows one of these neurons which was excited when the external temperature was lowered to 10°C, but not changed by external heating. It was thought that the receptors were on an exposed surface such as the nose where a concentration of thermal receptors has been described (HENSEL and WURSTER, 1970). The fact that neurons can be found in the anterior hypothalamus which are affected by ambient temperature is not evidence that these neurons are concerned

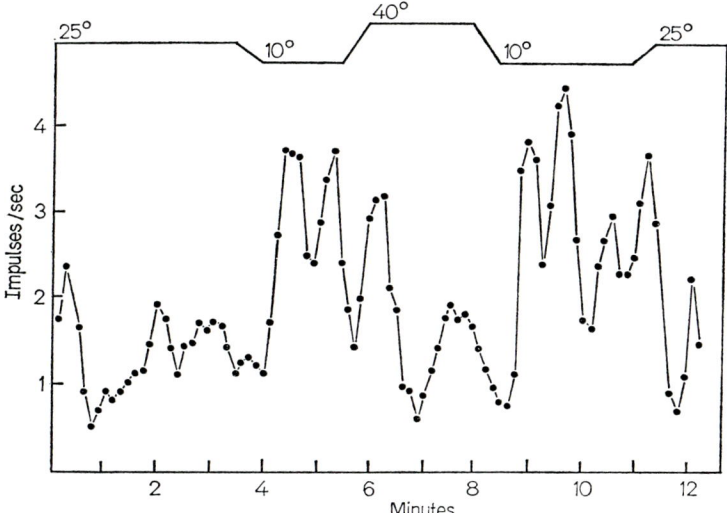

Fig. 10. Responses of a hypothalamic neuron in a rabbit (urethane) during changes of ambient temperature. The top of the frame shows ambient temperature. (From HELLON, 1970a)

with temperature regulation, since comparable responses have been seen in the lateral and posterior hypothalamus of rats (CROSS and SILVER, 1963). However, the additional observation has been made that cells which responded to ambient temperature also responded to hypothalamic temperature and this makes it more probable that these neurons are involved in temperature control. WIT and WANG (1968a) observed this convergence of peripheral and local temperature signals on the same neuron. By prolonged heating they stimulated first the peripheral receptors and then after about thirty minutes, brain temperature began to rise. Five neurons were found which responded in an additive fashion to these two stimuli. In some later experiments (HELLON, 1970a) this same convergence was seen in the anterior hypothalamus of rabbits, but, because there was independent control of ambient and hypothalamic temperatures, a greater complexity of firing patterns was found. Six units showing this dual sensitivity were seen and each showed a different response pattern. However, with one exception, all the cells were affected in the same general direction by ambient and brain temperatures.

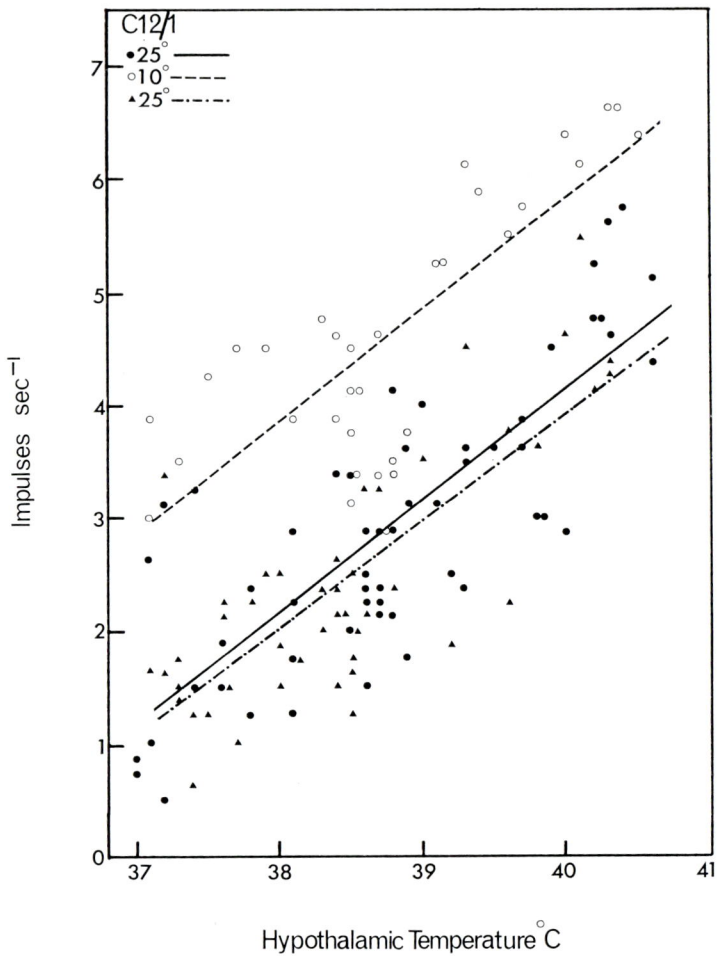

Fig. 11. Responses of a hypothalamic neuron in a cat (urethane) to hypothalamic temperature tested at ambient temperatures of 25° C, 10° C, and again at 25° C. Regression lines fitted by least squares. The intercept of the 10° C line on the temperature axis is significantly different from the other two lines. The slopes are not significantly different. (From HELLON, 1970b)

If ambient cooling was excitatory, then brain cooling was also excitatory. More recordings of this nature are needed to categorize the various responses.

In some comparable experiments NAKAYAMA and HARDY (1969) made recordings in the midbrain while changing the entire surface temperature of shaved, anaesthetized rabbits by means of a water bath. As in the hypothalamus, if units responded to the water bath, they did so by accelerating when the bath was cooled, but showed no response or inhibition to bath warming. The same convergence of peripheral and local temperature sensitivity was found in the midbrain. One third of the neurons which responded to a fall in bath temperature by increased activity were similarly influenced by a reduction in local midbrain temperature.

These recent recordings have revealed a complex variety of neurons in the anterior hypothalamus and midbrain which show a high degree of sensitivity to temperature in their own immediate surroundings and sometimes also to the temperature of the skin or air. Doubtless future investigations, which will be laborious, will add to our knowledge of the various categories of response and the proportions of cells in these categories. These categorizations will not however provide any tests of the various hypotheses of temperature regulation discussed in Section 5.

Nevertheless, it may be possible to test the validity of these hypotheses by single cell recordings, although there are obvious limitations to such an approach. Take for instance the question of how the temperature controller provides graded thermal responses to changes in ambient temperature in the absence of changes in hypothalamic temperature. HAMMEL (1965) has suggested that ambient temperature causes a shift in the set-point without a change in hypothalamic sensitivity to its own temperature. In neuronal terms it might be expected that responses such as those in Figs. 5 and 7 would show a shift to the right or left without any change in slope. On the other hand, the equations of STOLWIJK and HARDY (1966) predict that these curves would change in slope but not in position. In some recent preliminary experiments on cats and rabbits (HELLON, 1970b) the sensitivity of hypothalamic neurons to their own temperature has been tested first with the ambient temperature kept at 25°C and then after this had been lowered to 10°C or raised to 35°C. Out of 22 cells, none was changed at 35°C but four showed a change in their properties at 10°C. All four were warm-sensitive with a linear response over the range 37–41°C. In two of them there was a shift in the line relating hypothalamic temperature and firing rate without any change in the slope of this line. The results from one of these cells are in Fig. 11, which shows that cooling the exterior caused a higher firing rate to be associated with any given brain temperature. These results are similar to those HAMMEL's model would predict. The other two neurons behaved differently and showed a change in the slope of their lines. In one the slope was increased so that the cell was more sensitive to brain temperature and in the other there was a reversible loss of sensitivity to brain temperature. These results might be predicted on the STOLWIJK-HARDY equations.

(d) Spinal Cord Recordings

In view of the thermosensitivity of the spinal cord, it was to be anticipated that neurons would be found there which had the properties of thermodetectors. These recordings in the cord have only begun recently and there is not so much data available as there is for the brain stem.

In guinea-pigs anaesthetized with Nembutal, WÜNNENBERG and BRÜCK (1968) made recordings from the ventromedial part of the pons and a similar region of the cord at C5. The lower cervical and upper thoracic parts of the cord were heated with radio-frequency current. Similar responses were found in the pons (9 cells) and in the cord (2 cells). The units were warm-sensitive and had linear curves which were like those in Fig. 5. No cold-sensitive neurons were seen. These results indicate that the suppression of shivering in guinea pigs, which is brought about by warming from the overlying interscapular brown fat, may be due to the excitation of these warm receptors.

Unit recordings have also been made in the spinal cord of cats, under Nembutal anaesthesia (SIMON and IRIKI, 1970). The microelectrodes were inserted into the ventrolateral column at the level of C2–C3 and a U-shaped thermode was inserted extradurally to extend from the upper thoracic region to the sacrum. Twenty-five sensitive neurons were found and eighteen of them were excited by warming as in the guinea-pig; the other seven were excited by cooling. The exact sensitivities of these neurons could not be determined since the positions of the actual receptors and hence their temperatures were unknown. Similar unit recordings were made in animals with the cord transected at C1 and this indicates the activity was in afferent fibres.

These recent results provide an explanation for the responses which can be obtained by heating and cooling the spinal cord in conscious animals (Section 3). The cord must be regarded as a source of afferent information which is passed rostrally at least as far as the pons and which can evoke a fully-coordinated response to heating or cooling.

(e) Transmitter Substances and Temperature Sensitivity

The hypothalamus contains high concentrations of the monoamines, noradrenaline (NA) and 5-hydroxytryptamine (5-HT). It has been suggested by FELDBERG and his colleagues (see FELDBERG, 1968) that these monoamines may act as transmitter substances between neurons in the hypothalamus concerned with temperature regulation. The considerable literature will not be reviewed here and it will suffice to say that the two monoamines produce opposite effects on body temperature when given by intraventricular injection or by microinjection into the hypothalamus. There are however curious, but apparently genuine, species differences. In cats, dogs, and monkeys, NA causes hypothermia and 5-HT causes hyperthermia, but in rabbits, sheep and rats these two effects are reversed. There is also evidence that acetylcholine (ACh) may act like NA in the latter three species and cause hyperthermia (BLIGH and MASKREY, 1969).

An attempt to correlate the thermal properties of hypothalamic neurons with their responses to 5-HT and NA was made by CUNNINGHAM et al. (1967). The amines were given either by intraventricular injection or intravenously. The general findings were that both amines suppressed the activity of temperature-sensitive and temperature-insensitive units. This generalized depression by both substances does not conform with their antagonistic action on body temperature.

A more direct approach has been adopted by BECKMAN and EISENMAN (1970) who administered the drugs directly on to the neurons by microelectrophoresis from multi-barrelled micropipettes. Similar results were found in both rats and cats. Certain warm-sensitive cells did not respond to NA, 5-HT or ACh but were, affected by current flow from the pipette. These neurons may be true thermodetectors and may even lack a synaptic input. Other warm-sensitive cells, which were classed as interneurons rather than thermodetectors had their firing rate accelerated by ACh and inhibited by NA. Cold-sensitive interneurons were accelerated by NA but inhibited by 5-HT. These results are in agreement with the concept of the transmitter role of these substances in thermoregulation. In the rat the findings are what might be expected from the hyperthermic action of NA or ACh and the hypothermic action of 5-HT. However in cats these substances have opposite actions on body temperature, but the responses of their neurons are not reversed. This is a problem which has yet to be resolved.

(f) Fever

Finally we must consider how hypothalamic neurons may be concerned in the elevation of body temperature following bacterial or viral invasion.

It is commonly stated that fever is caused by a raising of the set-point so that body temperature is regulated at a higher level. There is evidence from MACPHERSON (1959) this is so. He studied a young man in a hot chamber and found that exercise caused the same changes in sweat rate and rectal temperature during a febrile illness as before it. A similar conclusion was reached by COOPER et al. (1964) on the basis of vasomotor responses to body heating during a fever.

It is now known that bacteria produce a pyrogen which reacts with the host's leucocytes and causes them to release a leucocyte or endogenous pyrogen. This endogenous pyrogen is the substance which acts on the hypothalamus to generate a fever. Microinjections of pyrogens in minute doses cause a rapid fever in cats and rabbits when given into the preoptic region, but not into other regions of the brain (VILLABLANCA and MYERS, 1965; JACKSON, 1967; COOPER et al., 1967). Since this is also the thermally-sensitive region, it was likely that the pyrogens were acting on the neurons we have been discussing.

There have been three papers describing how temperature-sensitive cells react when the animal is given a fever by intravenous injection of pyrogens (WIT and WANG, 1968b; CABANAC et al., 1968; EISENMAN, 1969). All three agree that the sensitivity of warm-sensitive cells is reduced by the pyrogen. Thus curves such as those in Fig. 5 showed a reduced slope or even became horizontal. The cold-sensitive ones become more sensitive to their own temperature. Assuming that the warm- and cold-sensitive neurons ultimately control heat-loss and heat-gain mechanisms respectively, these reactions are what might be expected to elevate body temperature. It has already been suggested (Section 6a) that the set-point of the thermoregulator may derive in part from the stable temperature-insensitive cells (Fig. 4), so the action of pyrogen on these units has also been tested. Out of a total of thirteen cells mentioned in the three papers, one has been found to react and its activity was reduced to zero within 18 min after the pyrogen injection (CABANAC et al., 1968). Thus the part played by these cells cannot yet be assessed.

There is at the moment no way of determining whether one or all of the various types of neuron are sensitive to the action of pyrogen. It is also worth emphasizing the great technical difficulties in these fever experiments when one cell has to be observed for long periods and only one pyrogen injection can be made in each animal.

WIT and WANG (1968b) have also made the interesting observation that the sensitivity of a warm neuron which had been suppressed by pyrogen could be restored by the injection of acetylsalicylate. Since the anti-pyretic action of salicylate appears to be largely within the CNS (CRANSTON et al., 1970), it is likely that there is some interaction between pyrogens and salicylate on neurons in the hypothalamus.

7. Summary and Conclusions

The preoptic and anterior hypothalamic regions of the brain stem have been warmed or cooled in conscious animals and this thermal stimulation evokes responses for heat loss or heat gain which are a combination of behavioural and physiological mechanisms. Exactly the same responses can be evoked by isolated thermal stimulation of the spinal cord and since these complex responses must involve supra-spinal control, it follows that the cord is probably passing information about its temperature to the brain stem.

By conducting experiments in calorimeters, it can be shown that there is a quantitative relation between the displacement of hypothalamic or cord temperature and the magnitude of the animal's responses. These quantitative experiments have given rise to various mathematical models which explain some features of the controlling system in terms of temperatures within the CNS and at the skin. All the models contain terms representing intangible set-point temperatures, which make them difficult to verify with our present techniques.

The recording of single neuron activity has confirmed the expected presence of cells in the brain stem and spinal cord which act as if they were thermal receptors. Some of these are activated by warming and others by cooling; their responses are in some cases linear and in others non-linear. However there is still no clear agreement as to what part this central thermosensitivity plays in normal temperature regulation. In particular, there is the contrast between the lack of thermoregulatory responses accompanying natural spontaneous fluctuations in hypothalamic temperature and those responses which can be generated when similar fluctuations are imposed artificially. The cutaneous thermoreceptors are likely to combine with those in the CNS, but the exact mechanisms have yet to be worked out. The only role for the central thermodetectors which can be stated with some degree of certainty is that they are concerned in the development of fever by bacterial pyrogens.

References

Abrams, R., Hammel, H. T.: Hypothalamic temperature in unanesthetized albino rats during feeding and sleeping. Amer. J. Physiol. 206, 641–646 (1964).
— — Cyclic variations in hypothalamic temperature in unanesthetized rats. Amer. J. Physiol. 208, 698–703 (1965).
Adair, E. R.: Control of thermoregulatory behavior by brief displacements of hypothalamic temperature. Psychon. Sci. 20, 11–13 (1970).
Adams, T.: Hypothalamic temperature in the cat during feeding and sleep. Science, N.Y. 139, 609–610 (1963).
Anand, B. K., Banerjee, M. G., Chhina, G. S.: Single neurone activity of hypothalamic feeding centres: effects of local heating. Brain Res. 1, 269–278 (1966).
Andersen, H. T., Andersson, B., Gale, C.: Central control of cold defense mechanisms and the release of "endopyrogen" in the goat. Acta physiol. scand. 54, 159–174 (1962).
Andersson, B., Brook, A. H., Ekman, L.: Further studies of the thyroidal response to local cooling of the "heat loss center". Acta physiol. scand. 63, 186–192 (1965).
— Ekman, L., Gale, C. C., Sundsten, J. W.: Control of thyrotrophic hormone (TSH) secretion by the "heat loss center". Acta physiol. scand. 59, 12–33 (1963).
— Gale, C. C., Hökfelt, B., Ohga, A.: Relation of preoptic temperatures to the function of the sympathico-adrenomedullary system and the adrenal cortex. Acta physiol. scand. 61, 182–191 (1964).

ANDERSSON, B., LARSON, B.: Influence of local temperature changes in the preoptic area and rostral hypothalamus on the regulation of food and water intake. Acta physiol. scand. **52**, 75–89 (1961).
BALDWIN, B. A., INGRAM, D. L.: The effect of heating and cooling the hypothalamus on behavioural thermoregulation in the pig. J. Physiol. (Lond.) **191**, 375–392 (1967).
— — The influence of hypothalamic temperature and ambient temperature on thermoregulatory mechanisms in the pig. J. Physiol. (Lond.) **198**, 517–529 (1968).
— — LEBLANC, J.: The effects of environmental temperature and hypothalamic temperature on excretion of catecholamines in the urine of pigs. Brain Res. **16**, 511–515 (1969).
BARBOUR, H. G.: Die Wirkung unmittelbarer Erwärmung und Abkühlung der Wärmezentra auf die Körpertemperatur. Naunyn-Schmiedebergs Arch. exp. Path. Pharmak. **70**, 1–26 (1912).
BECKMAN, A. L., EISENMAN, J. S.: Microelectrophoresis of biogenic amines on hypothalamic thermosensitive cells. Science, N.Y. **170**, 334–336 (1970).
BENZINGER, T. H.: Heat regulation: Homeostasis of central temperature in man. Physiol. Rev. **49**, 671–759 (1969).
BLIGH, J.: The thermosensitivity of the hypothalamus and thermoregulation in mammals. Biol. Rev. **41**, 317–367 (1966).
— MASKREY, M.: A possible role of acetylcholine in the central control of body temperature in sheep. J. Physiol. (Lond.) **203**, 55–57 P (1969).
BRÜCK, K., WÜNNENBERG, W.: Beziehung zwischen Thermogenese im „braunen" Fettgewebe, Temperatur im cervicalen Anteil des Vertebralkanals und Kältezittern. Pflügers Arch. ges. Physiol. **290**, 167–183 (1966).
— — Eine kälteadaptive Modifikation: Senkung der Schwellentemperaturen für Kältezittern. Pflügers Arch. ges. Physiol. **293**, 226–235 (1967).
— — "Meshed" control of two effector systems: non-shivering and shivering thermogenesis. In: Physiological and behavioral temperature regulation, ed. J. D. HARDY, A. P. GAGGE and J. A. J. STOLWIJK. Springfield, Ill.: Charles Thomas 1970.
CABANAC, M., HAMMEL, H. T., HARDY, J. D.: Tiliqua scincoides: Temperature-sensitive units in lizard brain. Science N.Y. **158**, 1050–1051 (1967).
— HARDY, J. D.: Réponses unitaires et thermorégulatrices lors de réchauffements et refroidissements localisés de la région préoptique et du mésencéphale chez le lapin. J. Physiol. (Paris) **61**, 331–347 (1969).
— STOLWIJK, J. A. J., HARDY, J. D.: Effect of temperature and pyrogen on single unit activity in the rabbit's brain stem. J. appl. Physiol. **24**, 645–652 (1968).
CARLISLE, H. J.: Behavioral significance of hypothalamic temperature-sensitive cells. Nature (Lond.) **209**, 1324–1325 (1966).
CHAI, C. Y., MU, J. Y., BROBECK, J. R.: Cardiovascular and respiratory responses from local heating of the medulla oblongata. Amer. J. Physiol. **209**, 301–306 (1965).
COOPER, K. E., CRANSTON, W. I., HONOUR, A. J.: Observations on the site and mode of action of pyrogens in the rabbit brain stem. J. Physiol. (Lond.) **191**, 325–337 (1967).
— — SNELL, E. S.: Temperature regulation during fever in man. Clin. Sci. **27**, 345–356 (1964).
CORBIT, J. D.: Behavioral regulation of hypothalamic temperature. Science, N.Y. **166**, 256–258 (1969).
CRANSTON, W. I., LUFF, R. H., RAWLINS, M. D., ROSENDORFF, C.: The effects of salicylate on temperature regulation in the rabbit. J. Physiol. (Lond.) **208**, 251–259 (1970).
CROSS, B. A., DYER, R. G.: Response of hypothalamic neurones to anaesthetics. J. Physiol. (Lond.) **210**, 67 P (1970).
— GREEN, J. D.: Activity of single neurones in the hypothalamus: effect of osmotic and other stimuli. J. Physiol. (Lond.) **148**, 554–569 (1959).
— SILVER, I. A.: Unit activity in the hypothalamus and the sympathetic response to hypoxia and hypercapnia. Exp. Neurol. **7**, 375–393 (1963).
CUNNINGHAM, D. J., STOLWIJK, J. A. J., MURAKAMI, N., HARDY, J. D.: Responses of neurons in the preoptic area to temperature, serotonin, and epinephrine. Amer. J. Physiol. **213**, 1570–1581 (1967).
DOWNEY, J. A., MOTTRAM, R. F., PICKERING, G. W.: The location by regional cooling of central temperature receptors in the conscious rabbit. J. Physiol. (Lond.) **170**, 415–441 (1964).

EISENMAN, J. S.: Pyrogen-induced changes in the thermosensitivity of septal and preoptic neurons. Amer. J. Physiol. **216**, 330–334 (1969).
— JACKSON, D. C.: Thermal response patterns of septal and preoptic neurons in cats. Exp. Neurol. **19**, 33–45 (1967).
EULER, C. VON: Slow "temperature potentials" in the hypothalamus. J. cell. comp. Physiol. **36**, 333–350 (1950).
— Physiology and pharmacology of temperature regulation. Pharmacol. Rev. **13**, 361–398 (1961).
— The gain of the hypothalamic temperature regulation mechanisms. In: Progress in brain research, vol. 5, ed. W. BARGMANN and J. P. SCHADÉ. Amsterdam: Elsevier 1964.
FELDBERG, W.: The monoamines of the hypothalamus as mediators of temperature responses. In: Recent advances in pharmacology, 4th ed., ed. J. M. ROBSON and R. S. STACEY. London: J. and A. Churchill 1968.
FORSTER, R. E., FERGUSON, T. B.: Relationship between hypothalamic temperature and thermoregulatory effectors in unanesthetized cat. Amer. J. Physiol. **169**, 255–261 (1952).
FREEMAN, W. J., DAVIS, D. D.: Effect on cats of conductive hypothalamic cooling. Amer. J. Physiol. **197**, 145–148 (1959).
FUSCO, M. M., HARDY, J. D., HAMMEL, H. T.: Interaction of central and peripheral factors in physiological temperature regulation. Amer. J. Physiol. **200**, 572–580 (1961).
GALE, C. C., MATHEWS, M., YOUNG, J.: Behavioral thermoregulatory responses to hypothalamic cooling and warming in baboons. Physiol. Behav. **5**, 1–6 (1970).
HALES, J. R. S., JESSEN, C.: Increase of cutaneous moisture loss caused by local heating of the spinal cord in the ox. J. Physiol. (Lond.) **204**, 40–42P (1969).
HAMILTON, C. L.: Hypothalamic temperature records of a monkey. Proc. Soc. exp. Biol. (N.Y.) **112**, 55–57 (1963).
HAMMEL, H. T.: Neurones and temperature regulation. In: Physiological controls and regulations, ed. W. S. YAMAMOTO and J. R. BROBECK. London: Saunders 1965.
— Regulation of internal body temperature. Ann. Rev. Physiol. **30**, 641–710 (1968).
— HARDY, J. D., FUSCO, M. M.: Thermoregulatory responses to hypothalamic cooling in unanesthetized dogs. Amer. J. Physiol. **198**, 481–486 (1960).
— JACKSON, D. C., STOLWIJK, J. A. J., HARDY, J. D., STRØMME, S. B.: Temperature regulation by hypothalamic proportional control with an adjustable setpoint. J. appl. Physiol. **18**, 1146–1154 (1963a).
— STRØMME, S. B., CORNEW, R. W.: Proportionality constant for hypothalamic proportional control of metabolism in unanesthetized dog. Life Sci. **12**, 933–947 (1963b).
HARDY, J. D.: Physiology of temperature regulation. Physiol. Rev. **41**, 521–606 (1961).
— Brain sensors of temperature. Brody Memorial Lecture VIII. Univ. of Missouri. Special Report No 103 (1969a).
— Thermoregulatory responses to temperature changes in the midbrain of the rabbit. Fed. Proc. **28**, 713 (1969b).
— GAGGE, A. P., STOLWIJK, J. A. J.: Physiological and behavioural temperature regulation. Springfield, Ill.: Charles Thomas 1970.
— HELLON, R. F., SUTHERLAND, K.: Temperature-sensitive neurones in the dog's hypothalamus. J. Physiol. (Lond.) **175**, 242–253 (1964).
HASAMA, B.: Pharmakologische und physiologische Studien über die Schweißzentren. II. Über den Einfluß der direkten mechanischen, thermischen und elektrischen Reizung auf die Schweiß- sowie Wärmezentren. Naunyn-Schmiedebergs Arch. exp. Path. Pharmak. **146**, 129—161 (1929).
HAYWARD, J. N., BAKER, M. A.: A comparative study of the role of cerebral arterial blood in the regulation of brain temperature in five mammals. Brain Res. **16**, 417–440 (1969).
HELLON, R. F.: Thermal stimulation of hypothalamic neurones in unanaesthetized rabbits. J. Physiol. (Lond.) **193**, 381–395 (1967).
— The stimulation of hypothalamic neurones by changes in ambient temperature. Pflügers Arch. **321**, 56–66 (1970a).
— Interaction between peripheral temperature receptors and central neurones responding to brain temperature. J. Physiol. (Lond.) **210**, 161P (1970b).
HELLSTRØM, B., HAMMEL, H. T.: Some characteristics of temperature regulation in the unanesthetized dog. Amer. J. Physiol. **213**, 547–556 (1967).

HENSEL, H.: Electrophysiology of cutaneous thermoreceptors. In: The skin senses, ed. D. R. KENSHALO. Springfield, Ill.: Charles Thomas 1968.
— KRÜGER, F. J.: Hautdurchblutung der wachen Katze bei Kühlung des vorderen Hypothalamus. Pflügers Arch. ges. Physiol. **268**, 72 (1958).
— WURSTER, R. D.: Static properties of cold receptors in nasal area of cats. J. Neurophysiol. **33**, 271–275 (1970).
HOLMES, R. L., NEWMAN, P. P., WOLSTENCROFT, J. H.: A heat-sensitive region in the medulla. J. Physiol. (Lond.) **152**, 93–98 (1960).
INGRAM, D. L., MCLEAN, J. A., WHITTOW, G. C.: The effect of heating the hypothalamus and the skin on the rate of moisture vaporization from the skin of the ox. (Bos Taurus.) J. Physiol. (Lond.) **169**, 394–403 (1963).
— WHITTOW, G. C.: The effect of heating the hypothalamus on respiration in the ox. (Bos taurus.) J. Physiol. (Lond.) **163**, 200–210 (1962).
IRIKI, M.: Änderung der Hautdurchblutung bei unnarkotisierten Kaninchen durch isolierte Wärmung des Rückenmarkes. Pflügers Arch. ges. Physiol. **299**, 295–310 (1968).
JACKSON, D. L.: A hypothalamic region responsive to localized injection of pyrogens. J. Neurophysiol. **30**, 586–602 (1967).
JESSEN, C.: Auslösung von Hecheln durch isolierte Wärmung des Rückenmarks am wachen Hund. Pflügers Arch. ges. Physiol. **297**, 53–70 (1967).
— Hecheln bei isolierten Temperaturänderungen von Rückenmark und Hypothalamus am wachen Hund. Pflügers Arch. **316**, R 53–54 (1970).
— MEURER, K. A., SIMON, E.: Steigerung der Hautdurchblutung durch isolierte Wärmung des Rückenmarks am wachen Hund. Pflügers Arch. ges. Physiol. **297**, 35–52 (1967).
— SIMON, E., KULLMANN, R.: Interaction of spinal and hypothalamic thermodetectors in body temperature regulation of the conscious dog. Experientia (Basel) **24**, 694–695 (1968).
KAHN, R. H.: Über die Erwärmung des Carotidblutes. Arch. Anat. Physiol. (Suppl.) **28**, 81–134 (1904).
KRÜGER, F. J., KUNDT, H. W., HENSEL, H., BRÜCK, K.: Das Verhalten der Hautdurchblutung bei Hypothalamuskühlung an der wachen Katze. Pflügers Arch. ges. Physiol. **269**, 240–247 (1959).
KUNDT, H. W., BRÜCK, K., HENSEL, H.: Hypothalamustemperatur und Hautdurchblutung der nichtnarkotisierten Katze. Pflügers Arch. ges. Physiol. **264**, 97–106 (1957).
MACPHERSON, R. K.: The effect of fever on temperature regulation in man. Clin. Sci. **18**, 281–287 (1959).
MAGOUN, H. W., HARRISON, F., BROBECK, J. R., RANSON, S. W.: Activation of heat loss mechanisms by local heating of the brain. J. Neurophysiol. **1**, 101–114 (1938).
MOOREHOUSE, V. H. K.: Effect of increased temperature of the carotid blood. Amer. J. Physiol. **28**, 223–234 (1911).
MURAKAMI, N., STOLWIJK, J. A. J., HARDY, J. D.: Responses of preoptic neurons to anaesthetics and peripheral stimulation. Amer. J. Physiol. **213**, 1015–1024 (1967).
MURGATROYD, D.: Central and peripheral factors in behavioral thermoregulation in the rat. Physiologist **9**, 251 (1960).
NAKAYAMA, T., EISENMAN, J. S., HARDY, J. D.: Single unit activity of anterior hypothalamus during local heating. Science, N.Y. **134**, 560–561 (1961).
— HAMMEL, H. T., HARDY, J. D., EISENMAN, J. S.: Thermal stimulation of electrical activity of single units of the preoptic region. Amer. J. Physiol. **204**, 1122–1126 (1963).
— HARDY, J. D.: Unit responses in the rabbit's brain stem to changes in brain and cutaneous temperature. J. appl. Physiol. **27**, 848–857 (1969).
PEMBREY, M. S.: Animal heat. In: SCHÄFER, E. A. (ed.), Textbook of physiology, vol. 1, p. 865. Edinburgh: Pentland 1898.
PRINCE, A. L., HAHN, L. J.: The effect on the body temperature induced by thermal stimulation of the heat centre in the brain of the cat. Amer. J. Physiol. **46**, 412–415 (1918a).
— — The effect on the volume of the hind limb induced by heating or cooling the corpus striatum of the rabbit. Amer. J. Physiol. **46**, 416–419 (1918b).
RAUTENBERG, W.: Die Bedeutung der zentralnervösen Thermosensitivität für die Temperaturregulation der Taube. Z. vergl. Physiol. **62**, 235–266 (1969).

Rawson, R. O., Hammel, H. T.: Hypothalamic and tympanic membrane temperatures in Rhesus monkey. Fed. Proc. **22**, 283 (1963).
— Stolwijk, J. A. J., Graichen, H., Abrams, R.: Continuous radio telemetry of hypothalamic temperatures from unrestrained animals. J. appl. Physiol. **20**, 321–325 (1965).
Satinoff, E.: Behavioural thermoregulation in response to local cooling of the rat brain. Amer. J. Physiol. **206**, 1389–1394 (1964).
Serota, M. M.: Temperature changes in the cortex and hypothalamus during sleep. J. Neurophysiol. **2**, 42–47 (1939).
Sherrington, C. S.: Notes on temperature after spinal transection with some observations on shivering. J. Physiol. (Lond.) **58**, 405–424 (1924).
Simon, E., Iriki, M.: Ascending neurons of the spinal cord activated by cold. Experientia (Basel) **26**, 620–622 (1970).
Smith, R. E., Roberts, J. C.: Thermogenesis of brown adipose tissue in cold-acclimated rats. Amer. J. Physiol. **206**, 143–148 (1964).
Stolwijk, J. A. J., Hardy, J. D.: Temperature regulation in man—a theoretical study. Pflügers Arch. ges. Physiol. **291**, 129–162 (1966).
Ström, G.: Effect of hypothalamic cooling on cutaneous blood flow in the unanesthetized dog. Acta physiol. scand. **21**, 271–277 (1950a).
— Influence of local thermal stimulation of the hypothalamus of the cat on cutaneous blood flow and respiratory rate. Acta physiol. scand. **20** (Suppl. 70), 47–76 (1950b).
Thauer, R.: Wärmeregulation und Fieberfähigkeit nach operativen Eingriffen am Nervensystem homoiothermer Säugetiere. Pflügers Arch. ges. Physiol. **236**, 102–147 (1935).
Villablanca, J., Myers, R. D.: Fever produced by micro-injection of typhoid vaccine into hypothalamus of cats. Amer. J. Physiol. **208**, 703–707 (1965).
Wit, A., Wang, S. C.: Temperature-sensitive neurons in preoptic/anterior hypothalamic region: effects of increasing ambient temperature. Amer. J. Physiol. **215**, 1151–1159 (1968a).
— — Temperature-sensitive neurons in preoptic/anterior hypothalamic region: actions of pyrogen and acetylsalicylate. Amer. J. Physiol. **215**, 1160–1169 (1968b).
Wünnenberg, W., Brück, K.: Single unit activity evoked by thermal stimulation of the cervical spinal cord in the guinea pig. Nature (Lond.) **218**, 1268–1269 (1968).

Chapter 6

Receptors Subserving Hunger and Thirst

By

Bengt Andersson, Stockholm (Sweden)

With 8 Figures

Contents

Introduction 187
Brain-Barrier Systems and Central Enteroceptors 188
Satiety and Hunger 191
    I. Energy Balance 191
    II. Hypothalamus and Satiety 192
        1. Ventromedial Lesions and Stimulation 192
        2. Hypothetical "Satiety Receptors" 194
            a) The Glucostatic Theory 194
            b) The Lipostatic Theory 197
            c) The Thermostatic Theory 197
    III. Hypothalamus and Hunger 198
        1. Lateral Hypothalamic Lesions and Stimulation 198
        2. Hypothetical Hunger Receptors 200
    IV. Satiety-Hunger Interaction 200
Thirst 200
    I. Hypothalamus and Thirst 200
    II. Thirst as a Variable of Fluid Balance 203
    III. Hypothetical "Thirst Receptors" 205
        1. The Osmoreceptor Theory 205
        2. Volumetric Influences 208
        3. Specific Na^+ Ion Sensitivity 209
Conclusions 210
References 211

Introduction

"Without eating no life, without life no drinking." These obvious facts may explain the keen interest shown by psychologists and physiologists for the mechanisms behind the sensations of hunger and thirst. As subjective sensations hunger and thirst can only be usefully examined in ourselves. Nevertheless, psychological observations on these urges in human beings have given rise to fruitful studies in animals which have provided valuable information on the mechanisms regulating caloric and water intake. Literally, this chapter should deal only with brain receptors subserving hunger and thirst. No effort is made, therefore, to present a comprehensive picture of what is known about the complex regulations of food and water consumption. The role played by oropharyngeal, gastric, and hepatic receptors in these regulations is neglected. Still, if this chapter had to be a strict description of the nature of brain receptors subserving hunger and thirst, one, or at most two sentences would have been sufficient. These receptors have one feature in common with the Holy Ghost; they are often talked about but nobody appears to have seen them so far. This feature is no denial of the existence of either. Much evidence has been produced that central receptors involved in the regulation of eating and drinking reside in the hypothalamus and/or closely associated parts of the brain, although alimentary behaviour undoubtedly can also be modified from other parts of the brain (cf. STEVENSON, 1969). The general idea is that receptors in the hypothalamus are influenced by the composition of the blood, and that they serve to maintain energy and body water homeostasis. So far the evidence for hypothalamic receptors of these kinds is indirect, however. Induction of neuronal activity within certain parts of the hypothalamus by stimuli known to induce hunger, satiety, or the urge to drink, does not necessarily mean that receptor activity is recorded. Feeding, food-aversion, or drinking elicited by electrical or drug stimulation at different sites in the hypothalamus do not necessarily mean that "hunger", "satiety", or "thirst receptors" are stimulated. As well the recorded neuronal activity as the effects of the stimulations may reflect stimulation of secondary neuronal systems. The nature of brain receptors regulating the hunger and thirst drives can not be described adequately until answers have been given to at least some of the following hitherto unsolved questions: Where exactly in the hypothalamic region are these receptors located? Are they neurons, glial cells, modified ependymal cells, or cells of still another kind? What are their physiological stimuli? Are they stimulated from the outside or from the inside of the blood-brain barrier? Since precise information on any of these points is lacking, much of the following has to assume the character of rather theoretical speculation.

Brain-Barrier Systems and Central Enteroceptors

Brain mechanisms controlling energy and water balance appear to need receptors adapted to detect changes in the composition and the water content of their environment. As long as it is not known whether the sensoring part of such receptors is in direct contact with the blood, any discussion of their function must consider the possible influence of the blood-brain barrier and other brain barriers. It appears reasonable, therefore, to emphasize at this stage certain characteristics of these barriers which affect the composition of the brain extracellular fluid (ECF). A blood-brain barrier may be said to exist for any substance in the blood plasma which fails to attain equal concentration in the ECF of the brain, or else does so much more slowly than in other organs and tissues of the body. The question whether an extracellular space really exists in the brain tissue has been extensively debated (cf. VAN HERREVELD, 1966). Early electron microscopic studies indicated that the brain extracellular space might be very small. Later investigations, however, have shown that there is an appreciable extracellular space in the brain tissue which is filled with fluid. Therefore, it must be assumed that the neurons and the glial cells of the brain, like most other cells in the body, live in an environment of ECF. Neurons require a stable ionic environment for their adequate functioning. Small changes in the concentrations of major ions like Na^+, K^+, and

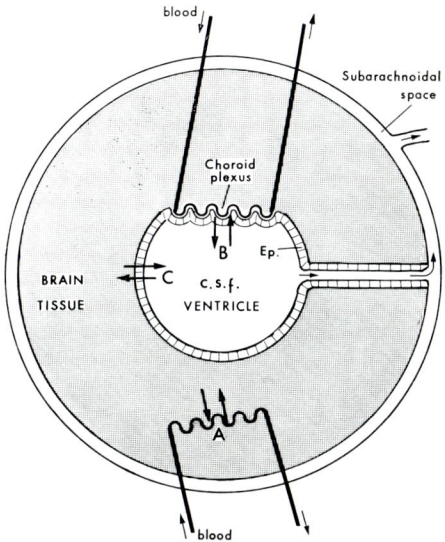

Fig. 1 A–C. Schematic illustration of the brain barrier systems which may be of importance for central receptor mechanisms. A The blood-brain barrier. The brain capillaries behave as if they were less permeable to solutes than other capillaries in the body. It prevents or delays the direct transfer of many substances and ions from blood plasma to brain extracellular fluid. B The blood-cerebrospinal fluid barrier. The cerebrospinal fluid (c.s.f.) is formed by a secretory-reabsorptive process in the epithelium of the choroid plexuses with the effect that the composition of the cerebrospinal fluid in many respects differs from that of a blood plasma ultrafiltrate. C The cerebrospinal fluid-brain barrier. The epithelial cell layer (the ependyma = $Ep.$) which covers the walls of the ventricles delays the exchange between the cerebrospinal fluid and the brain extracellular fluid. Yet, the ionic composition of the brain extracellular fluid is apparently almost identical to that of the cerebrospinal fluid

Ca^{++} in the ECF of the brain modify neuronal activity significantly. Such changes can occur rapidly in the blood plasma and in the ECF of most organs and tissues. The brain ECF, however, is protected by barrier systems which tend to modify the effects of variations in the composition of the blood plasma. The direct exchange of dissolved material between blood plasma and brain tissue across the walls of the blood capillaries is limited by the blood-brain barrier (Fig. 1, A). Another, more indirect, route by which the exchange between blood and brain takes place is via the cerebrospinal fluid (CSF). This fluid is formed by a secretory-reabsorptive process in the epithelium of the choroid plexuses. For this reason the composition of the CSF in many respects differs from that of an ultrafiltrate of the blood plasma. Hence, the epithelial cells of the choroid plexuses act as a barrier for the free exchange of solutes between the blood and the CSF (Fig. 1, B). During the bulk flow of CSF backwards through the ventricular system, an exchange of water and dissolved material takes place between the CSF and the brain ECF. To some extent this exchange is limited by the ependymal cell layer which thus forms a CSF-brain barrier (Fig. 1, C).

A carbonic anhydrase dependent reabsorptive process in the choroid plexus epithelium seems to determine the HCO_3^- concentration of the CSF, which normally is maintained somewhat lower than the blood plasma concentration of HCO_3^- (Davson and Luck, 1957). In its turn, the $p(HCO_3^-)$ and pH of the CSF influence the activity of the respiratory centre (Leusen, 1950; Mitchell et al., 1963). It is possible that brain barriers and CSF composition also play a role in central regulation of food and/or water intake. So far, this possibility has received relatively little attention.

Here will be mentioned briefly some features of the brain barrier systems which may be of importance in central control of energy, water and salt balances. A barrier exists for the free exchange of glucose between blood and brain tissue. The percentage extraction of glucose from blood into brain tissue decreases with rising blood glucose concentration (Crone, 1965). In man the same amount of glucose is extracted from the blood by the brain during the fed and the fasting state (Rowe et al., 1959), and the brain glucose uptake appears to remain practically unaltered at blood-glucose levels from 60 to 250 mg/100 ml (Erbslöh et al., 1958). An active reabsorption of glucose from the newly formed CSF takes place in the epithelium of the choroid plexuses (cf. Csáky and Rigor, 1968). In this manner a 2:1 proportion is maintained between blood and CSF concentrations of glucose. This proportion is preserved also during hyperglycemia. Glucose passes relatively easily in both directions between the CSF and the surrounding brain tissue. The transependymal transport is partly carrier-mediated but there is no evidence that this transport occurs against a concentration gradient (Bradbury and Davson, 1964).

The manner in which certain ions are transported within the central nervous system may be of importance in brain control of water and salt balances. A blood-brain barrier exists for many cations. Hence, the half-time for the exchange of Na^+ between blood and brain in the dog is about 8 hr (Brooks et al., 1970), and the corresponding half-time for K^+ is even longer (Katzman and Leiderman, 1953). In the choroidal plexuses K^+ and HCO_3^- are actively moved from the CSF to the blood plasma while an active transport of Na^+ takes place in the opposite direction (cf. Woodbury, 1968). Therefore, the concentrations of the Na^+ and Cl^- are maintained higher and the concentration of K^+ considerably lower in the CSF than in the blood plasma. Since ions pass the ependymal lining of the ventricular walls relatively freely, the ECF of the brain has an ion composition which is similar to that of the CSF and which consequently differs from that of the blood plasma (cf. van Herreveld, 1966). Therefore, it is evident that factors which interfere with the secretory and reabsorptive processes in the choroid plexuses may induce alterations in the ionic composition of the CSF. Under such circumstances the brain cells (especially those located near the ventricles) may be exposed to changes in their ionic environment which do not necessarily reflect corresponding changes in the blood plasma composition. On the other hand, rapidly induced changes in the ionic composition of the blood plasma are bound to affect the brain ECF composition more slowly than they affect the ECF in other organs and tissues in the body.

Certain parts of the hypothalamus belong to the most vascularized parts of the central nervous system which may facilitate the exposure of hypothalamic receptor systems to stimuli carried via the blood. However, the hypothalamus surrounds the ventral part of the 3rd ventricle (Fig. 2) which makes its ECF easily accessible to changes in the composition of the CSF. At some sites hypothalamic neurons even make direct contact with the CSF by protruding through the ependymal cell layer (LEONHARDT, 1968). Nothing seems to be known about the physiological significance of these protrusions. It is evident, however, that suitable physiological and anatomical conditions exists for stimuli to reach hypothalamic enteroceptors via the CSF.

Satiety and Hunger
I. Energy Balance

Energy balance is maintained when the caloric intake equals the energy output as heat loss, muscular activity and chemical work. An excess of intake over output is converted into true growth or into fat. Man and animals normally adjust their food intake rather quickly according to changes in activity, in environmental temperature, and in the caloric value of the food. This short-term regulation of caloric intake was recognized already at the turn of the century by the French physiologist ANDRÉ MAYER (1901; cf. GASNIER and MAYER, 1939). He also demonstrated the existence of a long-term regulation by showing that adult animals which have free access to food maintain body weight and fat depots almost unchanged for long periods of time. Further, MAYER found that the energy stores are readjusted after temporary depletion by starvation. The existence of a long-term regulation can also be demonstrated in animals which have been force-fed to obesity. After cessation of force-feeding the obese animals eat much less in the *ad libitum* situation than controls do. The food intake rises slowly to normal values as the original body weight is regained (COHN and JOSEPH, 1962). The awareness of a short-term and a long-term regulation of caloric intake has raised the question of whether there exist two regulatory mechanisms, or only one mechanism which is sensitive to both short-term and long-term factors. Hypotheses regarding the nature of the changes which are produced as a result of feeding or starvation, and which signal to the regulating system that feeding should be stopped or initiated, have been reviewed recently (*cf.* STEVENSON, 1969; MORGANE and JACOBS, 1969). Evidence has been presented that circulating substances which are related to carbohydrate metabolism may be the signals for the short-term regulation. Long-term regulation, on the other hand, has been thought to be dependent on some blood-borne factor which is released from the fat depots and which induces satiety. It is well established that the hypothalamus is the principal site of the central regulation of food intake (*cf.* STEVENSON, 1969). The evidence is less conclusive that receptors responding to humoral satiety or hunger stimuli reside in the hypothalamus. The possibility remains that this part of the brain functions as an integrative centre for hunger and satiety signals from sensory cells elsewhere in the body or the central nervous system. The available evidence for hypothalamic receptors involved in the regulation of food intake will be reviewed in the following section. Most of the evidence emanates from studies in the rat, but observations made in larger experimental animals and man are mostly in agreement with the results obtained from the rat.

II. Hypothalamus and Satiety

1. Ventromedial Lesions and Stimulation

The demonstration by HETHERINGTON and RANSON in 1939 that obesity can be produced in animals by discrete, bilateral lesions in the ventromedial region of the hypothalamus opened the way for controlled and reproducible studies of central regulation of food intake. The same authors (HETHERINGTON and RANSON, 1942) also found that obesity does not develop after simply hypophysectomy, but that it can also be produced by ventromedial lesions in hypophysectomized animals. This shows that the obesity is not due to pituitary hypo- or hyperfunction. A few years later it was conclusively shown that this obesity is caused by hyperphagia and not by any metabolic disturbance (BROBECK et al., 1943). Since then, numerous studies have been performed to elucidate the nature of "hypothalamic" hyperphagia. These studies have been extensively reviewed recently (STEVENSON, 1969; MORGANE and JACOBS, 1969). Some controversy appears to remain as to the exact localization and the extent of hypothalamic damage needed to induce hyperphagia. Effective lesions apparently must be bilateral and involve some part of the ventromedial nucleus (Fig. 2), but moderate hyperphagia has also been seen occasionally as result of unilateral damage to this nucleus (MAYER and BARNETT, 1955). The degree of hyperphagia does not seem to be proportional to the extent of the damage made to the ventromedial nuclei (HOEBEL, 1965) and very small, almost thread-like, bilateral lesions may be effective (MASSOPUST, 1955). Temporary hyperphagia can be induced in the goat by injections of anaesthetics into the lateral ventricles (BAILE and MAYER, 1966). This effect has been interpreted as the result of a depression of normal activity in the ventromedial region of the hypothalamus.

In their pioneering studies of "hypothalamic hyperphagia", BROBECK and coworkers (BROBECK et al., 1943) showed that this disturbance in the regulation of food intake is characterized by two stages (Fig. 3). Immediately after the animals have recovered from the operation they start to eat voraciously, and the daily intake of food amounts to two or three times normal. Body weight is rapidly gained during this dynamic phase of hyperphagia or obesity. Following the dynamic phase, food intake gradually returns to almost normal level and the weight curve tends to plateau (= static phase of obesity). The static phase is not due to a recovery from the effects of the hypothalamic damage. If static obese animals are starved to normal body weight, they immediately overeat again when fed *ad libitum*. Hence, it appears that the hunger drive of the hyperphagic animals abates concomitantly with the accumulation of body fat. Some change arises which alters the reaction of the animals to the palatability of the food and they become more discriminative (MILLER et al., 1950; KENNEDY, 1953a and b; TEITEL-

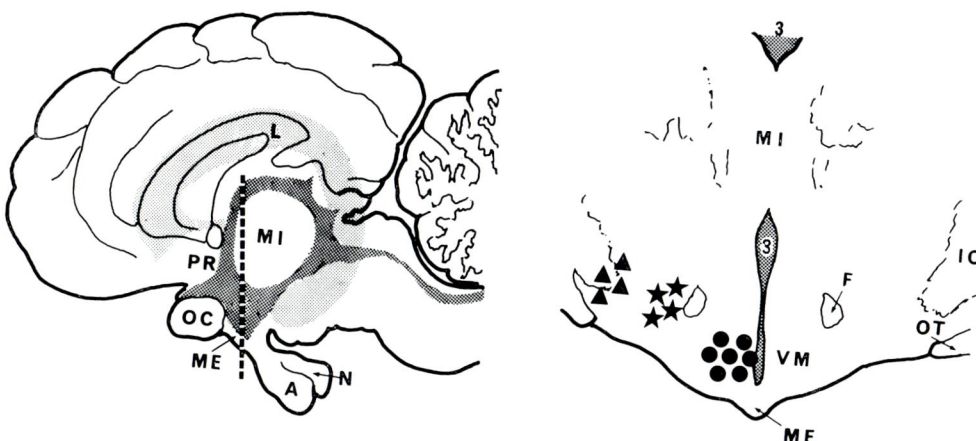

Fig. 2. Hypothalamic regions involved in the regulation of food intake. Left: A mid-sagittal section through a mammalian brain. The dashed line marks the transverse section to the right in the figure. Right: A transverse section through the hypothalamus and the thalamus. Black dots: The ventromedial region of the hypothalamus where bilateral lesions cause hyperphagia and obesity. Black triangles: The far lateral hypothalamus where bilateral lesions result in a refusal to eat and to drink (aphagia and adipsia). Black stars: The perifornical region where bilateral lesions also cause aphagia in the rat. *A* adenohypophysis, *F* descending column of the fornix, *IC* internal capsule, *L* lateral ventricle (light grey), *ME* median eminence, *MI* massa intermedia, *N* neurohypophysis, *OC* optic chiasma, *OT* optic tract, *PR* preoptic region, *VM* ventromedial nucleus, *3* 3rd ventricle (together with the sylvian aqueduct, dark grey)

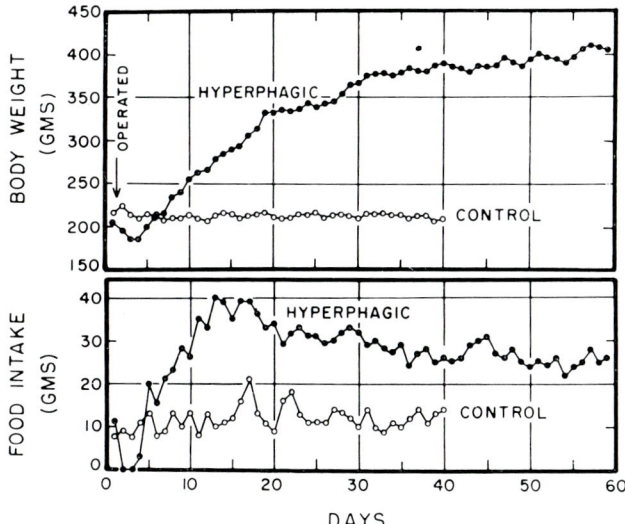

Fig. 3. Hypothalamic hyperphagia and obesity in the rat. Post-operative body weight and daily food intake of a rat after a bilateral lesion has been placed in the ventromedial hypothalamus. For comparison are shown the corresponding data from an unoperated control. After the initial "dynamic" phase the food intake returns towards normal level and the body weight tends to plateau ("static" phase of obesity). (From TEITELBAUM, 1955)

BAUM, 1955, 1957). The statically obese animals may refuse completely to accept distasteful or cellulose-adultered diets, although these diets are readily accepted by hungry normal rats and dynamically hyperphagic animals. The statically obese animals no longer "eat for calories" as normal rats do (ADOLPH, 1947). If they are put into situations in which they have to bar-press to receive food, they bar-press less and consequently obtain less food than normal or dynamically hyperphagic rats in the same situation, thus showing a lower than normal drive to gain calories (TEITELBAUM, 1957). On the other hand, if the statically obese animals are offered food which has been made more palatable than their usual diet, they increase their food intake and become even fatter (KENNEDY, 1953a; TEITELBAUM, 1955). It appears, therefore, that damage to the ventromedial hypothalamus both releases hunger from inhibition and allows discrimination to play a more important role in the regulation of food intake. That the ventromedial hypothalamus exerts an inhibitory influence on the hunger drive is also indicated by the effect of stimulation. Electrical stimulation in this region of the brain inhibits food conditioned movements and food intake in rats (HOEBEL and TEITELBAUM, 1962) and goats (WYRWICKA and DOBRZECKA, 1960).

The fact that ventromedial lesions may make animals more discriminative in their attitude towards food indicates that this part of the hypothalamus is involved in a facilitatory system for feeding. Most of the effects of hypothalamic lesions which have been mentioned, however, may be taken as an indication that the ventromedial hypothalamus primarily is the site of a sensory mechanism which responds directly to humoral satiety signals. The possible natures of these signals have been discussed widely. Some of the theories on the character of hypothalamic "satiety receptors", and some evidence for and against these theories are presented below.

2. Hypothetical "Satiety Receptors"

a) **The Glucostatic Theory.** The idea that metabolic signals are carried to "satiety receptors" in the ventromedial hypothalamus via the blood has received much attention. Since the glucose reserves in the body are small, the blood glucose level and the cell uptake of glucose (glucose "availability") decrease when food intake is unduly postponed. It appears likely, therefore, that the short-term regulation of food intake in some manner is related to the carbohydrate metabolism. For several reasons, however, it is improbable that the blood glucose level in itself determines the activity of central receptors regulating food intake. The blood glucose level and the hunger sensation are often not correlated (JANOWITZ and IVY, 1949), and the urge to eat is not inhibited by the hyperglycemia in diabetic subjects.

MAYER'S (1953) glucostatic theory implies that the availability of glucose for cell metabolism is an important component of satiety. This availability depends both on blood glucose concentration and on insulin which is necessary for the cellular uptake of glucose. The arterio-venous difference in glucose concentration gives a measure of the glucose availability and this difference has been shown to diminish during hunger in man (STUNKARD and WOLFF, 1954). The glucostatic theory assumes that there are glucoreceptors in the ventromedial hypothalamus, and that the response of these receptors waxes with increasing uptake of glucose.

Thus, a high glucose uptake would mean a strong receptor firing and the activation of the satiety mechanism. Experimental data support the glucostatic theory. The hunger sensation in man is depressed by the injection of glucagon, which raises the blood glucose concentration and increases cellular utilization of glucose (STUNKARD et al., 1955). A decreased glucose utilization can be obtained in most tissues and particularly in the brain by the administration of the glucose analogue 2-deoxy-D-glucose (WOODWARD and HUDSON, 1954). Injections of this substance into monkeys and rats have been found to increase food intake (SMITH and EPSTEIN, 1969). Also, the fact that injections of aurothioglucose in mice cause hyperphagia and obesity (BRECHER and WAXLER, 1949) has been taken as evidence for the presence of glucoreceptors in the hypothalamus. The aurothioglucose hyperphagic mice have lesions in the ventromedial hypothalamus (MARSHALL and MAYER, 1954), whereas mice treated with gold thiocarbohydrates of other kinds do not develop hypothalamic lesions or become hyperphagic (MAYER and MARSHALL, 1956). On these grounds, MAYER (1958) has suggested that destructive amounts of gold are dragged into cells in the ventromedial hypothalamus which have a peculiar affinity for glucose (= tentative glucoreceptors).

Electroencephalographic (EEG) and single cell activity have also been taken as support for the glucostatic theory. The EEG activity of the ventromedial hypothalamus of unanaesthetized monkeys has been found to decrease with time after feeding (ANAND et al., 1961b), and to increase after the intravenous injection of glucose (ANAND et al., 1961a). Recordings of single cell activity in the ventromedial hypothalamus also indicate the presence of neurons which increase their firing rate on intravenous (ANAND et al., 1964) or local (OOMURA et al., 1969) application of glucose (Fig. 4). By contrast, intravenous injections of insulin (ANAND et al., 1964) and intracarotid infusions of 2-deoxy-D-glucose (DESIRAJU et al., 1968) have been found to decrease the activity of single neurons in the ventromedial hypothalamus. The inhibitory effect of insulin was preceded by a transient increase in neuronal activity. This initial increase in activity has been explained as due to a temporary, insulin induced increase in the neuronal glucose uptake.

Remonstrances can be raised to most of the evidence produced for the presence of glucoreceptors in the ventromedial hypothalamic region. The uptake and metabolism of glucose in the brain is considered to be independent of insulin (PARK et al., 1957) and the brain uptake of glucose does not seem to change appreciably between the fed and fasting states in man (ROWE et al., 1959). It appears that hypothalamic glucoreceptors, in contrast to other cells of the brain, would require insulin for their carbohydrate metabolism. Otherwise diabetic subjects would fail to eat, rather than overeat. There are also studies which have failed to show correlation between the glucose utilization and the hunger sensation in man (cf. STEVENSON, 1969). Even if a glucoreceptor mechanism may reside in the hypothalamus of man and monogastric mammals, such a mechanism is not likely to participate in the regulation of food intake in ruminant animals. Neither blood glucose nor insulin levels change appreciably with feeding in these species (SIMKINS et al., 1965). Even severe hypoglycemia induced by insulin does not cause eating in ruminant animals, which on the whole appear to be rather independent of glucose for their intermediary metabolism (BAILE and MAYER, 1968).

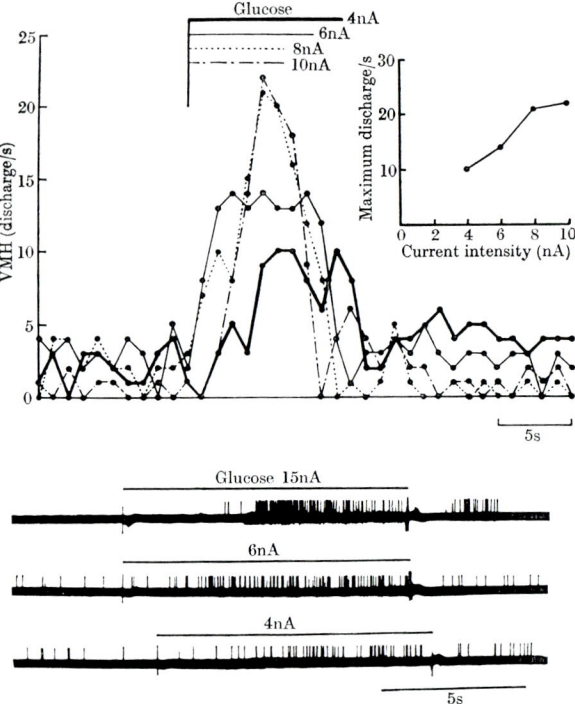

Fig. 4. Electrophysiological evidence for "glucoreceptors" in the hypothalamus of the rat. Above: Discharge rate of a neurone in the ventromedial hypothalamus before, during, and after electro-osmotic application of various amounts of glucose through a micropipette. Inset: Maximum discharge rate as a function of the amount of glucose applied. Below: The modulation of ongoing unit activity by the administration of glucose. Strength of glucose application expressed in nA because only the current which was passed through the electrode could be measured in each experiment. (From OOMURA et al., 1969)

Injections of glucose directly into the ventromedial hypothalamus does not depress the food intake of hungry rats as might have been expected if there are specific glucoreceptors in this part of the brain (EPSTEIN, 1960). Objections may also be raised against aurothioglucose hyperphagia as evidence for glucoreceptors in the ventromedial hypothalamus. Mice appear to be the only animals in which hyperphagia can be produced by aurothioglucose treatment, and even in this species the brain lesions induced by the substance are not restricted to the ventromedial nuclei of the hypothalamus (LIEBELT and PERRY, 1957). Usually the ependymal wall of the 3rd brain ventricle and certain extrahypothalamic structures are also damaged. BROBECK (1960a) suggests that the degree and localization of brain lesions induced by aurothioglucose may depend upon relative blood supply, metabolic rate or other nonspecific factors, rather than specific sensory receptors for glucose.

The evidence for glucoreceptors in the ventromedial hypothalamus obtained by recordings of electrical activity may seem highly suggestive. Nevertheless, they are not free from critizism. Changes in EEG and single neuron activity related to starvation, feeding and intravascular injections or infusions of glucose, insulin and 2-deoxy-D-glucose may as well be signs of reflex influences on hypothalamic activity. Neurons which seem to respond specifically to the local application of glucose have also been found in the lateral hypothalamus (OOMURA et al., 1969) where stimulation elicits feeding (see below). If this activity were recorded from receptors showing a specifically high glucose utilization, it might be expected that the lateral hypothalamus should also be damaged in the aurothioglucose hyperphagic mouse. Such is not the case, however (LIEBELT and PERRY, 1957).

Hence, the evidence presented so far for the existence of glucoreceptors in the ventromedial hypothalamus is suggestive but by no means conclusive.

b) The Lipostatic Theory. Even if glucoreceptors which monitor cellular glucose utilization may reside in the ventromedial hypothalamus, such receptors do not seem well adapted for the long-term control of energy balance. The body fat constitutes a substantial portion of the stored energy in the normal adult. It has been suggested that also some parameter of the lipid depot level may act as a feedback in the central regulation of food intake (KENNEDY, 1953b). This would explain why an animal which is made hyperphagic by a lesion in the ventromedial hypothalamus stops overeating when it has attained a particular weight, and why its food intake is adjusted to maintain this static level of obesity during increased or decreased energy expenditure. HERVEY's (1959) experiments with parabiotic rats support the possibility that some unidentified chemical compound is released into the circulation in proportion to the size of the fat depots, and that this compound acts as a satiety signal to the hypothalamus. HERVEY found that a lesion of the ventromedial hypothalamus in one member of a pair of rats joined in parabiosis causes hyperphagia and obesity in the lesioned animal. A concomitant gradual decrease of the fat depots takes place in the partner with the intact hypothalamus. This partner apparently reduces its food intake. HERVEY's experiments suggest that some humoral factor in the obese "twin" crosses the parabiotic union and serves as a satiety signal in the intact hypothalamus of the partner. HERVEY has also shown that a subsequent lesion made in the ventromedial hypothalamus of the thin parabiotic partner will induce hyperphagia and obesity also in this animal. It would mean that "receptors" which are sensitive to the postulated humoral factor reside in the ventromedial hypothalamus. Why, then, do animals which have this part of the brain destroyed stop overeating when their fat depots have reached a particular size? It may be concluded that there is, as yet, no direct evidence for the existence of hypothalamic receptors which monitor the blood level of fat metabolites.

c) The Thermostatic Theory. It has been realized for a long time that food consumption in homeotherms varies inversely with the environmental temperature (cf. HAMILTON, 1967). Because lesions in the hypothalamus cause abnormalities of regulation not only of food intake, body weight and activity, but also of body temperature, BROBECK (1945) has suggested that this part of the brain may function as an integrator of all the variables of energy balance. According to BROBECK, food intake is partly regulated as if it were a mechanism of temperature regulation (= the thermostatic theory). However, the influence of environmental temperature on food intake remains unaffected in animals made hyperphagic and obese by lesions in the ventromedial hypothalamus (KENNEDY, 1953b). Hence, temperature factors involved in the regulation of eating do not seem to operate via this part of the hypothalamus. Since the most anterior part of the hypothalamus and the preoptic region (Fig. 2) are involved in the regulation of body temperature (MAGOUN et al., 1938), it has been suggested that the thermostatic regulation of eating may be exerted from that region of the brain (BROBECK, 1960a). Central temperature receptors reside within the medial part of the anterior hypothalamus and the preoptic region. The nature of these receptors and their importance in temperature regulation are described by HELLON in the following chapter of this volume. Experiments which have involved ablation or local cooling or warming of the anterior hypothalamus and the preoptic region have been taken as indicating that the central temperature receptors in fact participate in the regulation of food intake. It has been shown that goats (ANDERSSON and LARSSON, 1961) and rats (HAMILTON and BROBECK, 1964; STEVENSON et al., 1964) with lesions medially in the anterior hypothalamus and the preoptic region continue to eat at body temperatures well above 40°C during heat stress. When control animals of the same species are subjected to heat stress to the point that they have this elevation of body temperature, they stop eating. The effects on food intake obtained by local temperature alterations in this region of the brain are equivocal, however. In the goat local cooling of the anterior hypothalamus and the preoptic region has been found to increase, and the corresponding warming to decrease food intake (ANDERSSON and LARSSON, 1961; ANDERSSON et al., 1962). On the other hand, temperature alterations more selectively confined to the preoptic region rather have the reverse effects on eating in the rat (SPECTOR et al., 1968) and in the pig (INGRAM, 1968). This suggests that the effects obtained by local alterations of brain temperature in the goat may have been accomplished by functional ablation (cooling) and nonspecific stimulation (heating) of the anterior part of the ventromedial hypothalamus. That lesions in the anterior hypothalamus and the preoptic region abolish thermal inhibition of food intake does not mean of necessity that central temperature receptors are involved in the regulation of food intake, since some afferent inflow from peripheral temperature receptors is relayed within the lesioned part of the brain (cf. HELLON, this volume).

It can be concluded that heat production and environmental temperature obviously exert an influence on food intake. The importance of central temperature receptors in the regulation of food intake remains obscure, however.

III. Hypothalamus and Hunger

1. Lateral Hypothalamic Lesions and Stimulation

Although it had been noticed in some previous studies that hypothalamic injury sometimes may cause a reduction or total inhibition of food intake (cf. STEVENSON, 1969), it was first clearly shown by ANAND and BROBECK (1951a and b) that small bilateral lesions in the lateral hypothalamus of cats and rats produce a failure to eat and death from starvation. This has later been confirmed in the same and other animal species. Lesions in the lateral hypothalamus produce aphagia also in animals which previously have been made hyperphagic by ventromedial hypothalamic lesions (ANAND et al., 1955). The vast majority of studies which have been made to elucidate the nature of hypothalamic aphagia have been carried out in the rat. Comprehensive reviews of this work have recently been published (Stevenson, 1969; Morgane and Jacobs, 1969). A reader of these reviews will find that the results obtained are somewhat conflicting and confusing. Bilateral lesions as well in the perifornical region as in the far-lateral region of the hypothalamus (Fig. 2) may induce aphagia in the rat. From studies of the behaviour of animals having one or the other type of hypothalamic damage, it has been claimed that the perifornical region exerts a "motivational" effect of feeding whereas activity in the far-lateral hypothalamus rather reflects a "metabolic" need for food (Morgane, 1961a and b). The lesions originally made by Anand and Brobeck (1951a) were located in the far-lateral hypothalamus at the transverse level of the ventromedial nuclei. The nature of aphagia caused by this type of lesions has been studied most extensively. Such lesions produce not only aphagia but also a failure to drink (adipsia) in the rat, and the question has been raised whether the alimentary effect of the lesions is primary aphagia or an aphagia which is secondary to adipsia (Montemurro and Stevenson, 1957). The aphagia and the adipsia due to lesions in the far-lateral hypothalamus are not irreversible. After a period of complete refusal to take either food or water, the rats begin to accept palatable liquid food and then slowly recover to the stage when they again drink water and eat ordinary dry food (Teitelbaum and Stellar, 1954; Teitelbaum and Epstein, 1962).

With the production of aphagia by hypothalamic lesions it became generally accepted as a working hypothesis that the lateral hypothalamus has an appetite function and controls the desire for food. This idea is supported by some studies of alimentary behaviour in rats during the recovery from hypothalamic aphagia. Rodgers et al. (1965) gave such animals the free choice between eating food of high palatability and bar-pressing to obtain injections of liquid food directly into the stomach. Feeding by mouth returned before bar-pressing and it was concluded that rats with lesions in the far-lateral hypothalamus "can eat but do not because they are not hungry". Rather conflicting results have been obtained by Bailly and Morrison (1963) in a similar study. They found that rats which have been trained to obtain their entire food supply by bar-pressing for intragastric injections of liquid food, continued to feed themselves by bar-pressing after lesions had been made in the far-lateral hypothalamus. However, when they were offered food and water by mouth the lesioned animals showed complete aphagia and adipsia. The conclusion reached in this study was that "the response to lesions in the lateral hypothalamus is predominantly and invariably a motor failure, i.e. true aphagia".

It is no easy task to integrate all the available data on hypothalamic aphagia into a plausible description of the importance of the lateral hypothalamic region in the regulation of food intake. This part of the brain has often been referred to as the "feeding" or "hunger centre", which may be an exaggeration. It seems safe to conclude, however, that the far-lateral hypothalamus contains a fundamental system for feeding which has both a motoric and a "motivational" component. This dualism is indicated also by the effects of electrical stimulation. Feeding in response to stimulation of the lateral hypothalamus is generally associated with futile masticatory and licking movements (Brügger, 1943; Larsson, 1954), but such stimulation also may induce active food-seeking behaviour (Miller, 1960).

2. Hypothetical Hunger Receptors

The designation "hunger centre" which has been used for the lateral hypothalamus, would imply the presence of receptors which respond to metabolic hunger signals and which participate in the control of energy balance. Local application of noradrenaline or adrenaline in the lateral hypothalamus elicits eating in the satiated rat. The effect can be eliminated by systemic administration of a ganglionic blocking agent (GROSSMAN, 1960, 1964). It appears more likely that this reflects adrenergic transmission at some synaptic relay in the hypothalamic feeding system than a catecholamine induced stimulation of hypothalamic hunger receptors. The EEG and single neuron activity in the lateral hypothalamus have been found to vary inversely to the activity recorded in the ventromedial hypothalamus (ANAND et al., 1961a and b, 1964; DESIRAJU et al., 1968). This might be expected if a ventromedial satiety mechanism exerts an inhibitory influence on spontaneous activity in a lateral feeding system (see below). It can be stated, therefore, that no suggestive evidence has been provided so far for the presence of specific "hunger" receptors in the lateral hypothalamus.

IV. Satiety-Hunger Interaction

The demonstration of two antagonistic feeding systems in the hypothalamus —a ventromedial inhibitory, "satiety" system, and a lateral, excitatory, "hunger" system—has led to the idea that the homeostatic regulation of food intake is achieved by the reciprocal action of the two systems (ANAND and BROBECK, 1951a and b). It appears that the lateral excitatory system has a more basic influence on food intake, and that the regulation primarily is exerted by the ventromedial inhibitory system. This conception is supported by the effects of lesions. When both the excitatory and the inhibitory system are destroyed, the ablation of the excitatory system dominates and aphagia develops (ANAND et al., 1955). Hyperphagia, on the other hand, can be induced in rats merely by cutting the neural connection between the ventromedial and the lateral system (GOLD, 1970). Hyperphagia seems to develop, therefore, in any situation when the lateral excitatory system is deprived of the inhibitory tonus which normally is exerted on it from the ventromedial "satiety" system. Hence, it appears that the hypothalamus primarily regulates the degree of satiety, rather than the intensity of the hunger sensation.

Thirst

I. Hypothalamus and Thirst

The theory that dehydration and elevated blood tonicity might elicit thirst by stimulating "thirst receptors" in the brain was advanced by ANDRÉ MAYER as early as 1900. Substantial evidence for this conception has been provided by later studies of water intake and thirst-motivated behaviour in experimental animals. Although other parts of the brain may be of importance for the development of the urge to drink, so far only the hypothalamic region has been clearly shown to play an essential role in the thirst mechanism. The importance of the hypothalamus

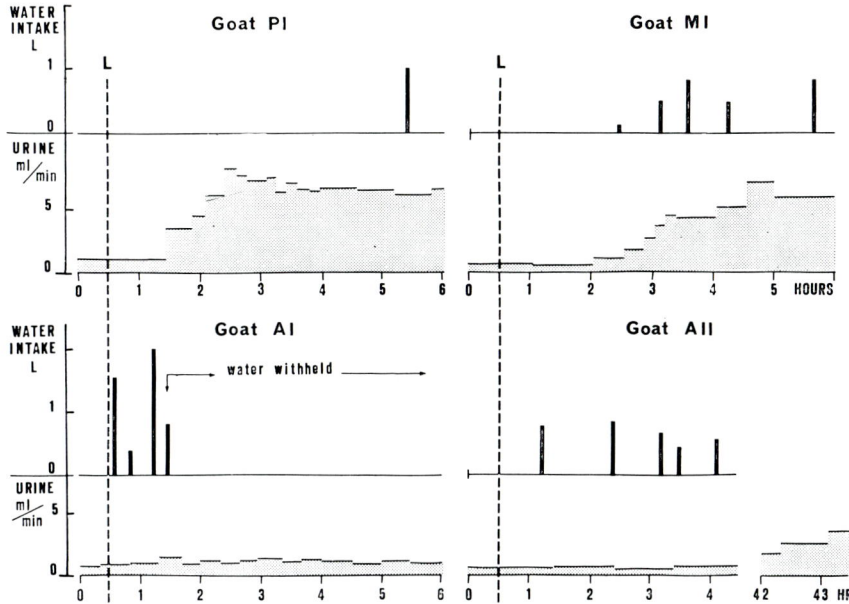

Fig. 5. Influence of the extent of the hypothalamic damage on the relation between polyuria and polydipsia during the temporary phase of diabetes insipidus in the goat. Above, left: Commencement of the temporary phase of diabetes insipidus in an animal with a lesion restricted to the median eminence. Polyuria precedes polydipsia. Above, right: Simultaneous onset of polyuria and polydipsia in a goat with a median eminence lesion extending into the mid-hypothalamic region. Below: Primary hyperdipsia with lack of temporary polyuria (left), or with much delayed polyuria (right) in two goats with median eminence lesions encroaching upon the anterior hypothalamus. The permanent phase of diabetes insipidus developed in all animals with polyuria preceding polydipsia. L production of lesion (radio-frequency heating between permanently implanted thermocouple electrodes). (From OLSSON, 1970)

and the hypothalamic-neurohypophyseal system in the homeostatic control of water balance has become more and more evident. The studies of diabetes insipidus in man first directed attention to the involvement of the hypothalamus and the pituitary gland in the control of renal water excretion (cf. VON HANN, 1919). The disease was early subjected to great attention because of its dramatic clinical manifestations, characterized by the excretion of very large amounts of dilute urine and intense thirst. The development of techniques for producing discrete hypothalamic lesions in animals offered new possibilities for studying the cause of diabetes insipidus. RANSON and coworkers (FISHER et al., 1938) showed that an interruption of the neural connections between the hypothalamus and the neurohypophysis has a typical triphasic effect on the water turnover. Within a few hours after a median eminence lesion is produced, transient polyuria and polydipsia occur (= the temporary phase of experimental diabetes insipidus). After a few days, the water exchange returns to normal or subnormal levels and remains low for about a week. During this interphase degenerative processes in the neurohypophysis result in an uncontrolled release of the water-saving antidiuretic hormone

(ADH), which reduces the renal loss of water to a minimum (O'Connor, 1952; Laszlo and de Wied, 1966). When the neurohypophyseal storage of ADH is emptied, the renal loss of water again becomes excessive and polydipsia develops (= the permanent phase of experimental diabetes insipidus).

In most instances the increased thirst in diabetes insipidus has been shown to be secondary to the extreme water losses. However, experiments have also been reported in which a water intake much in excess of the actual need has been observed during the temporary phase of diabetes insipidus (Bellows and van Wagenen, 1938; Smith and McCann, 1964). As an explanation for this excessive drinking, it has been suggested that the hypothalamic lesions may have knocked out a mechanism which normally inhibits thirst. Another explanation is provided by a recent study in goats subjected to radio-frequency lesions in the ventral hypothalamus (Olsson, 1970). The immediate post-lesion course of water turnover was studied without the interference of anaesthesia. It was found that the increase in water intake which takes place during the temporary (Fig. 5, left above) and permanent phases of diabetes insipidus is clearly secondary to an excessive renal loss of water as long as the lesions are restricted to the median eminence. Body fluid homeostasis is maintained in these animals, indicating that their thirst mechanism reacts in a normal manner to the increased demand for water. However, when the median eminence lesions are extended to encroach upon parts of the hypothalamus where electrical stimulation causes drinking in this species (the anterior medial hypothalamus) (Andersson and McCann, 1955), excessive drinking (hyperdipsia) takes place during the first hours after the lesion has been produced (Fig. 5, below). In spite of the initial overhydration, no transient polyuria, or only a much delayed temporary increase in urine flow are observed. The absence of transient polyuria in goats with median eminence lesions which encroach upon the anterior medial hypothalamus indicates an immediate post-lesion release of antidiuretic material from damaged parts of the hypothalamic neurosecretory system. The most likely explanation for the hyperdipsia seems to be that irritative processes develop in the periphery of the hypothalamic lesion and act as a temporary, non-specific stimulus to the thirst mechanism. A more permanent effect might have been expected if the excessive drinking had been due to the ablation of a system which normally inhibits thirst. In a review on the regulation of feeding and drinking, Brobeck (1960b) stressed that a state of persistent overdrinking, analogous to hypothalamic hyperphagia, does not appear to have been discovered. This statement is obviously still valid. The evidence for the presence in the hypothalamus of an excitatory system for drinking is abundant, however.

It was originally shown by Stevenson (Stevenson, 1949; Stevenson et al., 1950) that medially placed hypothalamic lesions cause reduced water intake (hypodipsia) and a significant increase in serum Na^+ concentration in rats. Such lesions have later been shown to have the similar effects in the dog (Andersson and McCann, 1956). If the brain damage is extended in anterior direction to involve the medial part of the preoptic region, dogs (Witt et al., 1952; Keller, 1959) and goats (Andersson and Larsson, 1961; Andersson et al., 1965) show a complete and persistent loss of the urge to drink water (permanent adipsia). The animals retain the urge to eat and to drink liquid food but refuse to take any water even when severely dehydrated. The effects of stimulation provide additional evidence that the anterio-medial hypothalamus is involved in the regulation of water intake. Injections of minute amounts of hypertonic saline (ANDERSSON, 1952, 1953) into this part of the hypothalamus, and its electrical stimulation (ANDERSSON and McCANN, 1955) elicits excessive drinking in the goat. A sensation of thirst appears to be the cause of this drinking since the stimulation may induce an animal to solve simple problems to obtain water (ANDERSSON et al., 1960), or may elicit a thirst motivated conditioned motor reaction in hydrated animals (ANDERSSON and WYRWICKA, 1957).

Also the lateral hypothalamus is obviously involved in the regulation of water intake. Far-lateral hypothalamic lesions (Fig. 2) cause both adipsia and aphagia in the rat (MONTEMURRO and STEVENSON, 1960; EPSTEIN and TEITELBAUM, 1964) and both eating and drinking are seen as effect of electrical stimulation of the lateral hypothalamus of this species (cf. STEVENSON, 1969). A significant minority of the rats with far-lateral hypothalamic lesions become permanently adipsic. Those that recover water intake, however, do not drink in response to simple water deprivation. They drink only when they eat dry food. These observations have led EPSTEIN and TEITELBAUM (1964) to suggest that rats with lateral hypothalamic lesions may retain a neurological mechanism that mediates the thirst of alimentation (postprandial drinking) but that they have lost the mechanism which initiates drinking in response to dehydration.

Hence, the effects of ablation and stimulation indicate that central "thirst receptors" may be found in the anterio-medial and/or the far-lateral parts of the hypothalamus. Some evidence will be presented below that primary receptors for thirst reside medially in the hypothalamic region near the wall of the 3rd brain ventricle. Before discussing the hypothetical nature of these receptors it appears useful, however, to consider briefly the importance of thirst in the control of body fluid homeostasis.

II. Thirst as a Variable of Fluid Balance

The body water is distributed in two main compartments, the extracellular fluid (ECF) and the intracellular fluid (ICF). The principal cation of the ECF is Na^+, whereas K^+ dominates inside the cells. Water is freely diffusible between the two fluid compartments but the cell membranes behave as if they were relatively impermeable to Na^+ ions. This makes the relation between intracellular and extracellular fluid volumes dependent on the amount of sodium present in the ECF. Deviations from the normal balance may occur as shown diagrammatically (Fig. 6). Such deviations are known to stimulate thirst and/or other compensatory mechanisms, all of them aimed at restoring normal water and salt balances.

An uncompensated simple water loss results in absolute dehydration (Fig. 6B). During absolute dehydration the solute concentration of the body fluid is increased. Both intra- and extracellular fluid volumes are reduced without any obvious change in the proportion between the two. Absolute dehydration stimulates the thirst mechanism and the release of ADH from the neurohypophysis. Due to the intensified action of this hormone the renal reabsorption of water

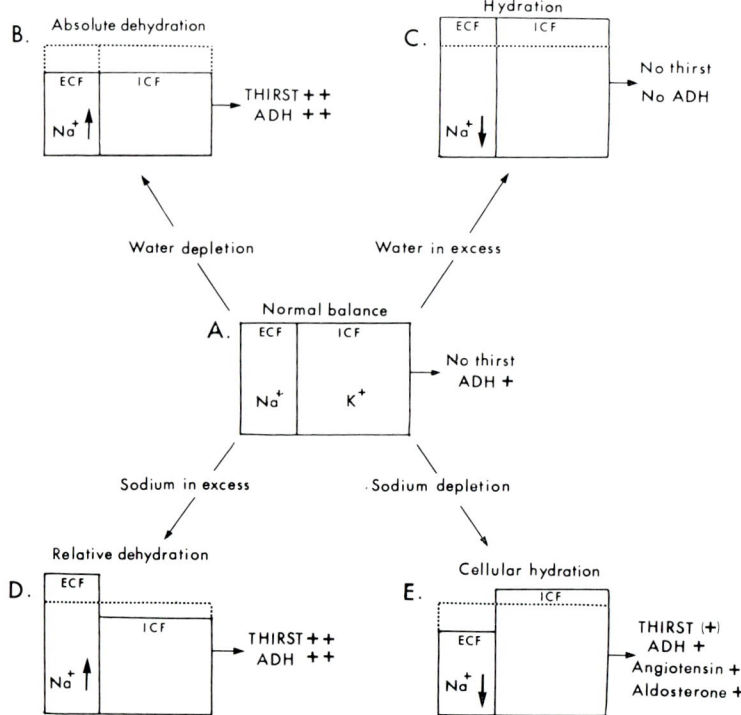

Fig. 6. Diagrams of changes in body fluid distribution and composition during negative and positive water and salt balances. To the right of the diagrams are indicated compensatory factors aimed to restore normal balance. *ECF* Extracellular fluid compartment; *ICF* Intracellular fluid compartment; *ADH* Antidiuretic hormone

becomes maximal and the inevitable water loss via the kidney is kept at a minimum. A normal proportion between the ECF and the ICF volumes is preserved also after an excessive intake of water. Both fluid volumes become expanded with reduced solute concentration (hydration) (Fig. 6C). Hydration inhibits thirst and the release of ADH and large volumes of almost solute free urine are formed. In this manner the body can eliminate the surplus water and restore normal water balance.

A high intake of salt raises the Na^+ concentration of the ECF which in turn causes a shift of water in the direction of ICF to ECF (relative dehydration) (Fig. 6D). Relative dehydration is compensated for in two ways. (1) Thirst and increased release of ADH work to dilute the extracellular Na^+ concentration to normal. (2) The Na^+ excretion via the kidney becomes augmented. The excessive salt is eliminated gradually with the urine and the ECF concomitantly returns to normal volume. During sodium depletion, on the other hand, the Na^+ concentration of the ECF becomes subnormal. Consequently water is shifted in the direction of ECF to ICF (Fig. 6E). The intracellular fluid under such circumstances may expand above normal volume (cellular hydration). A diminished amount of ECF means reduced blood volume and less effective blood supply to various organs.

A diminished blood supply to the kidney activates the renin-angiotensin system (cf. PEART, 1965). Angiotensin II (in the following referred to as just angiotensin) appears in increased amounts in the blood and helps to maintain normal blood pressure in spite of the reduced blood volume. But angiotensin also works to compensate for sodium depletion in at least one other way. It stimulates the secretion of the sodium retaining hormone aldosterone from the adrenal cortex (cf. DENTON, 1965). It is of particular interest that sodium depleted animals may continue to drink water and to release ADH in spite of the fact that their extracellular Na^+ concentration is subnormal and their intracellular fluid volume may be expanded (CIZEK et al., 1951; HOLMES and CIZEK, 1951).

Hence, we find that two deviations from normal water and salt balance elicit thirst and increased release of ADH, viz. absolute and relative dehydration. Both are characterized by reduced ICF volume (cellular dehydration) and by an elevated Na^+ concentration of the ECF (Fig. 6B and D). It may be expected, therefore, that either cellular dehydration or an elevated extracellular Na^+ concentration in some manner stimulates the thirst and ADH releasing mechanism. For reasons to be presented below, the most espoused view is that cellular dehydration is the crucial factor, and that it is the reduced volume of certain brain cells ("osmoreceptors") that is the ultimate cause of thirst and accelerated ADH release during absolute and relative dehydration. However, neither an elevated extracellular Na^+ concentration, nor "osmoreceptor" stimulation can be the cause of drinking and ADH release seen in sodium depleted animals. Here the extracellular Na^+ concentration is subnormal and the cell volume is often expanded, whereas the ECF volume is reduced. Since sodium depleted animals continue to drink water and to secrete ADH, it has been suggested that a "volumetric" regulation of thirst and ADH release also exists. The "volumetric" regulation is thought to control the ECF volume and to work in parallel with an "osmometric" regulation.

III. Hypothetical "Thirst Receptors"

1. The Osmoreceptor Theory

With the production of diabetes insipidus in experimental animals (FISHER et al., 1938), the interest became focused on mechanisms underlying the nervous regulation of the ADH release from the neurohypophysis. The fact that the kidney starts to produce large volumes of dilute urine (water diuresis) during hydration indicated that changes in the concentration of certain solutes or in the osmolality of the blood might influence the release of ADH. With this in mind, VERNEY (1947) in the 1940 ies performed a series of ingenious experiments in the dog which form the basis for the conception of an "osmometric" regulation of the ADH release. VERNEY found that a rise in the tonicity of the blood which flows through the hypothalamus and adjacent parts of the brain may cause a release of ADH in hydrated animals. This happened when hypertonic solutions of sodium salts, sucrose, and d-glucose were used to increase blood tonicity. However, when a hypertonic urea solution was employed for the same purpose, no ADH release was obtained. The lack of response to urea was explained as due to the high diffusibility of the substance. On the basis of these experiments VERNEY postulated that there are "osmoreceptors" in the brain, probably in the anterior hypothalamus, which regulate the release of ADH. It was assumed that these receptors are not stimulated by a rise in body fluid tonicity as such, but rather by any change in the composition of the ECF which causes cellular dehydration and by this a reduction in the volume of the "osmoreceptors". It would explain why urea, which easily diffuses into the cells, was not found to stimulate the receptors. Since small amounts of ADH are released continuously during normal water balance (Fig. 6A), it was surmised that some stimulation of the "osmoreceptors" takes place also at normal volume of the receptors, but that the activity of the "osmoreceptors" ceases completely when their volume expands above normal. This would explain why ADH is no longer released during hydration.

The antidiuretic hormone and drinking serve one purpose in common, that is the restoration of normal water balance during absolute and relative dehydration. Therefore, it has been surmised that the release of ADH and thirst are regulated by the the same or rather similar mechanisms (WOLF, 1950). When it was found that injections of small amounts of hypertonic NaCl solution into the anteriomedial hypothalamus elicit drinking in the goat, it was taken as an indication that "osmoreceptors" in the hypothalamus also participate in the regulation of water intake (ANDERSSON, 1953).

A number of electrophysiological studies since have indicated the presence of neurons in the anterior hypothalamus (VON EULER, 1953; SAWYER and GERNANDT, 1956; HOLLAND et al., 1959; CROSS and GREEN, 1959; VINCENT and HAYWARD, 1970) and in other parts of the brain (CLEMENTE et al., 1957; SUNDSTEN and SAWYER, 1959, and others) which respond specifically to an increased osmolality of the blood. Strongly hypertonic NaCl solution has been used to raise blood osmolality in most of these studies which, with one exception (VINCENT and HAYWARD, 1970), have been carried out in anaesthetized preparations. To the reviewer's mind none of these studies provide conclusive evidence that "osmoreceptors" in VERNEY's sense are present in those parts of the brain where the neuronal activity has been recorded.

A hypothalamic "osmoreceptor" mechanism which regulates both thirst and the release of ADH may seem well adapted to its purposes. Yet, several aspects of the current idea of an "osmometric" control of water balance can be criticized. A rise in blood tonicity obtained by injections of hypertonic urea solution is a much weaker stimulus to thirst than the same rise in blood tonicity elicited by injections of sodium salts (HOLMES and GREGERSEN, 1950), and VERNEY obtained no ADH response to intracarotid injections of hypertonic urea solution in hydrated dogs (1947). However, because of the blood-brain barrier, hypertonic urea solution injected into the blood causes marked dehydration of the brain at least in man (cf. KLEEMAN and CUTLER, 1963). This ought to cause appreciable shrinkage of the brain cells and thereby act as a powerful stimulus to postulated "osmoreceptors". Another observation worthy of note is that hypertonic sucrose solution, which elicits thirst and ADH release when applied in the blood, has none of these effects when infused into the 3rd brain ventricle (OLSSON, 1969). This appears incompatible with the osmoreceptor theory. Since the sucrose molecules do not penetrate into the cells, one would expect that hypertonic sucrose solution should also stimulate brain osmoreceptors effectively when applied inside the blood-brain barrier. VERNEY's later work (JEWELL and VERNEY, 1957) also confuses

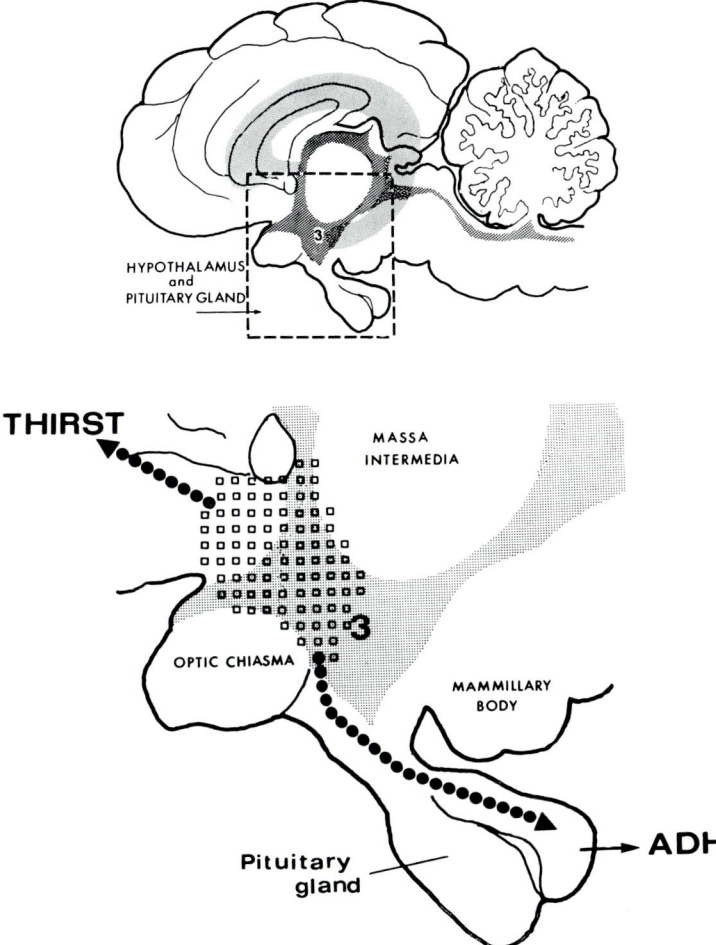

Fig. 7. Probable localization of central receptors regulating thirst and the release of antidiuretic hormone (ADH). Above: A mid-sagittal section through a mammalian (goat's) brain. Parts of the thalamus, the hypothalamus and the pituitary gland are enclosed by the hatched lines. Below: A magnification of the region of the brain enclosed in the upper figure. Square dotted area = The part of the hypothalamus in which by all probability receptors are located which regulate thirst and the release of ADH from the neurohypophysis. 3 Ventral part of the 3rd brain ventricle

somewhat the picture of osmoreceptors in the anterior hypothalamus which are stimulated directly over the blood-brain barrier by an elevated blood tonicity. The ADH release in response to unilateral carotic infusions of hypertonic NaCl was studied in a large number of dogs. In these animals JEWELL and VERNEY had selectively tied off different branches of the internal carotid artery. By elimination they came to the conclusion that osmoreceptors are located in the anterior part of the hypothalamus and/or the preoptic region (Fig. 7). The blood supply to the choroid plexuses was not considered in these experiments. JEWELL and VERNEY were successful in exclusively supplying the anterior hypothalamus and the preoptic region with hypertonic blood in two of the dogs only. It is notable that these two dogs did *not* respond to the infusions with ADH release. The hypertonic blood apparently did not reach the choroid plexuses in the two dogs, and for this reason probably did not affect the composition of the CSF. It may be, therefore, that an ADH releasing mechanism in the hypothalamus is sensitive to changes in the ionic composition of the CSF rather than to osmotic stimuli acting directly over the blood-brain barrier. Some support for this idea will be presented in the following.

2. Volumetric Influences

Thirst may develop in situations when the fluid volume of the body is reduced without appreciable change in the extracellular Na^+ concentration or tonicity (*cf.* FITZSIMONS, 1961). An example of this is the thirst which may follow more severe bleeding. It has also been suggested that drinking occurring after the ingestion of food (postprandial drinking) may be the consequence of an isosmotic reduction of the body fluids elicited by intensified secretion of gastric and intestinal juices (GREGERSEN, 1932). The fact that thirst may develop without any rise in the Na^+ concentration or in the tonicity of the ECF has been taken as indicative of a "volumetric" regulation of water intake working independently of an "osmometric" regulation. Still, osmoreceptor stimulation can not be excluded if the isosmotic reduction involves not only the ECF but also the ICF. An "osmometric" control of water balance, however, can not explain thirst and ADH release in sodium depleted subjects, having if anything, an expanded ICF volume (Fig. 6E). Here it appears difficult to understand the urge to drink water and the release of a water-saving hormone without suggesting some kind of regulatory mechanism which responds to changes in the ECF volume. It has been proposed that there are volume receptors in the heart and larger blood vessels which influence the release of ADH reflexly via the vagal nerve (GAUER *et al.*, 1951, 1954). More convincing are the experimental evidences that the kidney participates in a "volumetric" control of the water balance. It was originally shown by LINAZASORO and coworkers (LINAZASORO *et al.*, 1954; JIMENEZ DIAS *et al.*, 1959) that removal of the kidneys reduces the water intake in rats, and that drinking under such circumstances may be restored by the injection of extracts of kidney tissue. The importance of a renal factor in the regulation of water intake has been further elucidated by FITZSIMONS. He has shown convincingly that an activation of the renin-angiotensin system stimulates the thirst mechanism. Restriction of the renal blood flow causes the liberation of renin into the blood and also elicits thirst in the rat, although there is no change in over-all water and salt balance (FITZSIMONS, 1969). Accordingly, the injection of angiotensin into the blood causes inappropriate drinking, and the effects on drinking of angiotensin and hypertonic NaCl solution appear to be additive when both are administered intravenously in the rat (FITZSIMONS and SIMONS, 1969). The effect of angiotensin on thirst appears to be central since injections of only a few ng of the substance into the anterior hypothalamus, the preoptic region and the septum elicit drinking in rats in normal water balance (EPSTEIN *et al.*, 1969, 1970). For these reasons, FITZSIMONS has suggested that angiotensin either stimulates thirst receptors in the brain directly or makes these receptors over-sensitive to other thirst stimuli (FITZSIMONS, 1969). Of particular interest is that angiotensin also enhances the release of ADH from the neurohypophysis (BONJOUR and MALVIN, 1970), apparently by affecting the brain control of the liberation of this hormone (ANDERSSON and WESTBYE, 1970).

Hence, the renin-angiotensin system provides an additional regulation of thirst and ADH release. It seems to have its main physiological importance as a guard against fatal reduction of the ECF in situations like sodium depletion when the extracellular tonicity or Na^+ concentration cannot act effectively as stimuli to restore normal fluid balance. Yet, the central effects of angiotensin seem to be highly dependent on the Na^+ concentration of the CSF of the brain ECF. Evidence for this is presented below.

3. Specific Na^+ Ion Sensitivity

VERNEY's (1947) fundamental studies have made the idea widely accepted that brain osmoreceptors (reacting to changes in their own volume) play an essential role in the regulation of ADH release and thirst. An alternative would be that the activity of brain receptors involved in the control of water balance is influenced by variations in the environmental Na^+ ion concentration rather than by osmotically induced changes in the receptor volume. This possibility has received relatively little consideration, but is supported by experiments in which the thirst and ADH releasing mechanisms have been approached *via* the CSF of the 3rd brain ventricle in the goat. Slow infusions (10 μl/min) of hypertonic sodium salts into the 3rd ventricle elicit cumulative drinking in goats in normal water balance and cause a liberation of ADH in hydrated animals (ANDERSSON et al., 1967, 1969). The goats may start to drink within a minute or two after the onset of the infusion, at which time less than a mg of salt has been infused. For this reason it may be assumed that thirst receptors are located close to the wall of the 3rd ventricle. Certain facts speak against the possibility that these effects of sodium salts are due to stimulation of osmoreceptors in VERNEY's sense. Strongly hypertonic solutions of d-glucose (ANDERSSON et al., 1967) and of sucrose (OLSSON, 1969) do not elicit thirst or ADH release when applied into the 3rd ventricle. Yet, it appears likely that the increased tonicity of the ventricular fluid under these circumstances reduces the volume of adjacent brain cells by dehydration.

In similar infusion experiments a striking interaction between angiotensin and Na^+ has been observed. This interaction provides additional support for the idea that the Na^+ ion concentration as such, and not the ECF tonicity, influences the activity of central receptors which are involved in the regulation of thirst and the ADH release (ANDERSSON and WESTBYE, 1970; ANDERSSON et al., 1970). Like the infusions of hypertonic NaCl solution, the infusion into the 3rd ventricle of angiotensin (dissolved in slightly hypotonic NaCl solution) stimulates thirst and ADH release in the goat. A conspicuous potentiation of these effects occurs when angiotensin is infused together with hypertonic NaCl solution (Fig. 8). On the other hand, if angiotensin is administered in iso- or hypertonic solutions of non-electrolytes, these solutions reduce the Na^+ concentration of the CSF, and little or no drinking or ADH release is seen. It appears, therefore, that the effect of angiotensin on central receptors which regulate thirst and the release of ADH in some manner is mediated by Na$^+$ ions. There seem to be at least three possibilities for such an angiotensin-Na$^+$ interaction. (1) Angiotensin may facilitate the transport of Na$^+$ ions from the CSF into the brain tissue. This would mean that the nearby brain cells become exposed to a rise in their environmental Na$^+$ concentration. (2) Angiotensin may make receptors for thirst and ADH release more sensitive to the existing Na$^+$ concentration of the brain ECF. (3) Angiotensin may facilitate passage of Na$^+$ ions into the receptor cells. The latter would mean, however, that the intracellular Na$^+$ concentration, rather than the extracellular concentration of this ion, determines the activity of the receptors.

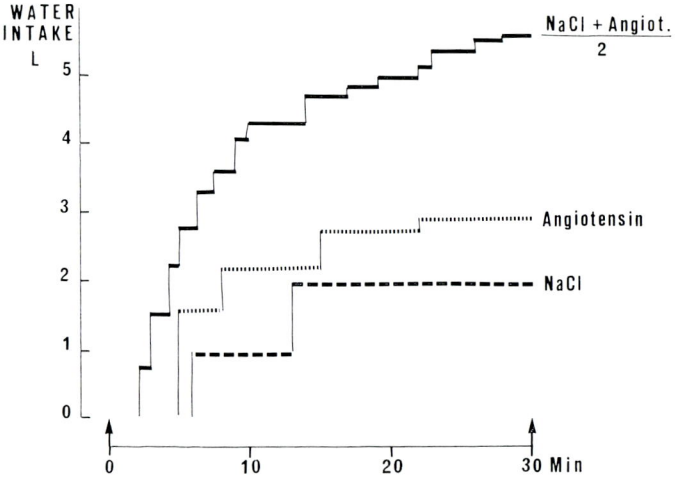

Fig. 8. Synergistic action of Na$^+$ and angiotensin on central regulation of thirst $\frac{\text{"NaCl+Angiot."}}{2}$ = The water consumption of a goat during 30 min infusion into 3rd brain ventricle of angiotensin (2 ng/kg min) dissolved in hypertonic (0.33 M) NaCl. "Angiotensin" = The drinking of the same animal during the corresponding infusion of the double amount of angiotensin dissolved in slightly hypotonic saline. "NaCl" = The water consumption of the animal during 30 min intraventricular infusion of 0.5 M NaCl. Rate of infusion 10 µl/min. Infusion periods between arrows. Interval between experiments 3 days. (From ANDERSSON and WESTBYE, 1970)

Conclusions

It is now well established that the hypothalamus is the principal site of the central control of energy balance. The demonstration of two antagonistic feeding systems—a ventromedial, inhibitory ("satiety") system, and a lateral, excitatory ("hunger") system—indicates that the homeostatic regulation of food intake is achieved by the reciprocal action of the two systems. The excitatory system seemingly has a more basic influence on food intake and the regulation is exerted mainly by the inhibitory system. Hence, it appears that the hypothalamus primarily regulates the degree of satiety, rather than the intensity of the hunger sensation. Suggestive evidence has been provided that "satiety" receptors, which respond to humoral metabolic signals reside in the ventromedial hypothalamus.

The possibility remains, however, that the importance of this part of the brain in the control of energy balance is merely as an integrative centre for hunger and satiety signals from sensory cells elsewhere in the body or the central nervous system.

At the present stage it may seem presumptuous to put forward speculations which diverge from the current idea of an "osmometric" and a "volumetric" regulation of thirst and the release of ADH. Strong evidence has obviously been produced that the water balance principally is controlled by the two, which appear complementary to each other in their common endeavour to maintain body fluid homeostasis. Nevertheless, some of the above-mentioned studies of brain mechanisms regulating thirst and the release of ADH may be an excuse for the following "unorthodox" ideas.

A possible alternative to hypothalamic osmoreceptors in VERNEY'S (1947) sense would be receptors which are located near the 3rd brain ventricle and which are influenced by the Na^+ ion concentration of the cerebrospinal fluid. The observed angiotensin-Na^+ interaction indicates that angiotensin either make such Na^+ receptors more sensitive to the environmental Na^+ concentration, or in some manner increase the exposure of the receptors to Na^+. Therefore, a "volumetric" regulation of thirst and ADH release which is mediated by the renin-angiotensin system also may have its final link in such Na^+ receptors.

References

ADOLPH, E. F.: The urge to eat and drink in rats. Amer. J. Physiol. **151**, 110–125 (1947).
ANAND, B. K., BROBECK, J. R.: Localization of a "feeding center" in the hypothalamus of the rat. Proc. Soc. exp. Biol. (N.Y.) **77**, 323–324 (1951a).
— — Hypothalamic control of food intake in rats and cats. Yale J. Biol. Med. **24**, 123–140 (1951b).
— CHHINA, G. S., SHARMA, K. N., DUA, S., SINGH, B.: Activity of single neurons in the hypothalamic feeding centres: effect of glucose. Amer. J. Physiol. **207**, 1146–1154 (1964).
— DUA, S., SHOENBERG, K.: Hypothalamic control of food intake in cats and monkeys. J. Physiol. (Lond.) **127**, 143–152 (1955).
— — SINGH, B.: Electrical activity of the hypothalamic "feeding centers" under the effect of changes in blood chemistry. Electroenceph. clin. Neurophysiol. **10**, 54–59 (1961a).
— SUBBERWAL, U., MANCHANDA, S. K., SINGH, B.: Glucoreceptor mechanism in the hypothalamic feeding centres. Indian. J. med. Res. **49**, 717–724 (1961b).
ANDERSSON, B.: Polydipsia caused by intrahypothalamic injections of hypertonic NaCl-solutions. Experientia (Basel) **8**, 157–159 (1952).
— The effect of injections of hypertonic NaCl-solutions into different parts of the hypothalamus of goats. Acta physiol. scand. **18**, 188–201 (1953).
— DALLMAN, M. F., OLSSON, K.: Observations on central control of drinking and of the release of antidiuretic hormone (ADH). Life Sci. **8**, 425–432 (1969).
— ERIKSSON, L., OLTNER, R.: Further evidence for angiotensin-sodium interaction in central control of fluid balance. Life Sci. **9**, 1091–1096 (1970b).
— GALE, C. C., HÖKFELT, B., LARSSON, B.: Acute and chronic effects of preoptic lesions. Acta physiol. scand. **65**, 45–60 (1965).
— — SUNDSTEN, J. W.: Effects of chronic central cooling on alimentation and thermoregulation. Acta physiol. scand. **55**, 177–188 (1962).
— LARSSON, B.: Influence of local temperature changes in the preoptic area and rostral hypothalamus on the regulation of food and water intake. Acta physiol. scand. **52**, 75–89 (1961).
— — PERSSON, N.: Some characteristics of the hypothalamic "drinking centre" in the goat as shown by the use of permanent electrodes. Acta physiol. scand. **50**, 140–152 (1960).

ANDERSSON, B., MCCANN, S. M.: Drinking, antidiuresis and milk-ejection from electrical stimulation within the hypothalamus of the goat. Acta physiol. scand. **35**, 191–201 (1955).
— — The effect of hypothalamic lesions on the water intake of the dog. Acta physiol. scand. **35**, 312–320 (1956).
— OLSSON, K., WARNER, R. G.: Dissimilarities between the central control of thirst and the release of antidiuretic hormone (ADH). Acta physiol. scand. **71**, 57–64 (1967).
— WESTBYE, O.: Synergistic action of sodium and angiotensin on brain mechanisms controlling fluid balance. Life Sci. **9**, 601–608 (1970a).
— WYRWICKA, W.: The elicitation of a drinking motor conditioned reaction by electrical stimulation of the hypothalamic "drinking area" in the goat. Acta physiol scand. **41**, 194–198 (1957).
BAILE, C. A., MAYER, J.: Hyperphagia in ruminants induced by a depressant. Science **151**, 458–459 (1966).
— — Effect of insulin induced hypoglycemia and hypoacetoemia on eating behavior in goats. J. Dairy Sci. **51**, 1495–1499 (1968).
BAILLY, P., MORRISON, S. D.: The nature of the suppression of food intake by lateral hypothalamic lesions in rats. J. Physiol. (Lond.) **165**, 227–245 (1963).
BELLOWS, R. T., WAGENEN, W. P. VAN: The relationship of polydipsia and polyuria in diabetes insipidus. A study of experimental diabetes insipidus in dogs with and without esophageal fistulae. J. nerv. ment. Dis. **88**, 417–473 (1938).
BONJOUR, J. P., MALVIN, R. L.: Stimulation of ADH release by the renin-angiotensin system. Amer. J. Physiol. **218**, 1555–1559 (1970).
BRADBURY, M. W. B., DAVSON, H.: The transport of urea, creatinine, and certain monosacharides between blood and fluid perfusing the cerebral ventricular system of rabbits. J. Physiol. (Lond.) **170**, 195–211 (1964).
BRECHER, G., WAXLER, S.: Obesity in albino mice due to a single injection of goldthioglucose. Proc. Soc. exp. Biol. (N.Y.) **70**, 498–503 (1949).
BROBECK, J. R.: Effects of variations in activity, food intake and environmental temperature on weight gain in albino rat. Amer. J. Physiol. **143**, 1–5 (1945).
— Food and temperature. Recent Progr. Hormone Res. **16**, 439–466 (1960a).
— Regulation of feeding and drinking. In: Handbook of physiology, vol. II, sect. 1, Neurophysiology. Washington D.C.: Amer. Physiol. Soc. 1960b.
— TEPPERMAN, J., LONG, C. N. H.: Experimental hypothalamic hyperphagia in the albino rat. Yale J. Biol. Med. **15**, 831–853 (1943).
BROOKS, C. F., KOCH, A., WANG, J.: Kinetics of sodium transfer from blood to brain of the dog. Amer. J. Physiol. **218**, 693–702 (1970).
BRÜGGER, M.: Freßtrieb als hypothalamisches Symptom. Helv. physiol. Pharmacol. Acta **1**, 183–198 (1943).
CIZEK, L. J., SEMPLE, R. E., HUANG, K. C., GREGERSEN, M. I.: Effect of extracellular electrolyte depletion on water intake in dogs. Amer. J. Physiol. **164**, 415–422 (1951).
CLEMENTE, C. D., SUTIN, J., SILVERSTONE, J. T.: Changes in the electrical activity of the medulla on the intravenous injection of hypertonic solutions. Amer. J. Physiol. **188**, 193–198 (1957).
COHN, C., JOSEPH, D.: Influence of body weight and body fat on appetite of "normal" lean and obese rats. Yale J. Biol. Med. **34**, 598–607 (1962).
CRONE, C.: Facilitated transfer of glucose from blood into brain tissue. J. Physiol. (Lond.) **181**, 103–113 (1965).
CROSS, B. A., GREEN, J. D.: Activity of single neurons in the hypothalamus: Effect of osmotic and other stimuli. J. Physiol. (Lond.) **148**, 554–569 (1959).
CSAKY, T. Z., RIGOR, B. M.: The choroid plexus as a glucose barrier. In: Brain barrier systems. Progr. Brain Res. **29**, 147–158 (1968).
DAVSON, H., LUCK, C. P.: The effect of acetazoleamide on the chemical composition of the aqueous humour and cerebrospinal fluid of some mammalian species and on the rate of turnover of ^{24}Na in these fluids. J. Physiol. (Lond.) **137**, 279–293 (1957).
DENTON, D. A.: Evolutionary aspects of the emergence of aldosterone secretion and salt appetite. Physiol. Rev. **45**, 245–295 (1965).
DESIRAJU, T., BANERJEE, M. G., ANAND, B. K.: Activity of single neurons in the feeding centres: effect of 2-deoxy-d-glucose. Physiol. Behav. **3**, 757–760 (1968).

EPSTEIN, A. N.: Reciprocal changes in feeding behavior produced by intrahypothalamic chemical injections. Amer. J. Physiol. **199**, 969–974 (1960).
— FITZSIMONS, J. T., ROLLS, B. J. (née SIMONS): Drinking induced by injection of angiotensin into the brain of the rat. J. Physiol. (Lond.) **210**, 457–474 (1970).
— — SIMONS, B. J.: Drinking caused by the intracranial injection of angiotensin into the rat. J. Physiol. (Lond.) **200**, 98—100P (1969).
— TEITELBAUM, P.: Severe and persistent deficits in thirst produced by lateral hypothalamic damage. In: Thirst. Oxford: Pergamon Press 1964.
ERBSLÖH, F., KLÄRNER, P., BERNSMEIER, A.: Über die Bilanz des cerebralen Zuckerstoffwechsels. Klin. Wschr. **36**, 849–852 (1958).
EULER, C. VON: A preliminary note on slow hypothalamic "osmiopotentials". Acta physiol. scand. **29**, 133–136 (1953).
FISHER, C., INGRAM, W. R., RANSON, S. W.: Diabetes insipidus and the neuro-hormonal control of water balance. Ann Arbor, Mich.: Edwards Bros. 1938.
FITZSIMONS, J. T.: Drinking by rats depleted of body fluids without increase in osmotic pressure. J. Physiol. (Lond.) **159**, 297–309 (1961).
— The role of a renal thirst factor in drinking induced by extracellular stimuli. J. Physiol. (Lond.) **201**, 349–368 (1969).
— SIMONS, B. J.: The effect on drinking in the rat of intravenous infusion of angiotensin, given alone or in combination with other thirst stimuli. J. Physiol. (Lond.) **203**, 45–57 (1969).
GASNIER, A., MAYER, A.: Recherches sur la régulation de la nutrition. Ann. Physiol. Physiochim. biol. **15**, 145–214 (1939).
GAUER, O. H., HENRY, J., SIEKER, J. P., WENDT, W. E.: Heart and lungs as a receptor region controlling blood volume. Amer. J. Physiol. **167**, 786–787 (1951).
— HENRY, J. P., SIEKER, H. O., WENDT, W. E.: The effect of negative pressure breathing on urine flow. J. clin. Invest. **33**, 287–296 (1954).
GOLD, R. M.: Hypothalamic hyperphagia produced by parasagittal knife cuts. Physiol. Behav. **5**, 23–26 (1970).
GREGERSEN, M. I.: The physiological mechanism of thirst. Amer. J. Physiol. **101**, 44–45 (1932).
GROSSMAN, S. P.: Eating and drinking elicited by direct adrenergic or cholinergic stimulation of hypothalamus. Science **132**, 301–302 (1960).
— Behavioral effects of direct chemical stimulation of central nervous system structures. Int. J. Neuropharmacol. **3**, 45–58 (1964).
HAMILTON, C. L.: Food and temperature. In: Handbook of physiology, vol. I, sect. 6, Alimentary canal. Washington, D.C.: Amer. Physiol. Soc. 1967.
— BROBECK, J. R.: Food intake and temperature regulation in rats with rostral hypothalamic lesions. Amer. J. Physiol. **207**, 291–297 (1964).
HANN, F. VON: Über die Bedeutung der Hypophysenveränderungen bei Diabetes insipidus. Frankfurt. Z. Path. **21**, 337–365 (1918).
HERREVELD, A. VAN: Brain tissue electrolytes. Molecular Biol. and Med. Series. London: Butterworths 1966.
HERVEY, G. R.: The effects of lesions in the hypothalamus in parabiotic rats. J. Physiol. (Lond.) **145**, 336–352 (1959).
HETHERINGTON, A. W., RANSON, S. W.: Experimental hypothalamico-hypophyseal obesity in the rat. Proc. Soc. exp. Biol. (N.Y.) **41**, 465–466 (1939).
— — The spontaneous activity and food intake of rats with hypothalamic lesions. Amer. J. Physiol. **136**, 609–617 (1942).
HOEBEL, B. G., TEITELBAUM, P.: Hypothalamic control of feeding and selfstimulation. Science **135**, 375–377 (1962).
— Hypothalamic lesions by electrocauterization. Disinhibition of feeding and self-stimulation. Science **149**, 452–453 (1965).
HOLLAND, R. C., CROSS, B. A., SAWYER, C. H.: EEG correlates of osmotic activation of the milk-ejection mechanism. Amer. J. Physiol. **196**, 796–802 (1959).
HOLMES, J. H., CIZEK, L. J.: Observations on sodium chloride depletion in the dog. Amer. J. Physiol. **164**, 407–414 (1951).
— GREGERSEN, M. I.: Observation on drinking induced by hypertonic solutions. Amer. J. Physiol. **162**, 326–337 (1950).

Ingram, D. L.: Effects of heating and cooling the hypothalamus on food intake in the pig. Brain Res. 11, 714–716 (1968).
Janowitz, H. D., Ivy, A. C.: Role of blood sugar levels in spontaneous and insulin-induced hunger in man. J. appl. Physiol. 1, 643–650 (1949).
Jewell, P. A., Verney, E. B.: An experimental attempt to determine the site of the neurohypophysial osmoreceptors in the dog. Phil. Trans. B 240, 197–324 (1957).
Jiménez Dias, C., Linazasoro, J. M., Castro Mendoza, H.: Further study of the part played by the kidneys in the regulation of thirst. Bull. Inst. med. Res. (Madr.) 12, 60–67 (1959).
Katzman, R., Leiderman, P. H.: Brain potassium exchange in normal adult and immature rats. Amer. J. Physiol. 175, 263–270 (1953).
Keller, A. D.: Neurologically induced physiological resistence to hypothermia. Ann. N.Y. Acad. Sci. 80, 457–474 (1959).
Kennedy, G. C.: The effect of lesions in the hypothalamus on appetite. Proc. Nutr. Soc. 12, 160–165 (1953a).
— The role of depot fat in the hypothalamic control of food intake in the rat. Proc. roy. Soc. 140, 578–592 (1953b).
Kleeman, C. R., Cutler, R. E.: The neurohypophysis. Ann. Rev. Physiol. 25, 385–432 (1963).
Larsson, S.: On the hypothalamic organization of the nervous mechanism regulating food intake. Acta physiol. scand. 32 (Suppl. 115), 1–63 (1954).
Laszlo, F. A., Wied, D. de: Antidiuretic hormone content of the hypothalamo-neurohypophyseal system and urinary excretion of antidiuretic hormone in rats during the development of diabetes insipidus after lesions in the pituitary stalk. J. Endocr. 36, 125–137 (1966).
Leonhardt, H.: Bukettförmige Strukturen im Ependym der Regio Hypothalamica des Ventrikels beim Kaninchen. Z. Zellforsch. 88, 297–317 (1968).
Leusen, I.: Influence du pH du liquide céphalo-rachidien sur la respiration. Experientia (Basel) 6, 272 (1950).
Liebelt, R. A., Perry, J. H.: Hypothalamic lesions associated with goldthioglucose-induced obesity. Proc. Soc. exp. Biol. (N.Y.) 95, 774–780 (1957).
Linazasoro, J. M., Jiménez Dias, C., Castro Mendoza, H.: The kidney and thirst regulation. Bull. Inst. med. Res. (Madr.) 7, 53–61 (1954).
Magoun, H. W., Harrison, F., Brobeck, J. R., Ranson, S. W.: Activation of heat loss mechanisms by local heating of the brain. J. Neurophysiol. 1, 101–114 (1938).
Marshall, N. B., Mayer, J.: Energy balance in goldthioglucose obesity. Amer. J. Physiol. 178, 271–276 (1954).
Massopust, L. C., Jr.: The hypothalamic syndrome in rats with experimental lesions. Neurology (Minneap.) 5, 472–478 (1955).
Mayer, A.: Essai sur la soif: Ses causes et son mécanisme. Thesis. Paris: Jouve & Boyer 1900.
Mayer, J.: Genetic, traumatic, and environmental factors in the ethology of obesity. Physiol. Rev. 33, 472–508 (1953).
— Physiological and nutritional aspects of obesity. Rev. Nutr. Res. 19, 35–55 (1958).
— Barnett, R.: Unilateral lesions of N. ventromedialis and obesity in rat. Science 121, 599 (1955).
— Marshall, N. B.: Specificity of goldthioglucose for ventromedial hypothalamic lesions and hyperphagia. Nature (Lond.) 178, 1399–1400 (1956).
Miller, N. E.: Some motivational effects of brain stimulation and drugs. Fed. Proc. 19, 846 (1960).
— Bailey, C. J., Stevenson, J. A. F.: Decreased "hunger" but increased food intake resulting from hypothalamic lesions. Science 112, 256–259 (1950).
Mitchell, R. A., Loeschcke, H. H., Massion, W. H., Severinghaus, J. W.: Respiratory responses mediated through superficial chemisensitive areas on the medulla. J. appl. Physiol. 18, 523–533 (1963).
Montemurro, D. G., Stevenson, J. A. F.: Adipsia produced by hypothalamic lesions in the rat. Canad. Biochem. Physiol. 35, 31–37 (1957).
Morgane, J. P.: Medial forebrain bundle and "feeding centers" of the hypothalamus. J. comp. Neurol. 117, 1–26 (1961b).
— Jacobs, H. L.: Hunger and satiety. Wld Rev. Nutr. Diet. 10, 100–213 (1969).

Morgane, P. J.: Evidence of a "hunger motivational" system in the lateral hypothalamus of the rat. Nature (Lond.) **191**, 672–674 (1961a).
O'Connor, W. J.: The normal interphase in the polyuria which follows section of the supra-optico-hypophysial tracts in the dog. Quart. J. exp. Physiol. **37**, 1–10 (1952).
Olsson, K.: Studies on central regulation of secretion of antidiuretic hormone (ADH) in the goat. Acta physiol. scand. **77**, 465–474 (1969).
— Relations of polyuria and polydipsia in experimental diabetes insipidus. Acta physiol. scand. **78**, 20–27 (1970).
Oomura, Y., Ono, T., Ooyama, H., Wayner, M. J.: Glucose and osmosensitive neurones of the rat hypothalamus. Nature (Lond.) **222**, 282–284 (1969).
Park, C. R., Johnson, L. H., Wright, J. H., Batsel, H.: Effect of insulin on transport of several hexoses and pentoses into cells of muscle and brain. Amer. J. Physiol. **191**, 13–18 (1957).
Peart, W. S.: The renin-angiotensin system. Pharmacol. Rev. **17**, 143–182 (1965).
Rodgers, W. L., Epstein, A. N., Teitelbaum, P.: Lateral hypothalamic aphagia: motor failure or motivational deficit? Amer. J. Physiol. **208**, 334–342 (1965).
Rowe, G. G., Maxwell, G. M., Castillo, C. A., Crumption, C. W.: A study in man of cerebral blood flow and cerebral glucose, lactate, and pyruvate metabolism before and after eating. J. clin. Invest. **38**, 2154–2158 (1959).
Sawyer, C. H., Gernandt, B. E.: Effects of intracarotid and intraventricular injections of hypertonic solutions on electrical activity of the rabbit brain. Amer. J. Physiol. **185**, 209–216 (1956).
Simkins, K. L., Suttie, J. W., Baumgardt, B. R.: Regulation of food intake in ruminants. 3. Variation in blood and rumen metabolites in relation to food intake. J. Dairy Sci. **48**, 1629–1634 (1965).
Smith, G. P., Epstein, A. N.: Increased feeding in response to decreased glucose utilization in the rat and monkey. Amer. J. Physiol. **217**, 1083–1087 (1969).
Smith, R. W., McCann, S. M.: Increased and decreased water intake in the rat with hypothalamic lesions. Proc. 1st int. Symp. Thirst, p. 381–392. Oxford: Pergamon Press 1964.
Spector, N. H., Brobeck, J. R., Hamilton, C. L.: Feeding and core temperature in albino rats: Changes induced by preoptic heating and cooling. Science **161**, 286–288 (1968).
Stevenson, J. A. F.: Effects of hypothalamic lesions on water and energy metabolism in the rat. Recent Progr. Hormone Res. **4**, 363–394 (1949).
— Neural control of food and water intake. In: Hypothalamus, pp. 524–621. Springfield, Ill.: C. C. Thomas 1969.
— Box, B. M., Montemurro, D. G.: Evidence of possible association pathways for the regulation of food and water intake in the rat. Canad. J. Physiol. Pharmacol. **42**, 855–860 (1964).
— Welt, L. G., Orloff, J.: Abnormalities of water and electrolyte metabolism in rats with hypothalamic lesions. Amer. J. Physiol. **161**, 35–39 (1950).
Stunkard, A. J., Itallie, B. van, Reis, B. B.: The mechanism of satiety: effect of glucagon on gastric hunger contractions in man. Proc. Soc. exp. Biol. (N.Y.) **89**, 258–261 (1955).
— Wolff, H. G.: Correlation of arteriovenous glucose differences, gastric hunger contractions and the experience of hunger in man. Fed. Proc. **13**, 147 (1954).
Sundsten, J. W., Sawyer, C. H.: Electro-encephalographic evidence of osmosensitive elements in olfactory bulb of dog brain. Proc. Soc. exp. Biol. (N.Y.) **101**, 524–527 (1959).
Teitelbaum, P.: Sensory control of hypothalamic hyperphagia. J. comp. physiol. Psychol. **48**, 156–163 (1955).
— Random and food-directed activity in hyperphagic and normal rats. J. comp. physiol. Psychol. **50**, 486–490 (1957).
— Epstein, A. N.: The lateral hypothalamic syndrome: recovery of feeding and drinking after lateral hypothalamic lesions. Psychol. Rev. **69**, 74–90 (1962).
— Stellar, E.: Recovery from the failure to eat, produced by hypothalamic lesions. Science **120**, 894–895 (1954).
Verney, E. B.: The antidiuretic hormone and factors which determine its release. Proc. roy. Soc. B **135**, 25–106 (1947).

Vincent, J. D., Hayward, J. N.: Activity of single cells in the osmoreceptor-supraoptic nuclear complex in the hypothalamus of the waking rhesus monkey. Brain Res. 23, 105–108 (1970).

Witt, D. M., Keller, A. D., Batsel, H. L., Lynch, J. R.: Absence of thirst and resultant syndrome associated with anterior hypothalamectomy in the dog. Amer. J. Physiol. 171, 780 (1952).

Wolf, A. V.: Osmometric analysis of thirst in man and dog. Amer. J. Physiol. 161, 75–86 (1950).

Woodbury, D. M.: Distribution of nonelectrolytes and electrolytes in the brain as effected by alterations in cerebrospinal fluid secretion. In: Brain barrier systems, pp. 297–314. Amsterdam: Elsevier publ. comp. 1968.

Woodward, G. E., Hudson, M. T.: The effect of 2-deoxy-D-glucose on glycolysis and respiration of tumor and normal tissues. Cancer Res. 14, 599–605 (1954).

Wyrwicka, W., Dobrzecka, C.: Relationship between feeding and satiation centers of the hypothalamus. Science 132, 805–806 (1960).

Author Index

Page numbers in italics refer to the bibliography

Abbott, C. P., Daly, M. De B., Howe, A. 51, 53, *76*
— Howe, A., Joels, N. 53
Abrams, R., Hammel, H. T. 167, *182*
— see Rawson, R. O. 167, *186*
Adair, E. R. 164, *182*
Adams, T. 167, *182*
Adams, W. E., see Hirsch, E. F. 104, *110*
Adolph, E. F. 194, *211*
Adrian, E. D. 15, 25, 38, *40*; 99, 101, 107, *109*; 115, *156*
Agostini, E., Chinnock, J. E., Burgh Daly, M. De, Murray, J. G. 114, 115, 126, *156*
Agostoni, E., see Campbell, E. J. M. 106, *109*
Al-Lami, F., Murray, R. G. 53, 57, *76*
Alvarez-Buylla, R., Ramirez De, Arellano, J. 1, *40*
Amadon, R. S., see Schalk, A. F. 144, *160*
Amann, A., Schaefer, H. 36, *40*
Anand, B. K., Banerjee, M. G., Chhina, G. S. 175, *182*
— Brobeck, J. R. 198, 199, 200, *211*
— Chhina, G. S., Sharma, K. N., Dua, S., Singh, B. 195, *211*
— Dua, S., Shoenberg, K. 198, 200, *211*
— — Singh, B. 195, 200, *211*
— Pillai, R. V. 145, *156*
— Subberwal, U., Manchanda, S. K., Singh, B. 195, 200, *211*
— see Desiraju, T. 195, 200, *212*

Andersen, H. T., Andersson, B., Gale, C. 164, *182*
Andersson, B. 144, *156*; 203, 206, *211*
— Brook, A. H., Ekman, L. 164, *182*
— Dallman, M. F., Olsson, K. 209, *211*
— Ekman, L., Gale, C. C., Sundsten, J. W. 164, *182*
— Eriksson, L., Oltner, R. 209, *211*
— Gale, C. C., Hökfelt, B., Larsson, B. 202, *211*
— — — Ohga, A. 164, *182*
— — Sundsten, J. W. 198, *211*
— Larson, B. 164, *183*; 198, 202, *211*
— — Persson, N. 203, *211*
— McCann, S. M. 202, 203, *212*
— Olsson, K., Warner, R. G. 209, *212*
— Westbye, O. 209, 210, *212*
— Wyrwicka, W. 203, *212*
— see Andersen, H. T. 164, *182*
Andersson, S., Olbe, L. 147, *156*
Andrew, B. L. 91, *109*; 118, 134, *156*
Andrews, W. H. H., Stratman, C. J. 149, *157*
Angell James, J. E. 10, 12, 13, *40*
— Daly, M. De B. 12, *40*
Anichkov, S. V., Belen'Kii, M. L. 66, 67, 72, *76*
— Zakusov, V. V., Kuznetzov, A. I., Polyakov, N. G. 72, *76*
— see Krylov, S. S. 66, *79*
Armitage, A. K., Dean, A. C. B. 146, *157*

Arndt, J. O., Brambring, P., Hindorf, K., Röhnelt, M. 16, 17, 29, 30, *40*
— see Brambring, P. 17, 29, 30, *41*
Ash, R. W. 144, 149, *157*
— Kay, R. N. B. 124, 140, 141, 143, 144, *157*
Asteroth, H., see Haus, W. H. 8, *42*
Aviado, D. M., Li, T. H., Kalow, W., Schmidt, C. F., Turnbull, G. L., Peskin, G. W., Hess, M. E., Weiss, A. J. 95, *109*
— see Letona, J. M. L. De 95, *111*

Baile, C. A. 149, *157*
— Mayer, J. 192, 195, *212*
— Pfander, W. M. 149, *157*
Bailey, C. J., see Miller, N. E. 192, *214*
Bailly, P., Morrison, S. D. 199, *212*
Baker, M. A., see Hayward, J. N. 167, *184*
Balch, C. C., see Freer, M. 144, *158*
Baldwin, B. A., Ingram, D. L. 164, 167, *183*
— — Leblanc, J. 164, *183*
Band, D. M., Cameron, I. R., Semple, S. J. G. 75, *76*
Banerjee, M. G., see Anand, B. K. 175, *182*
— see Desiraju, T. 195, 200, *212*
Banister, J., Fegler, Hebb, C. 97, *109*
Barbour, H. G. 162, *183*
Barner, H. B., see Hirsch, E. F. 104, *110*
Barnett, R., see Mayer, J. 192, *214*
Batsel, H., see Park, C. R. 195, *215*

Batsel, H. L., see Witt, D. M. 202, *216*
Baumgardt, B. R., see Simkins, K. L. 195, *215*
Baumgarten, R. von, see Gupta, P. D. 17, 23, *42*
Bayliss, W. M., Starling, E. H. 139, *157*
Beckman, A. L., Eisenman, J. S. 180, *183*
Beghelli, V., Borgatti, G., Parmeggiani, P. L. 128, *157*
Belen'Kii, M. L., see Anichkov, S. V. 66, 67, 72, *76*
Beleslin, D., Varagic, V. 139, *157*
Bellows, R. T., Wagenen, W. P. van 202, *212*
Bensch, K. G., Gordon, G. B., Miller, L. R. 90, *109*
Benzinger, T. H. 161, *183*
Bernsmeier, A., see Erbslöh, F. 190, *213*
Bessou, P., Perl, E. R. 130, 131, *157*
Bevan, J. A., Kinnison, G. L. 39, *41*
Bianconi, R., Green, J. H. 10, 15, 27, 39, *41*
Biscoe, T. J. 58, 74, *76*
— Bradley, G. W., Purves, M. J. 58, 62, 63, 65, 71, 75, *76*
— Lall, A., Sampson, S. R. 57, 58, 62, 63, 75, *76*
— Purves, M. J. 58, 63, 75, *76*
— — Sampson, S. R. 58, 59, 60, 62, 63, 67, 68, 75, *76*
— Sampson, S. R. 58, 69, *76*
— Stehbens, W. E. 57, 58, *76*
— Taylor, A. 58, 74, *76*
— see Sampson, S. R. 71, *80*
Black, A. M. S., Torrance, R. W. 75, *76*
Blair, E. L., Brown, J. G., Harper, A. A., Scratcherd, T. 145, 149, *157*
Bligh, J. 161, 167, *183*
— Maskrey, M. 180, *183*
Bliss, E. L., see Zapata, P. 72, 73, *80*
Bloor, C. M. 10, 12, *41*

Bogue, J. Y., Stella, G. 49, 74, *76*
Bond, H. E., see Dougherty, R. W. 143, *157*
Bonjour, J. P., Malvin, R. L. 209, *212*
Borgatti, G., see Beghelli, V. 128, *157*
Bosma, J. F., see Takagi, Y. 88, *112*
Boss, J., Green, J. H. 10, *41*
Bouckaert, J. J., see Heymans, C. 48, 67, 72, 75, *78*
Bower, E. A. 118, 130, *157*
Boyd, I. A. 5, 38, *41*
Boyd, J. D., see Lever, J. D. 51, 53, *79*
Box, B. M., see Stevenson, J. A. F. 198, *215*
Bradbury, M. W. B., Davson, H. 190, *212*
Bradley, G. W., see Biscoe, T. J. 58, 62, 63, 65, 71, 75, *76*
Brambring, P., Röhnelt, M., Hindorf, K., Arndt, J. O. 17, 29, 30, *41*
Brambring, P., see Arndt, J. O. 16, 17, 29, 30, *40*
Brecher, G., Waxler, S. 195, *212*
Brightman, M. W., see Rall, W. 57, *80*
Brobeck, J. R. 196, 197, 198, 202 *212*
— Tepperman, J., Long, C. N. H. 192, *212*
— see Anand, B. K. 198, 199, 200, *211*
— see Chai, C. Y. 175, *183*
— see Hamilton, C. L. 198, *213*
— see Magoun, H. W. 162, 163, *185*; 198, *214*
— see Spector, N. H. 198, *215*
Bronk, D. W., Stella, G. 8, 25, *41*
— see Gammon, G. D. 15, 16, *42*; 130, 131, *158*
Brook, A. H., see Andersson, B. 164, *182*
Brooks, C. McC., Hoffman, B. F., Suckling, E. E. 25, *41*
Brooks, C. F., Koch, A., Wang, J. 190, *212*

Brown, A. M. 35, 36, *41*
Brown, J. G., see Blair, E. L. 145, 149, *157*
Brück, K., see Kundt, H. W. 162, 167, *185*
Brück, K., Wünnenberg, W. 165, 166, 170, *183*
— see Krüger, F. J. 164, *185*
— see Wünnenberg, W. 166, 179, *186*
Brügger, M. 199, *212*
Bülbring, E., Crema, A. 139, *157*
— Lin, R. C. Y. 139, 140, *157*
— — Schofield, G. C. 139, *157*
— see Whitteridge, D. 15, *45*
Burgh Daly, M. De, see Agostini, E. 114, 115, 126, *156*
Burger, E. J., Macklem, P. 106, *109*
Burton, A. C. 10, *41*

Cabanac, M., Hammel, H. T., Hardy, J. D. 170, *183*
— Hardy, J. D. 175, *183*
— Stolwijk, J. A. J., Hardy, J. D. 181, *183*
Cameron, I. R., see Band, D. M. 75, *76*
Campbell, E. J. M., Agostoni, E., Davis, J. N. 106, *109*
Campling, R. C., see Freer, M. 144, *158*
Cannon, W. B., Washburn, A. 128, 153, *157*
Carlisle, H. J. 164, *183*
Castillo, C. A., see Rowe, G. G. 190, 195, *215*
Castro Mendoza, H., see Jiménez Dias, C. 208, *214*
— see Linazasoro, J. M. 208, *214*
Cauna, N. C., Ross, L. L. 57, *76*
Celestino Da Costa 50, 65, *76*
Chai, C. Y., Mu, J. Y., Brobeck, J. R. 175, *183*
Chapman, K. M., Pearce, J. W. 17, *41*
— see Pearce, J. W. 22, *44*

Author Index

Chen, I-Li., Yates, R. D. 53, *76*
— — Duncan, D. 51, 53, *76*
— see Yates, R. D. 53, 71, 73, *80*
Chhina, G. S., see Anand, B. K. 175, *182*; 195, *211*
Chinnock, J. E., see Agostini, E. 114, 115, 126, *156*
Chiocchio, S. R., see Morita, E. 51, *79*
Christie, R. V. 75, *76*
Cizek, L. J., Semple, R. E., Huang, K. C., Gregersen, M. I. 205, *212*
— see Holmes, J. H. 205, *213*
Clemente, C. D., Sutin, J., Silverstone, J. T. 206, *212*
Code, C. F., Watkinson, G. 148, *157*
Cohn, C., Joseph, D. 191, *212*
Colebatch, H. J. H., see DeKock, M. A. 95, *110*
Coleridge, H., Coleridge, J. C. G., Howe, A. 47, 49, 50, *76, 77*; 107, *109*
Coleridge, H. M., Coleridge, J. C. G., Kidd, C. 3, 7, 26, 27, 29, 31, 32, 33, 34, 35, 36, 37, 38, 39, *41*
— — Luck, J. G. 102, *109*
Coleridge, J. C. G., Hemingway, A., Holmes, R. L., Linden, R. J. 16, 17, 20, 22, 25, 26, 27, 29, *41*
— Kidd, C. 39, *41*
— — Sharp, J. A. 39, *41*
— — Linden, R. J. 31, *41*
— see Coleridge, H. 47, 49, 50, *76, 77*; 107, *109*
— see Coleridge, H. M. 3, 7, 26, 27, 29, 31, 32, 33, 34, 35, 36, 37, 38, 39, *41*; 102, *109*
Comroe, J. H., Jr. 47, *77*
—, Schmidt, C. F. 60, 73, 74, 75, *77*
— see Dawes, G. S. 105, *109*
Cook, R. D., King, A. S. 89, *109*
Cooper, K. E., Cranston, W. I., Honour, A. J. 181, *183*
— — Snell, E. S. 181, *183*

Cooper, T., see Hirsch, E. F. 104, *110*
Corbett, J. L., Kerr, J. A., Prys-Roberts, C., Crampton-Smith, A., Spalding, J. M. K. 88, *109*
Corbit, J. D. 164, *183*
Cornew, R. W., see Hammel, H. T. 167, 168, *184*
Cragg, B. G., Evans, D. H. L. 145, 146, *157*
Crampton-Smith, A., see Corbett, J. L. 88, *109*
Cranston, W. I., Luff, R. H., Rawlins, M. D., Rosendorff, C. 181, *183*
— see Cooper, K. E. 181, *183*
Crema, A., see Bülbring, E. 139, *157*
Crone, C. 190, *212*
Cross, B. A., Dyer, R. G. 174, *183*
— Green, J. D. 171, *183*; 206, *212*
— Silver, I. A. 177, *183*
— see Holland, R. C. 206, *213*
Cross, K. W. 95, *109*
Crumption, C. W., see Rowe, G. G. 190, 195, *215*
Csaky, T. Z., Rigor, B. M. 190, *212*
Culver, G. A., Rahn, H. 95, *109*
Cunningham, D. J., Stolwijk, J. A. J., Murakami, N., Hardy, J. D. 163, 180, *183*
Cunningham, D. J. C., see Lloyd, B. B. 68, *79*
Cutler, R. E., see Kleeman, C. R. 206, *214*

Dallman, M. F., see Andersson, B. 209, *211*
Daly, M. De B., Evans, D. H. L. 114, 115, *157*
— Lambertsen, C. J., Schweitzer, A. 61, 62, 74, *77*
— Scott, M. J. 101, *109*
— see Abbott, C. P. 51, 53, *76*
— see Angell James, J. E. 12, *40*
Daniel, E. E., Wiebe, G. E. 145, 146, 147, *157*

Davis, D. D., see Freeman, W. J. 164, *184*
Davis, H. L., Fowler, W. S., Lambert, E. H. 100, *109*
Davis, J. N., see Campbell, E. J. M. 106, *109*
Davson, H., Luck, C. P. 190, *212*
— see Bradbury, M. W. B. 190, *212*
Dawes, G. S., Comroe, J. H., Jr. 105, *109*
— Mott, J. C., Widdicombe, J. G. 105, *109*
Day, J. J., Komarov, S. H. 148, *157*
Dean, A. C. B., see Armitage, A. K. 146, *157*
Dearnaley, D. P., Fillenz, M., Woods, R. I. 53, *77*
De Boissezon, P. 50, 65, *77*
DeCastro, F. 5, 9, 10, *41*; 48, 49, 50, 51, 57, 58, 65, 71, *77*
— Rubio, M. 50, 51, 53, 57, 58, 65, *77*
De Kock, L. L. 50, *77*
— Dunn, A. E. G. 58, *77*
DeKock, M. A., Nadel, J. A., Zwi, S., Colebatch, H. J. H., Olsen, C. R. 95, *110*
DeLorenzo, A. J. 57, *77*
Dempsey, E. W., see Smith, C. A. 57, *77*
Denny-Brown, D., Robertson, E. G. 128, *157*
Denton, D. A. 205, *212*
De Robertis, E., Franchi, C. M. 57, *77*
Desiraju, T., Banerjee, M. G., Anand, B. K. 195, 200, *212*
Desphande, S. S., Devanandan, M. S. 105, *110*
Devanandan, M. S. 4, 5, *42*
Devanandan, M. S., see Dephande, S. S. 105, *110*
Diamond, J. 12, 13, 14, *42*
— see Gray, J. A. B. 130, *158*
Dickinson, C. J. 27, 32, *42*
Dijkstra, C. 104, *110*
Dillon, W. H., see Opdyke, D. F. 22, *44*
Dobrzecka, C., see Wyrwicka, W. 194, *216*

Dougherty, R.W. 143, *157*
— Habel, R. E., Bond, H. E. 143, *157*
— Meredith, C. D. 143, *158*
Douglas, W.W. 73, *77*
Downey, J. A., Mottram, R. F., Pickering, G.W. 164, *183*
Downie, H. G. 144, *158*
Downing, S. E. 95, *110*
Downman, C. B. B., McSwiney, B. A., Vass, C. C. N. 147, 154, *158*
Dragstedt, L. R., Oberhelman, H. A., Zubiran, J. M., Woodward, E. R. 146, *158*
— Woodward, E. R. 146, *158*
Dua, S., see Anand, B. K. 195, 198, 200, *211*
Dubois, F. S., Foley, J. O. 115, *158*
Duke, H. N., Green, J. H., Neil, E. 60, *77*
Duncan, D., see Chen, I-Li. 51, 53, *76*
— see Yates, R. D. 53, 71, 73, *80*
Duncan, D. L. 128, *158*
Dunn, A. E. G., see De Kock, L. L. 58, *77*
Duomarco, J., see Opdyke, D. F. 22, *44*
Dussardier, M. 141, *158*
Dyer, R. G., see Cross, B. A. 174, *183*

Ead, H.W., Green, J. H., Neil, E. 8, 12, *42*
Edvardsen, P. 150, *158*
Edwards, Mc I. W., see Mills, E. 60, 74, *79*
Edwards, M.W., Mills, E. 74, *77*
Eisele, J. H., see Guz, A. 101, 106, *110*
— see Noble, M. I. M. 106, *111*
Eisenman, J. S. 181, *184*
— Jackson, D. C. 171, 173, 174, *184*
— see Beckman, A. L. 180, *183*
— see Nakayama, T. 171, 172, 175, *185*
Ekman, L., see Andersson, B. 164, *182*

Elftman, A. G. 101, 107, *110*
Elias, E., Gibson, G. J., Greenwood, Linda F., Hunt, J. N., Tripp, J. H. 148, *158*
Engström, H., Wersäll, J. 57, *77*
Epstein, A. N. 196, *213*
— Fitzsimons, J.T., Rolls, B. J. 208, *213*
— — Simons, B. J. 208, *213*
— Teitelbaum, P. 203, *213*
— see Rodgers, W. L. 199, *215*
— see Smith, G. P. 195, *215*
— see Teitelbaum, P. 199, *215*
Erbslöh, F., Klärner, P., Bernsmeier, A. 190, *213*
Eriksson, L., see Andersson, B. 209, *211*
Erlanger, J., see Schoepfle, G. M. 13, *45*
Ernsting, J. 106, *110*
Euler, C. von 161, 164, 170, *184*; 206, *213*
Euler, U. S. von, Liljestrand, G., Zotterman, Y. 9, *42*; 49, 67, *77*
Evans, D. H. L., see Cragg, B. G. 145, 146, *157*
— see Daly, M. De B. 114, 115, *157*
Evans, J. P. 115, *158*
Eyzaguirre, C., Koyano, H. 72, *77*
— — Taylor, J.R. 72, *77*
— Lewin, J. 64, 69, *78*
— Uchizono, K. 5, *42*
— Zapata, P. 72, *78*
— see Sato, A. 7, 12, *45*
— see Zapata, P. 72, 73, *80*

Fahim, M., Gupta, P. D. 16, 17, 20, 22, 25, *42*
Farber, S., see Heymans, C. 72, *78*
Fedde, M. R., Peterson, D. F. 107, 108, *110*
Fegler, see Banister, J. 97, *109*
Feldberg, W. 180, *184*
Fenn, W. O., see Otis, A. B. 101, *111*
Ferguson, T. B., see Forster, R. E. 167, *184*

Ferrer, P., Koller, E. A. 91, *110*
Feyrter, F. 90, *110*
Fex, J. 71, *78*
Fidone, S. J., Sato, A. 3, 4, 5, 6, 7, 8, 9, 12, 14, *42*
— see Sato, A. 7, 12, *45*
Fillenz, M. 53, *78*; 90, 103, *110*
— Woods, R. I. *78*; 82, 88, 89, 97, *110*
— see Dearnaley, D. P. 53, *77*
Fisher, A.W. F. 88, 97, *110*
Fisher, C., Ingram, W. R., Ranson, S.W. 201, 205, *213*
Fitzsimons, J.T. 208, *213*
— Simons, B. J. 208, *213*
— see Epstein, A. N. 208, *213*
Floyd, W. F., Neil, E. 10, *42*; 61, *78*
Foley, J. O. 114, *158*
— see Dubois, F. S. 115, *158*
Forster, R. E. 65, *78*
— Ferguson, T. B. 167, *184*
Fowler, W. S., see Davis, H. L. 100, *109*
Franchi, C. M., see De Robertis, E. 57, *77*
Franz, D. N., Iggo, A. 6, 7, *42*
Freeman, W. J., Davis, D. D. 164, *184*
Freer, M., Campling, R. C., Balch, C. C. 144, *158*
Fröhlich, F. 90, *110*
Fusco, M. M., Hardy, J. D., Hammel, H.T. 163, 165, *184*
— see Hammel, H.T. 165, *184*

Gagge, A. P., see Hardy, J. D. 161, *184*
Gale, C. C., Mathews, M., Young, J. 164, *184*
— see Andersson, B. 164, *182*; 198, 202, *211*
Gammon, G. D., Bronk, D.W. 15, 16, *42*; 130, 131, *158*
Garry, R. C., Roberts, T. D. M., Todd, J. K. 135, 150, 151, *158*

Gasser, H. S., Grundfest, H. 8, *42*; 131, *158*
Gasnier, A., Mayer, A. 191, *213*
Gauer, O. H., Henry, J. P. 22, *42*
— — Sieker, J. P., Wendt, W. E. 208, *213*
— see Henry, J. P. 22, *42*
Gaylor, J. B. 82, 88, 97, *110*
Gernandt, B., Zotterman, Y. 15, *42*; 115, 132, *158*
Gernandt, B. E., see Sawyer, C. H. 206, *215*
Gibson, G. J., see Elias, E. 148, *158*
Glazebrook, A. J., Welbourn, R. B. 155, *158*
Gold, R. M. 200, *213*
Goormaghtigh, N., Pannier, R. 50, *78*
Gordon, G. B., see Bensch, K. G. 90, *109*
Graichen, H., see Rawson, R. O. 167, *186*
Gray, E. G. 57, *78*
Gray, J. A. B. 130, *158*
— Diamond, J. 130, *158*
— Malcolm, J. L. 130, *158*
— Sato, M. 1, *42*; 130, *158*
Gregersen, M. I. 208, *213*
— see Cizek, L. J. 205, *212*
— see Holmes, J. H. 206, *213*
Green, J. D., see Cross, B. A. 171, *183*; 206, *212*
Green, J. H., see Bianconi, R. 10, 15, 27, 39, *41*
— see Boss, J. 10, *41*
— see Duke, H. N. 60, *77*
— see Ead, H. W. 8, 12, *42*
Greenwood, Linda F., see Elias, E. 148, *158*
Greer, A. P., see Irisawa, H. 22, *43*
Griffo, Z. J., see Hornbein, T. F. 68, *78*
Groat, W. C. De, Ryall, R. W. 150, 151, *158*
Grossman, S. P. 200, *213*
Grundfest, H., see Gasser, H. S. 8, *42*; 131, *158*
Gupta, P. D., Henry, J. P., Sinclair, R., Baumgarten, R. von 17, 23, *42*
— see Fahim, M. 16, 17, 20, 22, 25, *42*

Guz, A., Noble, M. I. M., Eisele, J. H., Trenchard, D. 101, 106, *110*
— — Widdicombe, J. G., Trenchard, D., Mushin, W. W. 106, *110*
— — — — — Makey, A. R. 106, *110*
— Trenchard, D. 84, 95, 105, *110*
— see Noble, M. I. M. 106, *111*
Guzman, F., see Potter, G. D. 152, *160*

Habel, R. E., see Dougherty, R. W. 143, *157*
Haferkorn, D., see Wellhöner, H. H. 17, 25, *45*
Hahn, L. J., see Prince, A. L. 162, *185*
Hakumäki, M. O. K. 17, 22, 31, *42*
Hales, J. R. S., Jessen, C. 165, *184*
Hammel, H. T. 161, 164, 168, 169, 170, 171, 173, 179, *184*
— Hardy, J. D., Fusco, M. M. 165, *184*
— Jackson, D. C., Stolwijk, J. A. J., Hardy, J. D., Strømme, S. B. 167, 168, *184*
— Strømme, S. B., Cornew, R. W. 167, 168, *184*
— see Abrams, R. 167, *182*
— see Cabanac, M. 170, *183*
— see Fusco, M. M. 163, 165, *184*
— see Hellstrøm, B. 168, 169, *184*
— see Nakayama, T. 171, 172, 175, *185*
— see Rawson, R. O. 167, *186*
Hamilton, C. L. 167, *184*; 197, *213*
— Brobeck, J. R. 198, *213*
— see Spector, N. H. 198, *215*
Hamouda, F., see Hirsch, E. F. 104, *110*
Hann, F. von 201, *213*
Harding, R., Leek, B. F. 128, 136, 141, *158*
Hardy, J. D. 161, 164, 172, 176, *184*

Hardy, J. D., Gagge, A. P., Stolwijk, J. A. J. 161, *184*
— Hellon, R. F., Sutherland, K. 171, 173, 176, *184*
— see Cabanac, M. 170, 175, 181, *183*
— see Cunningham, D. J. 163, 180, *183*
— see Fusco, M. M. 163, 165, *184*
— see Hammel, H. T. 165, 167, 168, *184*
— see Murakami, N. 174, 176, *185*
— see Nakayama, T. 171, 172, 174, 175, 176, 178, *185*
— see Stolwijk, J. A. J. 169, 170, 179, *186*
Harper, A. A., see Blair, E. L. 145, 149, *157*
Harrison, F., see Magoun, H. W. 162, 163, *185*; 198, *214*
Hasama, B. 162, *184*
Haus, W. H., Kreuziger, H., Asteroth, H. 8, *42*
Hayashi, S. 88, 97, *110*
Hayward, J. N., Baker, M. A. 167, *184*
— see Vincent, J. D. 206, *216*
Head, H. 95, *110*
Hebb, C., see Banister, J. 97, *109*
Hebb, C. O. 72, *78*
Hellon, R. F. 174, 175, 177, 178, 179, *184*; 198
— see Hardy, J. D. 171, 173, 176, *184*
Hellstrøm, B., Hammel, H. T. 168, 169, *184*
Hemingway, A., see Coleridge, J. C. G. 16, 17, 20, 22, 25, 26, 27, 29, *41*
— see Simmons, D. H. 95, *112*
Henry, J., see Gauer, O. H. 208, *213*
Henry, J. P., Gauer, O. H., Reeves, J. L. 22, *42*
— Pearce, J. W. 17, 22, *42*
— see Gauer, O. H. 22, *42*
— see Gupta, P. D. 17, 23, *42*
— see Pearce, J. W. 22, *44*

Hensel, H. 173, *185*
— Krüger, F. J. 162, *185*
— Wurster, R. D. 177, *185*
— see Krüger, F. J. 164, *185*
— see Kundt, H. W. 162, 167, *185*
Herreveld, A. van 189, 190, *213*
Hervey, G. R. 197, *213*
Hess, A. 53, *78*
— see Zapata, P. 72, 73, *80*
Hess, M. E., see Aviado, D. M. 95, *109*
Hetherington, A. W., Ranson, S. W. 192, *213*
Heymans, C., Bouckaert, J. J. 48, *78*
— — Farber, S., Hsu, F. J. 72, *78*
— — Regniers, P. 48, 67, 75, *78*
— Neil, E. 3, 9, 10, *42*; 50, 63, 67, 72, 74, *78*
— Rijlant, P. 49, *78*
— see Heymans, J. F. 47, *78*
Heymans, J. F., Heymans, C. 47, *78*
Hilgeson, M. D., see Reynolds, L. D. 97, *112*
Hindorf, K., see Arndt, J. O. 16, 17, 29, 30, *40*
— see Brambring, P. 17, 29, 30, *41*
Hirsch, E. F., Kaiser, G. C., Barner, H. B., Cooper, T., Raus, J. J. 104, *110*
— — — Nigro, S. L., Hamouda, F., Cooper, T., Adams, W. E. 104, *110*
Ho, A. K. S., see Schofield, G. C. 90, *112*
Hoebel, B. G. 192, *213*
— Teitelbaum, P. 194, *213*
Hökfelt, B., see Andersson, B. 164, *182*; 202, *211*
Hoelzel, F. 153, *158*
Hoffman, B. F., see Brooks, C. McC. 25, *41*
Hoglund, R. 51, 57, *78*
Holland, R. C., Cross, B. A., Sawyer, C. H. 206, *213*
Hollinshead, W. H. 50, 51, *78*
Holmes, J. H., Cizek, L. J. 205, *213*
— Gregersen, M. I. 206, *213*

Holmes, R., Torrance, R. W. 38, *42*
Holmes, R. L., Newman, P. P., Wolstencroft, J. H. 175, *185*
— see Coleridge, J. C. G. 16, 17, 20, 22, 25, 26, 27, 29, *41*
Homberger, A. C. 91, *110*
Homma, S., Suzuki, S. S. 29, *42*
Honjin, R. 88, 97, *110*
Honour, A. J., see Cooper, K. E. 181, *183*
Hopkins, A. 145, 146, *158*
Hornbein, T. F., Griffo, Z. J., Roos, A. 68, *78*
Howard, B. R. 128, *158*
Howe, A. 50, *78*
— see Abbott, C. P. 51, 53, *76*
— see Coleridge, H. 47, 49, 50, *76*, 77; 107, *109*
Hsu, F. J., see Heymans, C. 72, *78*
Huang, K. C., see Cizek, L. J. 205, *212*
Hudson, M. T., see Woodward, G. E. 195, *216*
Hunt, C. C. 15, *42*
— McIntyre, A. K. 15, *42*
Hunt, J. N. 148, *159*
— Knox, M. T. 148, *159*
— Pathak, J. D. 148, *159*
— see Elias, E. 148, *158*
Hursh, J. B. 38, *42*
Hurst, A. F. 114, 153, *159*

Iggo, A. 9, *42*; 116, 117, 118, 119, 120, 121, 122, 123, 124, 125, 126, 127, 128, 130, 133, 134, 135, 136, 137, 140, 141, 143, 144, 145, 146, 147, 150, 151, 153, *159*
— Leek, B. F. 116, 124, 129, 140, 141, 142, 143, 144, 150, *159*
— Muir, A. R. 90, *110*
— see Franz, D. N. 6, 7, *42*
Ingram, D. L. 198, *214*
— McLean, J. A., Whittow, G. C. 164, *185*
— Whittow, G. C. 164, *185*
— see Baldwin, B. A. 164, 167, *183*
Ingram, W. R., see Fisher, C. 201, 205, *213*

Inman, D. R., Peruzzi, P. 13, *43*
Iriki, M. 165, *185*
— see Simon, E. 166, 180, *186*
Irisawa, H., Greer, A. P., Rushmer, R. F. 22, *43*
Irwin, J. V., see Takagi, Y. 88, *112*
Ishiko, N., Loewenstein, W. R. 13, *43*
Itallie, B. van, see Stunkard, A. J. 195, *215*
Ivanco, I., Korpas, J. 85, *110*
— — Tomori, Z. 85, *111*
Ivy, A. C., see Janowitz, H. D. 194, *214*

Jackson, D. C., see Eisenman, J. S. 171, 173, 174, *184*
— see Hammel, H. T. 167, 168, *184*
Jackson, D. L. 181, *185*
Jacobs, H. L., see Morgane, J. P. 191, 192, *214*
Janowitz, H. D., Ivy, A. C. 194, *214*
Jansson, G. 146, *159*
Jarisch, A., Landgren, S., Neil, E., Zotterman, Y. 14, *43*
— Zotterman, Y. 31, 32, 36, *43*
Jessen, C. 165, 166, 169, *185*
— Meurer, K. A., Simon, E. 165, *185*
— Simon, E., Kullmann, R. 165, 166, *185*
— see Hales, J. R. S. 165, *184*
Jewell, P. A., Verney, E. B. 206, 207, *214*
Jiménez Dias, C., Linazasoro, J. M., Castro Mendoza, H. 208, *214*
— see Linazasoro, J. M. 208, *214*
Joels, N. 69, *78*
— Neil, E. 65, 66, 67, *79*
— see Abbott, C. P. 53
— see Neil, E. 17, 25, 32, 35, *44*; 68, 73, *79*
Johnson, L. H., see Park, C. R. 195, *215*

Author Index

Joseph, D., see Cohn, C. 191, *212*
Jukes, M. G. M., see Lloyd, B. B. 68, *79*

Kahn, R. H. 162, *185*
Kaiser, G. C., see Hirsch, E. F. 104, *110*
Kalow, W., see Aviado, D. M. 95, *109*
Kappagoda, C. T., Linden, R. J., Snow, H. M. 17, *43*
Karczewski, W., Widdicombe, J. G. 95, *111*
Katz, B. 1, 2, 8, *43*
Katzman, R., Leiderman, P. H. 190, *214*
Kay, R. N. B., Phillipson, A. T. 143, *159*
— see Ash, R. W. 124, 140, 141, 143, 144, *157*
Keller, A. D. 202, *214*
— see Witt, D. M. 202, *216*
Kennedy, G. C. 192, 194, 197, *214*
Kenney, R. A., Neil, E. 73, *79*
Kent, D. C., see Widdicombe J. G. 91, *112*
Kerr, J. A., see Corbett, J. L. 88, *109*
Kezdi, P. 10, *43*
Kidd, C., Ledsome, J. R., Linden, R. J. 17, *43*
— see Coleridge, H. M. 3, 7, 26, 27, 29, 31, 32, 33, 34, 35, 36, 37, 38, 39, *41*
— see Coleridge, J. C. G. 39, *41*
King, A. S., see Cook, R. D. 89, *109*
Kinnison, G. J., see Bevan, J. A. 39, *41*
Klärner, P., see Erbslöh, F. 190, *213*
Kleeman, C. R., Cutler, R. E. 206, *214*
Knowlton, G. C., Larrabee, M. G. 99, 100, *111*
— see Larrabee, M. G. 95, *111*
Knox, M. T., see Hunt, J. N. 148, *159*
Koch, A., see Brooks, C. F. 190, *212*
Kolatat, T., Kramer, K., Mühl, N. 31, 32, 34, 36, *43*

Koller, E. A. 95, *111*
— see Ferrer, P. 91, *110*
Komarov, S. H., see Day, J. J. 148, *157*
Korpas, J., see Ivanco, L. 85, *110, 111*
Kosterlitz, H. W., Lees, G. M. 139, *159*
Kottegoda, S. R. 139, 140, *159*
Koyano, H., see Eyzaguirre, C. 72, *77*
Krahl, V. E. 107, *111*
Kramer, K. 17, 22, 25, *43*
— see Kolatat, T. 31, 32, 34, 36, *43*
— see Mühl, N. 17, 22, *43*
Krayer, O. 35, *43*
Kreuziger, H., see Haus, W. H. 8, *42*
Krüger, F. J., Kundt, H. W., Hensel, H., Brück, K. 164, *185*
— see Hensel, H. 162, *185*
Krylow, S. S., Anichkov, S. V. 66, *79*
Kulaev, B. S. 36, *43*
Kullmann, R., see Jessen, C. 165, 166, *185*
Kundt, H. W., Brück, K., Hensel, H. 162, 167, *185*
— see Krüger, F. J. 164, *185*
Kuznetzov, A. I., see Anichkov, S. V. 72, *76*

Lall, A., see Biscoe, T. J. 57, 58, 62, 63, 75, *76*
Lambert, E. H., see Davis, H. L. 100, *109*
Lambertsen, C. J., see Daly, M. De B. 61, 74, *77*
Landgren, S. 8, 9, 10, 14, *43*
— Neil, E. 60, 62, 63, 67, 73, 74, *79*
— — Zotterman, Y. 9, 10, 14, *43*
— see Jarisch, A. 14, *43*
Langrehr, D. 16, 17, 25, 26, *43*; 99, *111*
Larrabee, M. G., see Knowlton, G. C. 95, 99, 100, *111*
Larsell, O. 88, 97, 107, *111*
Larson, B., see Andersson, B. 164, *183*; 198, 202, 203, *211*

Larsson, S. 199, *214*
Laszlo, F. A., Wied, D. De 202, *214*
Leblanc, J., see Baldwin, B. A. 164, *183*
Ledsome, J. R., Linden, R. J. 17, *43*
— see Kidd, C. 17, *43*
Lee, K. D., Mattenheimer, H. 67, *79*
— Mayou, R. A., Torrance, R. W. 61, 62, 63, 67, *79*
Leek, B. F. 117, 118, 119, 120, 121, 122, 123, 124, 125, 126, 130, 140, 141, 142, 143, *159*
— see Harding, R. 128, 136, 141, *158*
— see Iggo, A. 116, 124, 129, 140, 141, 142, 143, 144, 150, *159*
Lees, G. M., Kosterlitz, H. W. 139, *159*
Leiderman, P. H., see Katzman, R. 190, *214*
Leitner, L. M., Perl, E. R. 15, 16, *43*; 131, 132, *159*
Leof, D., see Spencer, H. 88, 97, *112*
Leonhardt, H. 191, *214*
Letona, J. M. L. De, Mata, R. C. De La, Aviado, D. M. 95, *111*
Leusen, I. 190, *214*
Lever, J. D., Lewis, P. R., Boyd, J. D. 51, 53, *79*
Lewin, J., see Eyzaguirre, C. 64, 69, *78*
Lewis, P. R., see Lever, J. D. 51, 53, *79*
Li, T. H., see Aviado, D. M. 95, *109*
Liebelt, R. A., Perry, J. H. 196, 197, *214*
Liljestrand, G., see Euler, U. S. von 9, *42*; 49, 67, *77*
Lim, R. K. S., see Potter, G. D. 152, *160*
Lin, R. C. Y., see Bülbring, E. 139, 140, *157*
Linazasoro, J. M., Jiménez Dias, C., Castro Mendoza, H. 208, *214*
— see Jiménez Dias, C. 208, *214*

Linden, R. J., see Coleridge, J. C. G. 16, 17, 20, 22, 25, 26, 27, 29, 31, *41*
— see Kappagoda, C. T. 17, *43*
— see Kidd, C. 17, *43*
— see Ledsome, J. R. 17, *43*
Little, R. C. 22, *43*
— see Opdyke, D. F. 22, *44*
Lloyd, B. B., Jukes, M. G. M., Cunningham, D. J. C. 68, *79*
Loeschcke, H. H., see Mitchell, R. A. 190, *214*
Loewenstein, W. R. 130
— Rathkamp, R. 1, *43*
— Skalak, R. 130, *159*
— see Ishiko, N. 13, *43*
Long, C. N. H., see Brobeck, J. R. 192, *212*
Longmuir, I. S. 66, *79*
Luck, C. P., see Davson, H. 190, *212*
Luck, J. G., see Coleridge, H. M. 102, *109*
Luff, R. H., see Cranston, W. I. 181, *183*
Lynch, J. R., see Witt, D. M. 202, *216*

Machella, T. E. 155, *159*
Macklem, P., see Burger, E. J. 106, *109*
MacPherson, R. K. 181, *185*
Magoun, H. W., Harrison, F., Brobeck, J. R., Ranson, S. W. 162, 163, *185*; 198, *214*
Makey, A. R., see Guz, A. 106, *110*
Malcolm, J. L., see Gray, J. A. B. 130, *158*
Malvin, R. L., see Bonjour, J. P. 209, *212*
Manchanda, S. K., see Anand, B. K. 195, 200, *211*
Marshall, A. B., see Steven, D. H. 123, *160*
Marshall, N. B., Mayer, J. 195, *214*
— see Mayer, J. 195, *214*
Maskrey, M., see Bligh, J. 180, *183*
Massopust, L. C., Jr. 192, *214*
Massion, W. H., see Mitchell, R. A. 190, *214*

Mata, R. C. De La, see Letona, J. M. L. De 95, *111*
Mathews, M., see Gale, C. C. 164, *184*
Matthews, B. H. C. 8, 13, *43*; 122, *159*
Mattenheimer, H., see Lee, K. D. 67, *79*
Maxwell, G. M., see Rowe, G. G. 190, 195, *215*
Mayer, A. 191, 200, *214*
— see Gasnier, A. 191, *213*
Mayer, J. 194, 195, *214*
— Barnett, R. 192, *214*
— Marshall, N. B. 195, *214*
— see Baile, C. A. 192, 195, *212*
— see Marshall, N. B. 195, *214*
Mayou, R. A., see Lee, K. D. 61, 62, 63, 67, *79*
McCann, S. M., see Anderson, B. 202, 203, *212*
— see Smith, R. W. 202, *215*
McCloskey, D. I. 67, 68, *79*
— Torrance, R. W. 67, *79*
McIntyre, A. K., see Hunt, C. C. 15, *42*
McLean, J. A., see Ingram, D. L. 164, *185*
McSwiney, B. A., see Downman, C. B. B. 147, 154, *158*
Mead, J. 101, *111*
Mellanby, J., Pratt, C. L. G. 128, 151, *159*
Menguy, R. 155, *159*
Meredith, C. D., see Dougherty, R. W. 143, *158*
Meurer, K. A., see Jessen, C. 165, *185*
Miller, N. E. 199, *214*
— Bailey, C. J., Stevenson, J. A. F. 192, *214*
Mills, E. 63, *79*
— Edwards, McI. W. 60, 74, *79*
— Sampson, S. R. 63, *79*
— see Edwards, M. W. 74, *77*
Mills, J., Sellick, H., Widdicombe, J. G. 91, 92, 93, 105, 106, 107, *111*
Miller, L. R., see Bensch, K. G. 90, *109*

Mitchell, R. A., Loeschcke, H. H., Massion, W. H., Severinghaus, J. W. 190, *214*
Moir, R. J. 119, *159*
Montemurro, D. G., Stevenson, J. A. F. 199, 203, *214*
— see Stevenson, J. A. F. 198, *215*
Mori, T., see Nakayama, S. 146, *160*
Moore, R. M. 152, *159*
Moorehouse, V. H. K. 162, *185*
Morgane, J. P. 199, *214*
— Jacobs, H. L. 191, 192, 199, *214*
Morita, E., Chiocchio, S. R., Tramezzani, J. H. 51, *79*
Morrison, S. D., see Bailly, P. 199, *212*
Mott, J. C., Paintal, A. S. 25, *43*
— see Dawes, G. S. 105, *109*
Mottram, R. F., see Downey, J. A. 164, *183*
Mu, J. Y., see Chai, C. Y. 175, *183*
Mühl, N., Scholderer, I., Kramer, K. 17, 22, *43*
— see Kolatat, T. 31, 32, 34, 36, *43*
Muir, A. R., see Iggo, A. 90, *110*
Murakami, N., Stolwijk, J. A. J., Hardy, J. D. 174, 176, *185*
— see Cunningham, D. J. 163, 180, *183*
Muratori, G. 48, 49, *79*
Murgatroyd, D. 164, *185*
Murray, J. G., see Agostini, E. 114, 115, 126, *156*
Murray, R. G., see Al-Lami, F. 53, 57, *76*
Mushin, W. W., see Guz, A. 106, *110*
Myers, R. D., see Villablanca, J. 181, *186*

Nadel, J. A., Tierney, D. F. 101, *111*
— see De Kock, M. A. 95, *110*
— see Widdicombe, J. G. 91, 101, *112*

Author Index

Nail, B. S., Sterling, G. M., Widdicombe, J. G. 84, 85, 87, 88, *111*
Nakayama, S., Mori, T. 146, *160*
Nakayama, T., Eisenman, J. S., Hardy, J. D. 171, *185*
— Hammel, H. T., Hardy, J. D., Eisenman, J. S. 171, 172, 175, *185*
— Hardy, J. D. 171, 174, 175, 176, 178, *185*
Nasset, E. S. 138, *160*
— see Sharma, K. N. 138, 149, *160*
Neil, E. 10, *44*; 73, 74, *79*
— Joels, N. 17, 25, 32, 35, *44*; 68, 73, *79*
— O'Regan, R. G. 65, 69, 70, 71, *79*
— see Duke, H. N. 60, *77*
— see Ead, H. W. 8, 12, *42*
— see Floyd, W. F. 10, *42*; 61, *78*
— see Heymans, C. 50, 63, 67, 72, 74, *78*
— see Heymans, J. P. 3, 9, 10, *42*
— see Jarisch, A. 14, *43*
— see Joels, N. 65, 66, 67, *79*
— see Kenney, R. A. 73, *79*
— see Landgren, S. 9, 10, 14, *43*; 60, 62, 63, 67, 73, 74, *79*
Newman, P. P., Paul, D. H. 146, *160*
— see Holmes, R. L. 175, *185*
Nielsen, M., Smith, H. 68, *80*
Nigro, S. L., see Hirsch, E. F. 104, *110*
Niijima, A. 130, 135, *160*
Noble, M. I. M., Eisele, J. H., Trenchard, D., Guz, A. 106, *111*
— see Guz, A. 101, 106, *110*
Nonidez, J. F. 27, *44*

Oberhelman, H. A., see Dragstedt, L. R. 146, *158*
O'Connor, W. J. 202, *215*
Ohga, A., see Andersson, B. 164, *182*
Olbe, L., see Andersson, S. 147, *156*

Olsen, C. R., see De Kock, M. A. 95, *110*
Olsson, K. 201, 202, 206, 209, *215*
— see Andersson, B. 209, 211, *212*
Oltner, R., see Andersson, B. 209, *211*
Ono, T., see Oomura, Y. 195, 196, 197, *215*
Oomura, Y., Ono, T., Ooyama, H., Wayner, M. J. 195, 196, 197, *215*
Ooyama, H., see Oomura, Y. 195, 196, 197, *215*
Opdyke, D. F., Duomarco, J., Dillon, W. H., Schreiber, H., Little, R. C., Seely, R. D. 22, *44*
O'Regan, R. G. 69, 70, *80*
— see Neil, E. 65, 69, 70, 71, *79*
Orloff, J., see Stevenson, J. A. F. 202, *215*
Otis, A. B., Fenn, W. O., Rahn, H. 101, *111*
Ottoson, D. 13, *44*

Paintal, A. S. 2, 3, 4, 5, 6, 7, 8, 10, 11, 12, 13, 14, 15, 16, 17, 19, 20, 21, 22, 23, 24, 25, 26, 27, 28, 29, 31, 32, 33, 34, 35, 36, 38, 39, 40, *44*; 60, 73, 74, *80*; 83, 100, 101, 102, 103, 105, 106, *111*; 115, 116, 117, 118, 119, 120, 121, 124, 125, 127, 130, 132, 133, 152, *160*
— Riley, R. L. 4, *44*
— see Mott, J. C. 25, *43*
Palme, F. 10, *44*
Pannier, R., see Goormaghtigh, N. 50, *78*
Park, C. R., Johnson, L. H., Wright, J. H., Batsel, H. 195, *215*
Parmeggiani, P. L., see Beghelli, V. 128, *157*
Pathak, J. D., see Hunt, J. N. 148, *159*
Paul, D. H., see Newman, P. P. 146, *160*
Pearce, J. W., Henry, J. P., Chapman, K. M. 22, *44*
— Whitteridge, D. 17, *44*
— see Chapman, K. M. 17, *41*
— see Henry, J. P. 17, 22, *42*

Peart, W. S. 205, *215*
Pease, D. C., Quilliam, T. A. 57, *80*
Pembrey, M. S. 162, *185*
Perl, E. R., see Bessou, P. 130, 131, *157*
— see Leitner, J.-M. 15, 16, *43*; 131, 132, *159*
Perry, J. H., see Liebelt, R. A. 196, 197, *214*
Persson, N., see Andersson, B. 203, *211*
Peruzzi, P., see Inman, D. R. 13, *43*
Peskin, G. W., see Aviado, D. M. 95, *109*
Peterson, D. F., see Fedde, M. R. 107, 108, *110*
Pfander, W. M., see Baile, C. A. 149, *157*
Phillipson, A. T. 144, *160*
— see Kay, R. N. B. 143, *159*
Pickering, G. W., see Downey, J. A. 164, *183*
Pillai, R. V., see Anand, B. K. 145, *156*
Polyakov, N. G., see Anichkov, S. V. 72, *76*
Potter, G. D., Guzman, F., Lim, R. K. S. 152, *160*
Pratt, C. L. G., see Mellanby, J. 128, 151, *159*
Price, H. L., Widdicombe, J. 14, 15, *44*
Prince, A. L., Hahn, L. J. 162, *185*
Prys-Roberts, C. 106, *112*
— see Corbett, J. L. 88, *109*
Purves, M. J. 62, 64, 68, 75, *80*
— see Biscoe, T. J. 58, 59, 60, 62, 63, 65, 67, 68, 71, 75, *76*

Quilliam, T. A., see Pease, D. C. 57, *80*

Rahn, H., see Culver, G. A. 95, *109*
— see Otis, A. B. 101, *111*
Rall, W., Shepherd, E. M., Reese, T. S., Brightman, M. W. 57, *80*
Ramirez De Arellano, J., see Alvarez-Buylla, R. 1, *40*

Ranson, S. W. 114, *160*
— see Fisher, C. 201, 205, *213*
— see Hetherington, A. W. 192, *213*
— see Magoun, H. W. 162, 163, *185*; 198, *214*
Ranson et al. 162, 201
Rathkamp, R., see Loewenstein, W. R. 1, *43*
Raus, J. J., see Hirsch, E. F. 104, *110*
Rautenberg, W. 165, 166, *185*
Rawlins, M. D., see Cranston, W. I. 181, *183*
Rawson, R. O., Hammel, H. T. 167, *186*
— Stolwijk, J. A. J., Graichen, H., Abrams, R. 167, *186*
Rees, P. M. 10, *44*; 98
Reese, T. S., see Rall, W. 57, *80*
Reeves, J. L., see Henry, J. P. 22, *42*
Regniers, P., see Heymans, C. 48, 67, 75, *78*
Reid, C. S. W., Titchen, D. A. 143, *160*
Reis, B. B., see Stunkard, A. J. 195, *215*
Reynolds, L. D., Hilgeson, M. D. 97, *112*
Ries, F. A., see Teitelbaum, H. A. 88, *112*
Rigor, B. M., see Csaky, T. Z. 190, *212*
Rijlant, P., see Heymans, C. 49, *78*
Riley, R. L., see Paintal, A. S. 4, *44*
Roberts, J. C., see Smith, R. E. 170, *186*
Roberts, T. D. M., see Garry, R. C. 135, 150, 151, *158*
Robertson, E. G., see Denny-Brown, D. 128, *157*
Robertson, J. D., Swan, A. A. B., Whitteridge, D. 14, 15, *45*
Rodgers, W. L., Epstein, A. N., Teitelbaum, P. 199, *215*
Röhnelt, M., see Arndt, J. O. 16, 17, 29, 30, *40*
— see Brambring, P. 17, 29, 30, *41*

Rolls, B. J., see Epstein, A. N. 208, *213*
Roos, A., see Hornbein, T. F. 68, *78*
Rosendorff, C., see Cranston, W. I. 181, *183*
Ross, L. L., see Cauna, N. C. 57, *76*
Rowe, G. G., Maxwell, G. M., Castillo, C. A. Crumption, C. W, 190, 195, *215*
Rubio, M., see De Castro, F. 50, 51, 53, 57, 58, 65, *77*
Rushmer, R. F., see Irisawa, H. 22, *43*
Ryall, R. W., see Groat, W. C. De 150, 151, *158*

Sampson, S. R., Biscoe, T. J. 71, *80*
— see Biscoe, T. J. 57, 58, 59, 60, 62, 63, 67, 68, 69, 75, *76*
— see Mills, E. 63, *79*
Satinoff, E. 164, *186*
Sato, A., Fidone, S., Eyzaguirre, C. 7, 12, *45*
— see Fidone, S. J. 3, 4, 5, 6, 7, 8, 9, 12, *42*
Sato, M. 15, *45*
— see Gray, J. A. B. 1, *42*; 130, *158*
Sawyer, C. H., Gernandt, B. E. 206, *215*
— see Holland, R. C. 206, *213*
— see Sundsten, J. W. 206, *215*
Schaefer, H. 32, *45*
— see Amann, A. 36, *40*
Schalk, A. F., Amadon, R. S. 144, *160*
Schmidt, C. F., see Aviado, D. M. 95, *109*
— see Comroe, J. H., Jr. 60, 73, 74, 75, *77*
Schmidt, E. M., Stromberg, M. W. 5, *45*
Schoepfle, G. M., Erlanger, J. 13, *45*
Schofield, G. C., Ho, A. K. S., Southwell, J. M. 90, *112*
— see Bülbring, E. 139, *157*
Scholderer, I., see Mühl, N. 17, 22, *43*
Schreiber, H., see Opdyke, D. F. 22, *44*

Schweitzer, A., Wright, S. 72, *80*
— see Daly, M. De B. 61, 74, *77*
Scott, M. J., see Daly, M. De B. 101, *109*
Scratcherd, T., see Blair, E. L. 145, 149, *157*
Seely, R. D., see Opdyke, D. F. 22, *44*
Sellers, A. F., Stevens, C. E. 126, 141 *160*
— see Stevens C. E. 119 141, *160*
Sellick, H., Widdicombe, J. G. 91, 93, 95, 96, 97, 102, 106, *112*
— see Mills, J. 91, 92, 93, 105, 106, 107, *111*
Semple, R. E., see Cizek, L. J. 205, *212*
Semple, S. J. G., see Band, D. M. 75, *76*
Serafini-Fracassini, A., Volpin, D. 50, *80*
Serota, M. M. 167, *186*
Severinghaus, J. W., see Mitchell, W. H. 190, *214*
Sharma, K. N. 116, 123, *160*
— Nasset, E. S. 138, 149, *160*
— see Anand, B. K. 195, *211*
Sharp, J. A., see Coleridge, J. C. G. 39, *41*
Shepherd, E. M., see Rall, W. 57, *80*
Sherrington, C. S. 146, *160*; 162, *186*
Shoenberg, K., see Anand, B. K. 198, 200, *211*
Sieker, H. O., see Gauer, O. H. 208, *213*
Sieker, J. P., see Gauer, O. H. 208, *213*
Silver, I. A., see Cross, B. A. 177, *186*
Silverstone, J. T., see Clemente, C. D. 206, *212*
Simkins, K. L., Suttie, J. W., Baumgardt, B. R. 195, *215*
Simon, E., Iriki, M. 166, 180, *186*
— see Jessen, C. 165, 166, *185*

Author Index

Simons, B. J., see Epstein, A. N. 208, *213*
— see Fitzsimons, J. T. 208, *213*
Simmons, D. H., Hemingway, A. 95, *112*
Sinclair, R., see Gupta, P. D. 17, 23, *42*
Singh, B., see Anand, B. K. 195, 200, *211*
Sircus, W. 148, *160*
Skalak, R., see Loewenstein, W. R. 130, *159*
Sleight, P., Widdicombe, J. G. 3, 4, 7, 17, 31, 32, 35, 36, 37, 38, *45*
Smith, C. A., Dempsey, E. W. 57, *80*
Smith, G. P., Epstein, A. N. 195, *215*
Smith, H., see Nielsen, M. 68, *80*
Smith, R. E., Roberts, J. C. 170, *186*
Smith, R. W., McCann, S. M. 202, *215*
Snell, E. S., see Cooper, K. E. 181, *183*
Snow, H. M., see Kappagoda, C. T. 17, *43*
Southwell, J. M., see Schofield, G. C. 90, *112*
Spalding, J. M. K., see Corbett, J. L. 88, *109*
Spector, N. H., Brobeck, J. R., Hamilton, C. L. 198, *215*
Spencer, H., Leof, D. 88, 97, *112*
Starling, E. H., see Bayliss, W. M. 139, *157*
Stehbens, W. E., see Biscoe, T. J. 57, 58, *76*
Stella, G., see Bogue, J. Y. 49, 74, *76*
— see Bronk, D. W. 8, 25, *41*
Stellar, E., see Teitelbaum, P. 199, *215*
Sterling, G. M., see Nail, B. S. 84, 85, 87, 88, *111*
Steven, D. H., Marshall, A. B. 123, *160*
Stevens, C. E., Sellers, A. F. 119, 141, *160*
— see Sellers, A. F. 126, 141, *160*

Stevenson, J. A. F. 188, 191, 192, 195, 198, 199, 202, 203, *215*
— Box, B. M., Montemurro, D. G. 198, *215*
— Welt, L. G., Orloff, J. 202, *215*
— see Miller, N. E. 192, *214*
— see Montemurro, D. G. 199, 203, *214*
Stolwijk, J. A. J., Hardy, J. D. 169, 170, 179, *186*
— see Cabanac, M. 181, *183*
— see Cunningham, D. J. 163, 180, *183*
— see Hammel, H. T. 167, 168, *184*
— see Hardy, J. D. 161, *184*
— see Murakami, N. 174, 176, *185*
— see Rawson, R. O. 167, *186*
Stratman, C. J., see Andrews, W. H. H. 149, *157*
Ström, G. 162, *186*
Stromberg, M. W., see Schmidt, E. M. 5, *45*
Strømme, S. B., see Hammel, H. T. 167, 168, *184*
Struppler, A., Struppler, E. 27, 28, *45*
Struppler, E., see Struppler, A. 27, 28, *45*
Stunkard, A. J., Itallie, B. van, Reis, B. B. 195, *215*
— Wolff, H. G. 194, *215*
Subberwal, U., see Anand, B. K. 195, 200, *211*
Suckling, E. E., see Brooks, C. McC. 25, *41*
Sunder-Plassmann, P. 97, *112*
Sundsten, J. W., Sawyer, C. H. 206, *215*
— see Andersson, B. 164, *182*; 198, *211*
Sutherland, K., see Hardy, J. D. 171, 173, 176, *184*
Sutin, J., see Clemente, C. D. 206, *212*
Suttie, J. W., see Simkins, K. L. 195, *215*
Suzuki, S. S., see Homma, S. 29, *42*
Swan, A. A. B., see Robertson, J. D. 14, 15, *45*
Sweet, W. H., see White, J. E. 152, *160*

Takagi, Y., Irwin, J. V., Bosma, J. F. 88, *112*
Talaat, M. 115, *160*
Taylor, A., see Biscoe, T. J. 58, 74, *76*
Taylor, J. R., see Eyzaguirre, C. 72, *77*
Teitelbaum, H. A., Ries, F. A. 88, *112*
Teitelbaum, P. 192, 193, 194, *215*
— Epstein, A. N. 199, *215*
— Stellar, E. 199, *215*
— see Epstein, A. N. 203, *213*
— see Hoebel, B. G. 194, *213*
— see Rodgers, W. L. 199, *215*
Tepperman, J., see Brobeck, J. R. 192, *212*
Thauer, R. 165, *186*
Thomas, J. E. 148, *160*
Tierney, D. F., see Nadel, J. A. 101, *111*
Titchen, D. A. 141, 144, *160*
— see Reid, C. S. W. 143, *160*
Todd, J. K. 130, 135, 150, *160*
— see Garry, R. C. 135, 150, 151, *158*
Tomori, Z. 85, *112*
— Widdicombe, J. G. 85, 87, 88, 94, 97, *112*
— see Ivanco, L. 85, *111*
Torrance, R. W. 60, 66, 72, 73, *80*
— Whitteridge, D. 7, *45*
— see Black, A. M. S. 75, *76*
— see Holmes, R. 38, *42*
— see Lee, K. D. 61, 62, 63, 67, *79*
— see McCloskey, D. I. 67, *79*
Toussaint, F. Y. P., Toussaint, F. Y. 97, *112*
Toussaint, F. Y., see Toussaint, F. Y. P. 97, *112*
Tower, S. S. 115, *160*
Tramezzani, J. H., see Morita, E. 51, *79*
Trenchard, D., see Guz, A. 84, 95, 101, 105, 106, *110*
— see Noble, M. I. M. 106, *111*

Author Index

Tripp, J. H., see Elias, E. 148, *158*
Turnbull, G. L., see Aviado, D. M. 95, *109*

Uchizono, K., see Eyzaguirre, C. 5, *42*

Varagic, V., see Beleslin, D. 139, *157*
Vass, C. C. N., see Downman, C. B. B. 147, 154, *158*
Verney, E. B. 205, 206, 209, 211, *215*
— see Jewell, P. A. 206, 207, *214*
Villablanca, J., Myers, R. D. 181, *186*
Vincent, J. D., Hayward, J. N. 206, *216*
Volpin, D., see Serafini-Fracassini, A. 50, *80*

Wagenen, W. P. van, see Bellows, R. T. 202, *212*
Wang, J., see Brooks, C. F. 190, *212*
Wang, S. C., see Wit, A. 171, 174, 176, 177, 181, *186*
Warner, R. G., see Andersson, B. 209, *212*
Washburn, A., see Cannon, W. B. 128, 153, *157*
Watkinson, G., see Code, C. F. 148, *157*
Waxler, S., see Brecher, G. 195, *212*
Wayner, M. J., see Oomura, Y. 195, 196, 197, *215*
Weiss, A. J., see Aviado, D. M. 95, *109*
Weiss, K. E. 119, 143, *160*
Welbourn, R. B., see Glazebrook, A. J. 155, *158*
Wellhöner, H. H., Haferkorn, D. 17, 25, *45*
Welt, L. G., see Stevenson, J. A. F. 202, *215*
Wendt, W. E., see Gauer, O. H. 208, *213*
Wersäll, J., see Engström, H. 57, *77*
Westbye, O., see Andersson, B. 209, 210, *212*

White, A. M. S. 84, 87,
White, J. E., Sweet, W. H. 152, *160*
Whitteridge, D. 7, 10, 11, 15, 17, 27, 32, *45*; 95, *112*; 151, *160*
— Bülbring, E. 15, *45*
— see Pearce, J. W. 17, *44*
— see Robertson, J. D. 14, 15, *45*
— see Torrance, R. W. 7, *45*
Whittow, G. C., see Ingram, D. L. 164, *185*
Widdicombe, J. G. 38, 91, 95, 97, 100, 101, 107, 108, *112*
— Kent, D. C., Nadel, J. A. 91, *112*
— Nadel, J. A. 101, *112*
— see Dawes, G. S. 105, *109*
— see Guz, A. 106, *110*
— see Karczewski, W. 95, *111*
— see Mills, J. 91, 92, 93, 105, 106, 107, *111*
— see Nail, B. S. 84, 85, 87, 88, *111*
— see Price, H. L. 14, 15, *44*
— see Sellick, H. 91, 93, 95, 96, 97, 102, 106, *112*
— see Sleight, P. 3, 4, 7, 17, 31, 32, 35, 36, 37, 38, *45*
— see Tomori, Z. 85, 87, 88, 94, 97, *112*
Wiebe, G. E., see Daniel, E. E. 145, 146, 147, *157*
Wied, D. De, see Laszlo, F. A. 202, *214*
Winder, C. V. 71, 72, *80*
Wit, A., Wang, S. C. 171, 174, 176, 177, 181, *186*
Witt, D. M., Keller, A. D., Batsel, H. L., Lynch, J. R. 202, *216*
Witzleb, E. 14, *45*
Wolf, A. V. 206, *216*
Wolf, S., Wolff, H. G. 154, *160*
Wolff, H. G., see Stunkard, A. J. 194, *215*
— see Wolf, S. 154, *160*

Wolstencroft, J. H., see Holmes, R. L. 175, *185*
Woodbury, D. M. 190, *216*
Woods, R. I. 90
— see Dearnaley, D. P. 53, *77*
— see Fillenz, M. *78*; 82, 88, 89, 97, *110*
Woodward, E. R., see Dragstedt, L. R. 146, *158*
Woodward, G. E., Hudson, M. T. 195, *216*
Wright, J. H., see Park, C. R. 195, *215*
Wright, S., see Schweitzer, A. 72, *80*
Wünnenberg, W., Brück, K. 166, 179, *186*
— see Brück, K. 165, 166, 170, *183*
Wurster, R. D., see Hensel, H. 177, *185*
Wyrwicka, W., Dobrzecka, C. 194, *216*
— see Andersson, B. 203, *212*

Yates, R. D., Chen, I-Li., Duncan, D. 53, 71, 73, *80*
— see Chen, I-Li. 51, 53, *76*
Young, J., see Gale, C. C. 164, *184*

Zakusov, V. V., see Anichkov, S. V. 72, *76*
Zapata, P., Hess, A., Bliss, E. L., Eyzaguirre, C. 72, 73, *80*
— see Eyzaguirre, C. 72, *78*
Zipf, H. F. 2, *45*
Zotterman, Y. 49, *80*
— see Gernandt, B. 15, *42*; 115, 132, *158*
— see Euler, U. S. von 9, *42*; 49, 67, *77*
— see Jarisch, A. 14, 31, 32, 36, *43*
— see Landgren, S. 9, 10, 14, *43*
Zubiran, J. M., see Dragstedt, L. R. 146, *158*
Zwi, S., see De Kock, M. A. 95, *110*

Subject Index

Ambient temperature effects, interaction with effects of hypothalamic warming 164, 165
A-wave of atrial pressure and A-receptor discharge 29
Abomasal reflexes, caused by abomasal distension 144
— —, caused by abomasal acidification 144, 145
Absolute refractory period (A.R.P.) 7
Absolute refractory period for the initiation of an impulse (A.R.P.) 7
Acetylcholine, lack of effect on endings of medullated fibres 4
—, non-medullated fibre endings 4
Acetylcholinesterase, histochemical evidence in airways epithelium 88
Acetyl salicylate, antipyretic action of 181
A receptors (atrium), discharge characteristics 27
— — —, function of 29
— — —, identification 27, 28
— — —, ratio of A/B 16–17
Acid receptors 147–148
Adaptation of carotid baroreceptors 8, 9
Adipsia 199
Adrenaline, topical application to carotid sinus 10–14
Aldosterone 205
Alveoli, innervation 103
Ammonia, effect on lung irritant receptors 93
Anaesthetics (volatile), sensitisation of baroreceptors 15
Anaphylaxis, effect on lung irritant receptors 94
Angiotensin and central thirst receptors 208, 209, 210
Anodal block of vagus 84
Anoxia, types 48
Anterior hypothalamus, effects of warming 163
— —, effects of cooling 163
Antidiuretic hormone (ADH) and water balance 205, 206, 207, 208, 209, 210, 211
Antipyretic action of acetyl salicylic acid 181
Aortic bodies, location 49
Aortic nerve 12
Aphagia 200

Arterial baroreceptors, aortic 10–15
— —, carotid 7–10
— —, pulmonary artery 39, 40
Arterial chemoreceptors, location 47, 48, 49
— —, histology 50–58
Aspiration reflex 85
Atelectasis 94
Atria, filling 22
—, compliance 22
—, pressures 23
—, volume 22
Atrial receptors 16–32
Atrial wall tension and Type A atrial receptor discharge 29
Atropine and abolition of reflex effects of irritant gases 97
Aurothioglucose and hyperphagia 195
A-V anastomoses in carotid body 65

B-receptors (atrium), localisation and identification 11–16, 17
— — —, function by natural stimulus 22
— — —, ratio A/B 16–17
Baroreceptors, pulmonary arterial 39
—, systemic arterial 7–15
—, pseudo 15
Behavioural experiments in thermoregularoty studies 164
Bezold-Jarisch reflex, sensory endings concerned with 35
Bladder reflexes, tension receptors 150
— —, flow receptors 150, 151
— —, reflex emptying of bladder 150
Bloat 155
Blocking temperatures, of cardiovascular afferent fibres 6–7
Blood, brain barrier 189
Blood flow through carotid body 61, 62, 63
Breath holding time, effects of vagal block on 106
Breathing, role of vagi 101
Bronchial epithelium 90
Bronchodilatation 101
Butyryl cholinesterase in bronchiolar nerve bundles 88

Calcium ions, reversal of veratratrum reflexes 26
Calculi, bile duct 156

Calculi, urethra 156
Caloric intake, regulation 191
Carbon dust, effect on lung irritant receptors 93
Cardiac vagal receptors, atrial 16–31
— — —, epicardial 36, 37, 38
— — —, pericardial 38
— — —, ventricular 31–36
Cardia, movements of during eructation 143
Cardia, role in eructation 143
Carotid body, site and blood supply 48
— —, histology (light micros) 49, 50
— —, EM histology 51–58
— —, innervation of, afferent 50
— —, efferent 69, 70, 71
Carotid sinus, site 48
— —, innervation 48
Catecholamine excretion, effects of hypothalamic cooling 164
Catecholamines 88
Central Thermoreceptors 161–182
Cerebrospinal fluid, ionic composition and ADH release 208
Chemical Transmitter hypothesis of chemoreceptor excitation 71–73
Chemoreceptors, aortic 47
—, carotid 47
—, gastric pH receptors 136, 137
—, intestinal 138
Chloroform, effect on baroreceptor discharge 15
Choroid plexuses 189
Cigarette smoke, lung irritant receptors 94
Cold block of sensory nerves 3
Cold, sensitive neurons of hypothalamus 171, 172, 173
Colic 155
Compliance of atrium 97
Conduction block, by anodal current 105
— —, cooling 6, 7
Conduction Velocity, aortic baroreceptors 4, 5
— —, aortic chemoreceptors 4, 5
— —, lung irritant fibres 94
— —, pulmonary stretch receptor fibres 100
— —, pulmonary J-receptors 103
— —, gastro intestinal vagal fibres 125
— —, mesenteric Pacinian corpuscle fibres 131
— —, serosal receptor fibres 131
"Coronary" receptors 35
Cough and irritant receptors 88, 94
Cud, chewing 144
Cutaneous blood flow, changes induced by hypothalamic cooling 164
Cyclopropane effect on baroreceptor discharge 15

D-C potentials in antero-medial hypothalamus 170
Deflation of lung 100
Dense cored vesicles in bronchial epithelial cells 90
De-oxyglucose 195
Desensitization of atrial receptors by veratrum alkaloids 25
Diabetes insipidus 201
Diaphragmatic motoneurones, reflex excitation by epipharyngeal stimulation 85
Differential conduction block 84
Drugs, action on atrial receptors 25
—, action on baroreceptors 15
—, action on lung stretch receptors 101
Duodenal motility following gastric distension 145
Duodenum, acid receptors 148

Ectopic contractions of atria or ventricles, effects on cardiac receptor discharge 23, 24
Energy-rich phosphates and carotid body discharge 66, 67
Epicardial receptors 36, 37
Epipharynx, receptors 82
Epithelial receptors of airways 88
Ether, effect on baroreceptor discharge 15
—, effect on lung irritant receptors 94
Eupnoea and chemoreceptor discharge 75
Evaporative heat loss 168–169

Fever and derangement of thermoregulation 181
Flow receptors of urethra 135
Fluid balance and thirst 203, 204, 205
Forestomach, tension receptors 119

Gasp reflex 95, 97
Gastric pH receptors 135
Gastric stretch receptors 121
Gastric tension receptors 119
Gastro, intestinal mechanoreceptor reflexes 145–146
—, — chemoreceptor reflexes 147
Generator potential, dynamic off-effect 8
Generator region of sensory terminal 2, 3
Germitrine, stimulation of atrial receptors 25
Glossopharyngeal afferents, carotid body 47
— —, carotid sinus 2
— —, epipharynx 84–85
Glossopharyngeal efferents to carotid body 69, 70, 71
Glucagon 195

Subject Index

Glucoreceptors in the hypothalamus 195, 196
Glucostatic theory of satiety 194, 195, 196

Head's paradoxical reflex 96
Heart-rate, reflex effects of Type A receptors 30
Heart, receptors of – (see Atrial, ventricular receptors) 16–38
"Heat centres" of hypothalamus 162, 163, 164
Heat production 168, 169
Heat, sensitive neurones of hypothalamus 171, 172, 173
Hering Breuer reflex 95, 99, 101
Histamine, bronchoconstriction 101
Hunger 191
— centres 199
— pangs 153
— receptors 200
Hydration 204
Hydroxytryptamine (5-HT), hypothalamic content 180
Hypercapnia, effect on chemoreceptor activity 67, 68
—, effect on glomus blood flow 67, 68
—, effect on glomus oxygen usage 68
Hyperphagia 200
Hypertonic solutions, effects of hypothalamic injection 205, 206, 207
Hypodipsia 202
Hypothalamus, thermal sensitivity of 162, 163, 164, 165
—, electrical activity 170
—, preoptic single units 171
—, cold-sensitive neurons 171, 173
—, thermal-insensitive neurons 171, 172
—, warm-sensitive neurons 171, 173
—, peripheral inputs 176, 177, 178, 179
Hypoxia, effect on chemoreceptor activity 58, 59, 60
—, effect on glomus blood flow 64
—, effect on glomus oxygen usage 64

Identification of rhythmic discharges cardiovascular mechanoreceptors 18
Impactions (gut) 156
In-series location of viscus receptors function 123, 124, 126
Insulin 194
Internal carotid artery, effects of cooling 164
Intestinal chemoreceptors 138
Intrapulmonary bronchi, receptors 91
Intrinsic myogenic contraction of gut wall, effect on discharge of mechanoreceptors 123, 124
Irritant receptors of lungs 92–93

Irritant conduction velocity 94
Isometric contraction of viscus and visceral afferent discharge 121

Juxta, pulmonary capillary receptors 102
—, (J-receptors) – characteristics 102, 103
—, conduction velociaty 103
—, histological site 103
—, staining reactions 103
—, reflex effects 105
—, bronchomuscular tone 105
—, pulmonary oedema 105
—, nociceptive function 105

Larynx, receptors of 90, 91
Lateral Hypothalamus and Hunger 198, 199
Left atrium, receptor discharge characteristics 18
Lipostatic theory of satiety 197
Local anaesthetics and block of sensory conduction 2
Localisation of atrial nerve endings 26–27
Lung deflation and irritant receptor discharge 95
Lung, deflation receptors 84
—, stretch receptors 83
—, irritation receptors 92, 93
Lung inflation, effects on aortic baroreceptor discharge 18
— —, — — atrial receptor discharge 18
— —, — — the gasp reflex 95

Mechanical properties of lungs and optimal patterns of breathing 101
Mechanoreceptors of arterial system 7–15
— — lung 97, 98, 99, 100
— — stomach 118
Median eminence 202
Mediastinal receptors 107
Medullated fibre endings, effect of Ach 2
Methylene blue staining of tracheobronchial nerve endings 82
Mid-brain, thermo sensitive cells 175
Mucosal receptors, gastro-intestinal 134
Muscularis mucosae, receptors 133
Myenteric reflex of the gastro intestinal tract 138

Nasopharyngeal epithelium 88
Nasopharyngeal receptors and the aspiration reflex 84
Neurohypophysis 201
New-born, aortic receptor discharge 12
Nicotine, effects of intrapericardial injection on epicardial receptors 36
Nociceptive endings in lung 107

Non-painful sensations from stomach and bladder 153
Noradrenaline in hypothalamus 180
— response of thermosensitive hypothalamic neurons 180

Obesity 192
Off-response of tracheal irritant receptors 92
Osmometric regulation, of fluid balance 205
Osmoreceptors and ADH 205
Osmoreceptors (stomach), gastric secretion 148
— — duodenal motility 148
— — gastric emptying 148
Oxygen usage of carotid body 62
— effect of sympathetic discharge 64

Pacinian corpuscle 1, 13
Painful visceral sensation 154, 155, 156
Panting, in response to hypothalamic warming 165
Parabiosis 197
Pericardial vagal receptors 38
Peripheral input and Hypothalamic Thermoregulation 176
Peristaltic reflex 138–140
pH, effects on carotid chemoreceptor discharge 67
Phenyl-diguanide, effect on non-medullated nerve endings 4
— —, effect on J receptors 102–103
— —, lack of effect on medullated nerve endings 25
Pneumothorax 95
Polydipsia 201
Polypnoea, produced by hypothalamic warming 164
Polyuria 201
Post-prandial drinking 203, 208
Prostatic enlargement, effect on urethral and vesical receptor discharge 156
Pseudo-baroreceptors 15, 16
Pulmonary arterial receptors 39, 40
Pulmonary chemoreceptors 107
Pulmonary congestion and irritant receptors of the lung 95
Pulmonary stretch receptors 97, 98, 99, 100
Punctate stimulation of epicardial receptors 37
Pyloric antral motility, distension of duodenum 147
Pyrogens, response of thermo sensitive neurons 181

Q_{10} (temperature coefficient) of baroreceptors 13

Rapidly adapting pulmonary stretch receptors 99
Rapidly adapting visceral mechanoreceptors, mesenteric Pacinian corpuscles 130
— — — —, serosal receptors 130
Ratio of Type A to Type B atrial receptors 16
Recurrent laryngeal neurones 82
"Reference" temperature 168
Reflex evidence of abdominal visceral effects 138
Reflex responses from the lungs 83
Regenerative region of baroreceptors 2, 15
Relative refractory period of baroreceptors 13
Renin liberation and thirst stimulation 208
Reserpine effects on granules of glomus Type I cells 53
Reticulo-ruminal tension receptors 126
Right atrial receptors, characteristics of discharge 18
Rumen, primary cycle movements 141
Rumen, secondary cycle movements 141
Ruminant forestomach mechanoreceptor reflexes 140, 141, 142, 143
Rumination 144

Salivation, mechanoreceptor reflexes 140
Satiation 153
Satiety and Hypothalamus 191, 192
Satiety, hunger interaction 200
Satiety receptors 194, 195, 196, 197, 198
Sensations mediated by respiratory receptors 106
Sensitization by volatile anaesthetics
— of arterial baroreceptors 15
— of lung stretch receptors 100
Serosal movement receptors 131
"Set point" temperature 168
Sham-feeding and gastric secretion 147
Shivering and thermoregulation 170
Silver staining of lung nerves 87
Sinus nerve, afferent fibres 2, 47
— —, efferent fibres 69
Sinus nerve efferent stimulation
— — effects on afferent discharge 69, 70
— — effects on Type I cells 53
Slowly adapting mechanoreceptors of the gastro intenstinal tract 116, 117
Slowly adapting mesenteric receptors 129
Small baroreceptor fibres 8, 9
Smooth muscle of airways 82
Smooth muscle of carotid sinus 15
Specific Na^+ Ion Sensitivity 209
Spinal cord, effects of cooling 165
— —, effects of warming 165
— —, thermosensitive units 180

Subject Index

Splanchnic nerve, afferent components 114
Steady pressure and baroreceptor discharge 8, 12
Stretch receptors, aortic 10–15
— —, carotid sinus 7–10
— —, gastro-intestinal 118
— —, lungs 97–100
Sweating response 169
Sympathetic innervation of carotid body, effect on O_2 usage of carotid body 64
Synaptic vesicles, carotid body nerves 89, 98

Tachypnoea, irritant receptors 95
Takayashu's disease 3
Temperature coefficient (Q_{10}) of baroreceptors 13
— effect on baroreceptor discharge 12, 13
Temperature regulation and control engineering theory 168
Temperature sensitivity and transmitter substances 180
Tension receptors in viscera 117, 118, 119
— — — impulse activity 119
— — — function 126
Thermosensitive neurones in hypothalamus 171, 172, 173
Thermostatic theory of satiety 197, 198
Third Ventricle, effects of infusions 209
Thirst and Hypothalamus 200, 201, 202, 203
Thirst, osmometric and volumetric regulatory mechanisms 205, 206, 207, 208
Trachea, parasympathetic motor innervation 82
—, receptors 91
Type A atrial receptors, characteristic discharge 27, 28, 29, 30
— — — —, function 29, 30
Type B atrial receptors 17–27
Type I glomus cells, histology of nerve terminals 49, 50, 51, 53, 57
Type II glomus cells, histology 50, 51

Unit recordings of temperature sensitive hypothalamus neuron 170–174

Unit recordings of temperature sensitive mid-brain neurons 175
— — of temperature sensitive spinal neurons 180
Urethral receptors 135
Urinary tract reflexes 149, 150, 151

V-wave of atrial pressure activity of Type B receptors 22
Vagal blockade, effect on dyspnoea 106
Vagotomy, effect on breathing 101
Vagus, fibre components 114, 115
Varicose axons in tracheal epithelium 88
Vasoconstriction, in response to lowering hypothalamic temperature 164
Venous return and atrial receptor discharge 11
Ventricular afferent fibres 31 et seq.
— — identification 32–33
— — natural stimulation 34
Ventromedial hypothalamic lesions, and hyperphagia
— — —, and obesity 192, 193
Ventromedial hypothalamic stimulation and inhibition of food intake 194
Veratrum alkaloids 7
— — effects on atrial receptors 25
— — — — "coronary receptors" 35
— — — — ventricular receptors 32, 37
Visceral (gastric) chemoreceptors 135
— — — effect of pH 135, 136
— — — role of 136
Visceral nociceptor Theory 152
Visceral receptor mechanism 115–116
— — — slowly-adapting mechanoreceptors 116
— — — receptors in series with smooth muscle 117
Volatile anaesthetics and baroreceptor discharge 14, 15
Volume receptors of heart 208
Volumetric regulation of fluid balance 208

Warm, sensitive neurones of hypothalamus 171, 172, 173, 174
— — — of mid-brain 175, 176
— — — of spinal cord 179, 180
Water diuresis 205

Northern Michigan University
3 1854 001 199 978
EZNO
QP351 H34 vol. 3, no. 1
Enteroceptors,